THE SOUTH ASIAN ENERGY CORRIDOR

Toward Sustainability and Security

THE SOUTH ASIAN ENERGY CORRIDOR

Toward Sustainability and Security

Edited by

Amritanshu Shukla, PhD
Atul Sharma, PhD
Saurabh Mishra, PhD

AAP | APPLE ACADEMIC PRESS

First edition published 2025

Apple Academic Press Inc.
1265 Goldenrod Circle, NE,
Palm Bay, FL 32905 USA

760 Laurentian Drive, Unit 19,
Burlington, ON L7N 0A4, CANADA

CRC Press
2385 NW Executive Center Drive,
Suite 320, Boca Raton FL 33431

4 Park Square, Milton Park,
Abingdon, Oxon, OX14 4RN UK

Library and Archives Canada Cataloguing in Publication

Title: The South Asian energy corridor : toward sustainability and security / edited by Amritanshu Shukla, PhD, Atul Sharma, PhD, Saurabh Mishra, PhD.
Names: Shukla, Amritanshu, editor. | Sharma, Atul (Professor of environmental studies), editor. | Mishra, Saurabh, editor.
Description: First edition. | Includes bibliographical references and index.
Identifiers: Canadiana (print) 20230599265 | Canadiana (ebook) 20230619304 | ISBN 9781774914786 (hardcover) | ISBN 9781774914793 (softcover) | ISBN 9781003459507 (ebook)
Subjects: LCSH: Energy policy—South Asia. | LCSH: Energy industries—South Asia. | LCSH: South Asia—Environmental conditions.
Classification: LCC HD9502.S613 S68 2024 | DDC 333.790954—dc23

Library of Congress Cataloging-in-Publication Data

CIP data on file with US Library of Congress

ISBN: 978-1-77491-478-6 (hbk)
ISBN: 978-1-77491-479-3 (pbk)
ISBN: 978-1-00345-950-7 (ebk)

About the Editors

Amritanshu Shukla, PhD
Professor in Physics, Lucknow University, Lucknow, India

Amritanshu Shukla, PhD, is currently working as a Professor in Physics at Lucknow University, Lucknow, India. He has also served as Head of the Division of Basic Sciences and Humanities and as Chief Vigilance Officer of the Rajiv Gandhi Institute of Petroleum Technology, Jais, Amethi, Uttar Pradesh, India. He did his masters in Physics and completed his PhD from IIT Kharagpur in January 2005. He did his postdoctoral research work at some of the premier institutes across the globe, namely at the Institute of Physics Bhubaneswar (Department of Atomic Energy, Government of India), the University of North Carolina Chapel Hill USA, University of Rome/Gran Sasso National Laboratory, Italy, and Physical Research Laboratory Ahmedabad (Department of Space, Government of India).

His research interests include nuclear physics and physics of renewable energy (RE) resources. He has published over 150 research papers in various international journals and international and national conferences. He has delivered invited talks at various national and international institutes. He has several national and international projects and active research collaborations from India and abroad on his research interests. He has published over 50 refereed papers in international journals and delivered invited talks at conferences and symposia. His recent book publications include *Low Carbon Energy Supply: Trends, Technology, Management* (Springer); *Sustainability through Energy-Efficient Buildings* (CRC Press/Taylor and Frances); *Energy Security and Sustainability* (CRC Press, Taylor and Francis); and *Latent Heat-Based Thermal Energy Storage Systems* (Apple Academic Press).

Atul Sharma, PhD

Associate Professor at Rajiv Gandhi Institute of Petroleum Technology, India.

Atul Sharma, PhD, is currently an Associate Professor at Rajiv Gandhi Institute of Petroleum Technology (RGIPT), which was set up by the Ministry of Petroleum and Natural Gas, Government of India, through an Act of Parliament (RGIPT Act 54/2007) along the lines of the IITs with an "Institute of National Importance" tag. He has worked as a scientific officer at Devi Ahilya University, India; as a research assistant at Korea Institute of Energy Research, Daejeon, South Korea; and as a visiting Professor at the Department of Mechanical Engineering, Kun Shan University, Tainan, Taiwan, R.O.C.

Dr. Sharma published several edited books from well-known international publishers and research papers in various international journals and conferences. Dr. Sharma has completed several projects as Principal Investigator (P.I.) from various funding agencies such as the Department of Science and Technology (DST), New Delhi, and Council of Science and Technology, U.P. He is working on the development and applications of PCMs, green buildings, solar water heating systems, solar air heating systems, and solar drying systems, etc. Further, he reviewed many national and international journals, project reports, and book chapters. Dr. Sharma completed his MPhil and PhD from School of Energy and Environmental Studies, Devi Ahilya University, India.

Saurabh Mishra, PhD

Associate Professor of Communication at Rajiv Gandhi Institute of Petroleum Technology, India

Saurabh Mishra, PhD, is currently working as an Associate Professor of Communication at Rajiv Gandhi Institute of Petroleum Technology (RGIPT) (set up through an Act of Parliament by the Ministry of Petroleum and Natural Gas as an "Institute of National Importance" on the lines of IITs).

He has a teaching/research experience for more than 15 years. He is actively involved in studies on the impact of communication and energy proliferation. He has to his credit research papers and book chapters on varied interests. His current research concerns include examining the role of crucial stakeholders in propagating national awareness and the development of ideologies. His recent publications include papers and book chapters on renewable energy awareness and education in India, energy awareness and challenges in developing low-carbon societies, energy-efficient technology to policy awareness, and energy sustainability communications.

Contents

Contributors

Abhishek Anand
Department of Basic Sciences, Non-Conventional Energy Laboratory,
Rajiv Gandhi Institute of Petroleum Technology, Jais, Amethi, Uttar Pradesh, India

Deepali Atheaya
Department of Mechanical Engineering, Bennett University, Greater Noida, Uttar Pradesh, India

Rohit Bansal
Rajiv Gandhi Institute of Petroleum Technology Jais, Amethi, Uttar Pradesh, India

Koushik Guha Biswas
Rajiv Gandhi Institute of Petroleum Technology, An Institute of National Importance, Jais,
Uttar Pradesh, India

Dharam Buddhi
Department of Research and Innovation, Lovely Professional University, Phagwara, Punjab, India

A. K. Chaturvedi
AVSM, VSM – Freelance Contributor (Retired)

Ajay Kumar Chaturvedi
AWHO Colony, Tyagi Vihar, Lucknow, Uttar Pradesh, India

Shivani Chaturvedi
Enzyme and Microbial Biochemistry Lab, Department of Chemistry, Indian Institute of Technology,
Delhi, India

Manish Gupta
Mittal School of Business, Lovely Professional University, Phagwara, Punjab, India

Shweta Gupta
Rajiv Gandhi Institute of Petroleum Technology Jais, Amethi, Uttar Pradesh, India

Pooja Jaiswal
Rajiv Gandhi Institute of Petroleum Technology, An Institute of National Importance, Jais,
Uttar Pradesh, India

Raveena Kargwal
Department of Processing and Food Engineering, College of Agricultural Engineering and Technology,
C.C.S. Haryana Agricultural University, Hisar, Haryana, India

Sunil Kumar Khare
Enzyme and Microbial Biochemistry Lab, Department of Chemistry, Indian Institute of Technology,
Delhi, India

Anil Kumar
Department of Mechanical Engineering, Delhi Technological University, Delhi, India;
Center for Energy and Environment, Delhi Technological University, Delhi, India

Yogendra Kumar
Rajiv Gandhi Institute of Petroleum Technology, An Institute of National Importance, Jais, Uttar Pradesh, India

Saurabh Mishra
Department of Science and Humanities, Rajiv Gandhi Institute of Petroleum Technology, Uttar Pradesh, India

Lata Nain
Division of Microbiology, ICAR–Indian Agricultural Research Institute, New Delhi, India

Prasanta Kumar Rout
Process Chemistry and Technology Department, Central Institute of Medicinal and Aromatic Plants, Lucknow, Uttar Pradesh, India

Atul Sharma
Department of Basic Sciences, Rajiv Gandhi Institute of Petroleum Technology, Jais, Amethi, Uttar Pradesh, India

Mukul Sharma
Department of Mechanical Engineering, Bennett University, Greater Noida, Uttar Pradesh, India

Amritanshu Shukla
Department of Physics, University of Lucknow, Lucknow, U.P., India

Raman Shukla
Rajiv Gandhi Institute of Petroleum Technology, An Institute of National Importance, Jais, Uttar Pradesh, India

Jyoti Singh
Division of Microbiology, ICAR–Indian Agricultural Research Institute, New Delhi, India

Ramkishore Singh
Department of Research and Innovation, Lovely Professional University, Phagwara, Punjab, India

Shailendra Singh
Non-Conventional Energy Laboratory, Rajiv Gandhi Institute of Petroleum Technology, Jais, Amethi, Uttar Pradesh, India

Ankit Srivastava
Department of Basic Sciences, Non-Conventional Energy Laboratory, Rajiv Gandhi Institute of Petroleum Technology, Jais, Amethi, Uttar Pradesh, India; Department of Mechanical Engineering, Feroze Gandhi Institute of Engineering and Technology, Raebareli, Uttar Pradesh, India

Manoj Kumar Srivastava
Department of Physics, Army Cadet College, Indian Military Academy, Dehradun, Uttarakhand, India

Ashutosh Verma
Department of Research and Innovation, Lovely Professional University, Phagwara, Punjab, India

Rakesh Verma
Rajiv Gandhi Institute of Petroleum Technology Jais, Amethi, Uttar Pradesh, India

Manish Vyas
School of Pharmaceutical Sciences, Lovely Professional University, Phagwara, Punjab, India

Abbreviations

AD	anaerobic digestion
ADB	Asian Development Bank
AEDB	Alternative Energy Development Board
AEH	Asian Energy Highway
BBIN	Bhutan–Bangladesh–India–Nepal
BIMSTEC	Bangladesh, India, Myanmar, Sri Lanka, Thailand Economic Cooperation
BPDB	Bangladesh Power Development Board
BWR	boiling water reactors
CAATSA	Countering America's Adversaries Through Sanctions Act
CAGR	combined annual growth rate
CARs	Central Asian Republics
CASA	Central Asia–South Asia
CASAREM	CAS regional electricity market
CBET	cross-border electricity trade
CBPE	cross-border power exchange
CDM	clean development mechanism
CEA	Central Electricity Authority
CEB	Ceylon Electricity Board
CEM	Clean Energy Ministerial
CF	capacity factor
CIS	Commonwealth of Independent States
CO2	carbon dioxide
CPC	Ceylon Petroleum Corporation
CRED	Chinese Center for Renewable Energy Development
CRISIL	Chennai and Risk and Infrastructure Solutions Limited
DSD	direct solar drying
EEZ	exclusive economic zone
EPTA	electric power trade agreement
ESD	education for sustainable development
EU	European Union
FAEE	fatty acid ethyl esters
FATA	Federally Administered Tribal Areas
FBC	fluidized bed combustors

FOWPI's	First Offshore Wind Project for India
GAIL	Gas Authority of India Limited
GBI	generation-based incentive
GDP	gross domestic product
GHGs	greenhouse gases
GHI	global horizontal irradiance
GSI	Geological Survey of India
GSPA	gas sale purchase agreement
GW	gigawatt
HBD	hybrid solar drying
HDC	hydrodecarboxylation
HDI	Human Development Index
HDO	hydrodeoxygenation
HPO	hydropower obligation
HRSG	heat recovery steam generator
HVDC	high voltage direct current
IAEA	International Atomic Energy Agency
IDCOL	Infrastructure Development Company Limited
IEA	International Energy Agency
INDC	India's Nationally Determined Contributions
IoT	internet of things
IRADe	Integrated Research and Action for Development
IREDA	Indian Renewable Energy Development Agency
ISD	indirect solar drying
ISH	integrated surface hourly
JNNSM	Jawaharlal Nehru National Solar Mission
JSC	Joint Steering Committee
JWG	Joint Working Group
KPK	Khyber Pakhtunkhwa
LNG	liquefied natural gas
LPG	liquefied petroleum gas
MBI	Myanmar–Bangladesh–India
MMT	million metric ton
MNRE	Ministry of New and Renewable Energy
MoP	Ministry of Power
NAPCC	National Action Policy on Climate Change
NDCs	nationally determined contributions
NDRC	National Growth and Reform Commission
NEA	National Energy Administration

NGRI	National Geophysical Research Institute of Hyderabad
NHPC	National Hydro Power Corporation
NIWE	National Institute of Wind Energy
NPL	National Physical Laboratory
NREL	National Renewable Energy Laboratory
NTPC	National Thermal Power Corporation
ONGC	Oil and Natural Gas Corporation
OPEC	Organization of the Petroleum Exporting Countries
OSD	open sun-drying
OTEC	ocean thermal energy conversion
PCMs	phase change materials
PMP	phased manufacturing program
POL	petroleum oil lubricants
R&D	research and development
RE	renewable energy
RECs	renewable electricity certificates
RES	renewable energy sources
RO	renewable obligation
RPOs	renewable purchase obligations
SAARC	South Asia Association for Regional Cooperation
SAC	South-Asian countries
SAGE	South Asia Gas Enterprise Pvt Ltd
SAPG	solar-assisted power generation
SASEC	South Asia Sub-Regional Economic Cooperation
SCO	single-cell oil
SDGs	sustainable development goals
SECI	Solar Energy Corporation of India
SEZ	special economic zones
SF	solar fraction
SHP	small hydro plants
SPP's	simple payback period
SPV	special purpose vehicle
SREDA	Sustainable and Renewable Energy Development Authority
SRF	senior research fellowship
SSHT	small-scale hydro technology
TAP	Turkmenistan–Afghanistan–Pakistan
TAPI	Turkmenistan–Afghanistan–Pakistan–India gas pipeline
TCF	trillion cubic feet
TERI	The Energy and Resources Institute

TUTAP	Turkmenistan–Uzbekistan–Tajikistan–Afghanistan–Pakistan
UNFCCC	United Nations Framework Convention on Climate Change
US EIA	US Energy Information Administration
USAID	United States Agency for International Development
VFA	volatile fatty acids
VRE	variable renewable electricity
WtE	waste-to-energy

Acknowledgments

Before starting this project in January 2020, we were sure that we would need at least one year to develop it. The reason we decided to bring together an edited book was to share the latest updates, knowledge, and development for the South Asian energy corridor. Therefore, we connected with several researchers worldwide and requested them to provide their recent cutting-edge research in the form of book chapters for this edited book.

Since we had completed several such types of research in the recent years, we had good experience with it and in the beginning, we felt that it would be easy to complete this work. However, we couldn't compile the work within the assigned time frame as some chapters with particular research topics related to the book's theme were quite time-consuming. After completing this work, we realized how demanding the entire work had been since the authors who contributed to this book are from the scientific and academic community and had prior commitments. However, we received remarkable encouragement and enthusiasm from our authors and reviewers. We are honored to have all the contributors who have been very supportive, dedicated, and responsive throughout our interactions. All of us are very thankful to all our passionate authors and reviewers!

We would like to express our great appreciation to many people for their valuable and constructive suggestions during the planning and development of this book. Their willingness to give their time so generously is very much appreciated. This book would never have taken off without the generous support of the authors of the book chapters. It would also never have been completed without the cooperation, administrative, and editorial assistance of Dr. A. K. Haghi, Apple Academic Press. Our heartfelt thanks to all these dedicated and cooperative individuals.

This book would not have been possible without the support and encouragement of our families. Words cannot express our gratitude to them for their kind support and encouragement. We also feel sorry about neglecting them on many occasions, especially during weekends. Finally, we thank our families for tolerating our incessant disappearances into our offices. We dedicate this book to them for all their patience and moral support.

We are very excited, enthusiastic, and gratified about the final outcome.

—Amritanshu Shukla, PhD
Atul Sharma, PhD
Saurabh Mishra, PhD
Rajiv Gandhi Institute of Petroleum Technology, Jais, Amethi, Uttar Pradesh, India

Preface

Energy consumption is directly related to the growth of humans around the globe. History reveals how it has influenced mankind at large, on the one hand, giving rise to enhanced comfort and lifestyle, better mobility, communication, and other essential needs. On the other hand, the competition to own more energy resources (slaves to coal wells to oil wells) has created conflicts taking exploration to places like Antarctica and interstellar destinations. It is creating war-like situations in the world.

Approximately a billion of the world population (i.e., approximately 13% of the world population) either do not have or have very limited access to electricity. On top of that, there are more than two orders of difference in the electricity consumption per capita of the world. South Asia is home to almost one-fourth of the world's population despite having only one-tenth of the world's landmass. In particular, the population of South Asia has immensely increased as it houses three of the 10 most populous countries of the world. This has put roaring pressure on the energy supply and energy security. South Asia region seems to face several insecurities, including energy shortage.

One of Asia's biggest rising economy countries, i.e., India, is experiencing about a 4% annual demand increase, with a 10% shortage during peak hours. Along similar lines, other South Asian densely populated countries, such as Pakistan, Bangladesh, etc., show a similar pattern for energy demand but with a bigger difference between demand and supply. Pakistan's electricity demand is expected to cross 50 GW by 2030. In Bangladesh and Nepal, access to electricity is limited to about one-third of the households, and they experience regular power cuts due to a lack of power production. This situation is entirely different compared to Western countries, where the main focus presently is on the supply of clean fuel.

Moreover, the crisis of energy supply shortage is hit by climate change requirements and international market fluctuations, which further constrain the region's growth and stability. In addition, South Asian nations have some common energy issues in the basket. The growth pattern of population growth with rapid urbanization and industrialization has put considerable stress on the existing energy infrastructure. This problem gets multiplied due to the existence of cross-border conflicts.

Countries in South Asia are currently facing a difficult path of balancing between development and environmental concerns. Suppose the region has to sustain the growth pace. In that case, there is no second choice but to increase the width of the renewable window as per local climatic conditions in the respective country's energy mix, with an overall objective to increase access to electricity to most of the population. Related long-term policies, rules, and regulations need to be prioritized, with renewable energy (RE) and energy cooperation as key agenda points. Energy is an integral ingredient for economic development, and overall growth can open up avenues for further mutual development, regional stability, and conflict resolution.

In this connection, the present book offers up-to-date coverage of the energy scenario in the South Asian sector. The proposed book will showcase the major challenges being faced in the South Asian Energy Corridor in the context of Energy Security vs. Clean Energy. The role of diversifying supplies for the future to gain energy security can lead to mapping joint ventures among conventional rival countries, especially due to the rapidly emerging economies of this region. By their location and demography, India, China, Japan, and Pakistan occupy a major role in this corridor, though majorly fed by fossil fuels running towards achieving clean energy security.

The book will examine the role of major technologies vis-a-vis dominant energy players in South Asia to put a check and balance on energy security and clean energy resources hand in hand. The economics and geopolitics of various pipelines, ports, and regional strategic relations strongly favor developing an "energy bond" among nations, with both technologies and markets available within the same region. The proposed book strongly suggests increasing strategic energy cooperation between the major energy players to maximize mutual interest and reduce carbon emissions for larger interests. We hope the book will be uniquely interesting for all class of readers, including graduate students and researchers from academia and industry, providing them with an overall current update on the South Asian energy sector.

Authors are highly grateful for all esteemed contributors and our academic collaborators, who always provide energy and inspiration to contribute to this extremely relevant field of research for mankind. We are short of words to thank Dr. A. K. Haghi and all his team members from Apple Academic Press for their kind support and great effort in bringing the book to fruition, especially during this pandemic. Last but certainly not the least, the editors are sincerely thankful to their family members, whose constant support and affection have given them the strength to complete this assignment successfully.

—Amritanshu Shukla, PhD
Atul Sharma, PhD
Saurabh Mishra, PhD

CHAPTER 1

Introduction to the South Asia Energy Corridor

A. K. CHATURVEDI

Senior Vice Chairman, Society for Transformative Research Integration and Veterans' Empowerment (STRIVE), India

ABSTRACT

Sustainable availability of Energy at affordable cost is the prerequisite of development. A survey of South Asian Countries brings out clearly that average consumption of energy per capita per year in region is quite low as compared to world average. Which brings up an important question, whether the region is deficient of energy bearing resources? The answer to this question is an obvious 'No'. Most of the countries have adequate resources in terms of solar energy, hydro energy, coal, biomass, and marine based resources. Some amount of gas is also available in the offshore region. Region is also located close to West Asia and Central Asian Republics which are rich in petroleum and Gas. Similarly, Bay of Bengal is rich in gas-based resources. Therefore, neither the region is deficient in resources, nor it is far away from resources which can be exploited to enhance the probability of sustainable energy availability. In this regard three aspects are very important. Firstly, it needs to be appreciated that the South Asia is a single geographical entity and notwithstanding the political boundaries between the countries of the region the resources are monolithic and need to be exploited collectively. Secondly a collective research and development programme will go a long way in addressing common technological issues pertaining to provision of energy within the region. Thirdly resources of external to region

The South Asian Energy Corridor: Toward Sustainability and Security. Amritanshu Shukla, PhD, Atul Sharma, PhD & Saurabh Mishra, PhD (Eds.)

can be exfoliated economically only if enough consensus on modalities is worked out to ensure its passage and security during the passage.

1.1 CENTRALITY OF ENERGY IN THE ECONOMIC GROWTH OF A NATION/REGION

Energy is essential for sustaining life because it is needed to improve the quality of life. It is also necessary for improving agricultural production, optimizing resources in manufacturing, and ensuring the effectiveness of the services sector. Thus, there is a direct correlation between energy availability and economic growth. No wonder a country's per capita energy consumption is regarded as an important indicator of the economic development of the country under consideration. The aspects related to its easy availability or its deniability by countries/groups holding the energy production resources make it a strategic resource that can be used to shape the international relations between suppliers and users, besides impacting their respective economies. Historically, with the advent of the wheel and later the Industrial Revolution, its centrality for the economic well-being of a nation kept increasing steadily. Currently, a stage has been reached where, with increased automation in production, energy as an input plays a very critical role in ensuring the efficiency of the system. Energy use with the GDP per capita would be quite clear from (Figure 1.1).

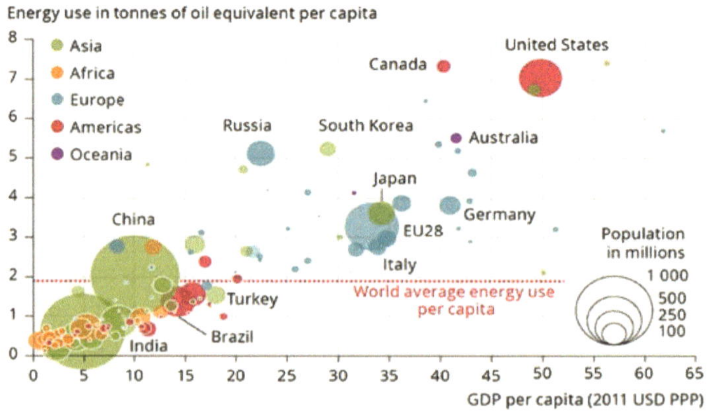

FIGURE 1.1 Correlation between energy use and GDP.
Source: Reprinted with permission from EEA, 2015 . https://creativecommons.org/licenses/by/4.0/

The above sketch explains the correlation between the economic growth and the energy consumption:

- The economic growth slowed down during 2019 due to a number of geopolitical and geo-economic reasons besides the effect of the Pandemic. In statistical terms, the global GDP growth fell from 3.6% in 2018 to 2.9% in 2019 (The Flagship Report, 2019). In advanced economies, the average fall of economic growth was as high as 25% between 2018 and 2019. However, in India, the economic growth dropped from 6.8% in 2018 to 4.8% in 2019. In China, economic growth dropped from 6.6% in 2018 to 6.1% in 2019 (The Flagship Report, 2019).
- One of the major impacts of slowing down of economic growth was the slowing down of the global energy demand during 2019 as compared to the rate of energy demand during 2018. The yearly energy consumption rate went down from 2.8% in 2018 to a mere 1.3%. This, in fact, was lower than the 10-year average rate of 1.6% per annum. The maximum drop in the energy demand was in the European Union (EU) at 2%. Even in the US, it fell by 0.8%. However, the drop in the world energy demand was somewhat checked due to an improvement in the energy demand in Southeast Asia, which was 3.4%.
- An analysis of the consumption pattern suggests that during this period, the consumption rate was quite low in North America, Europe, and the Commonwealth of Independent States (CIS). The US and Germany posted the largest decline. South and Central America displayed a below-average trend. However, the consumption pattern in the case of Africa, the Middle East, and Asia posted growth rates in line with historical averages. China was the biggest driver in the growth of primary energy. India and Indonesia were the next two contributors (BP Energy Review, 2020, p. 3).
- Impact of COVID-19 – There is a need to highlight the new realities on account of COVID-19. It is a fact that many of the developed nations used COVID-19 to arm-twist those developing/underdeveloped countries that were not in a position to fight the Pandemic on their own. India's response to such an approach of the developed countries is a case study. India enhanced indigenous capabilities to fight the Pandemic. One of the important lessons to enhance the resilience of the affected countries implied in the response of India shifted her focus from globalization to localization. The important contributory

factor was the sustained availability of energy. Thus, energy security assumes greater criticality for countries that are dependent substantially on the import of energy-bearing resources.

1.2 INDICATORS FOR FUTURE

- It is assumed that the standard of living of people in developing countries will continue to improve with the growing prosperity. The implication of such a phenomenon is that the energy demand in these countries would increase, though the inequalities in energy consumption and access to energy within the countries will continue to persist.
- The existing technologies to exploit conventional energy-bearing resources are reaching a plateau in development, and also, the prospected quantities of known energy-bearing resources are alarmingly becoming lesser. As such, they would increasingly become inadequate to meet the energy demand of the growing population of the world. This gap in demand and supply would result in the spiraling of the cost of energy. The denial regimes on account of geopolitical considerations will add to the degree of difficulty in getting access to energy by the affected countries. In years ahead, Climate change will further call for the identification of new low-carbon energy sources and the development of efficient and cost-effective technologies for their exploitation to meet the growing demand with a view to ensuring a smooth energy supply system. It may be noted that the energy supply system is undergoing transformation, and future systems will be such that fossil fuels will gradually be replaced by renewables. By an estimate, the use of renewables, less hydro, is likely to enhance 10 times, and that would be at the expense of hydrocarbons, whose share of primary energy is likely to decline from the datum of close to 85% in 2018 to around 40% by 2050. While the relevance of Natural Gas is unlikely to reduce (BP Energy Outlook, 2020, p. 77), the relevance of Coal, on the other hand, is likely to reduce due to its ill effects on the environment (BP Energy Outlook, 2020, p. 65). As far as hydroenergy is concerned, its growth as an energy source is likely to slow down. Two of the major factors for such a slowdown would be, firstly, the impact of climate change on the availability of hydro resources itself and, secondly, reduced land availability due to population growth to build large hydro projects (BP Energy Outlook, 2020,

p. 92). However, in the case of South Asia, cooperation in harnessing the hydro resources of trans-national channels in the region will help substantially in meeting the energy needs of the region.

- The world will see greater industrialization, resulting in more rural folks moving into urban slums in search of jobs. This new pattern of lifestyle will generate more quantities of waste, resulting in the contamination of water bodies and other related health issues. In recent years, technologies have been developed that not only help in generating substantial quantities of energy using waste but also help in waste management. It is felt that the conversion of waste to energy will play an important role in meeting the growing energy demand in the years ahead. In this connection, it is relevant to mention that India alone has the potential of 5,690 MW (An Overview by the Ministry of New and Renewable Energy). Similar would be the potential of the other countries of the Region, too, because of the higher density of the population and almost identical weather conditions prevailing in the region. Cooperation in this field of energy generation would be highly beneficial for the Region in more than one way.
- Hydrogen (BP Energy Outlook, 2020, p. 104) and nuclear conversion will be equally important (BP Energy Outlook, 2020, p. 92) in the energy matrix of the future. Indian resources (Monazite sands: Thorium) to generate nuclear energy will get a boost in India only if the second stage of nuclear conversion (fast breeder reactor-based system) gets operationalized.
- There will be more emphasis on electricity in the future for energy utilization, especially generated from a fully decarbonized (net negative) source of the power sector (BP Energy Outlook, 2020, p. 124). It is estimated that by 2050, electricity is likely to account for 50% of the consumption (BP Energy Outlook, 2020, p. 117). In a relative development, the demand for oil is likely to go down over the next 30 years. The pace of this decline and its scope would be determined by the relative growth of the use of electricity in the domain of power generation and road-based transportation. By an estimate, the number of electric vehicles on the road is projected to grow from 3,269,671 units in 2019 to 26,951,318 units by 2030 (Electric Vehicle Market by Vehicle), subject to the availability of an adequate quantity of Lithium, which is essential for manufacturing the battery for the electric cars. The criticality of lithium makes the countries like Afghanistan (Iain & Enda, 2021), Chile, Argentina, Australia, and China (https://

www.volkswagenag.com/en/news/stories/2020/03/lithium-mining-what-you-should-know-about-the-contentious-issue.html#) having adequate quantity of lithium very important in the supply chain of the batteries.

- A transition to a lower carbon energy system is likely to lead to a fundamental restructuring of the global energy system, with a more diverse energy mix, greater consumer choice, and more localized energy markets.
- In view of the above, an optimal energy management system will therefore be needed, which will be based on international cooperation and an integrated energy management system from exploration to utilization in respect of each of the countries.

1.3 SOUTH ASIA: A NEW GROWTH AREA

South Asia consists of Afghanistan, Pakistan, India, Nepal, Bhutan, Bangladesh, Maldives, and Sri Lanka. Some of the peculiarities of the region are as follows:

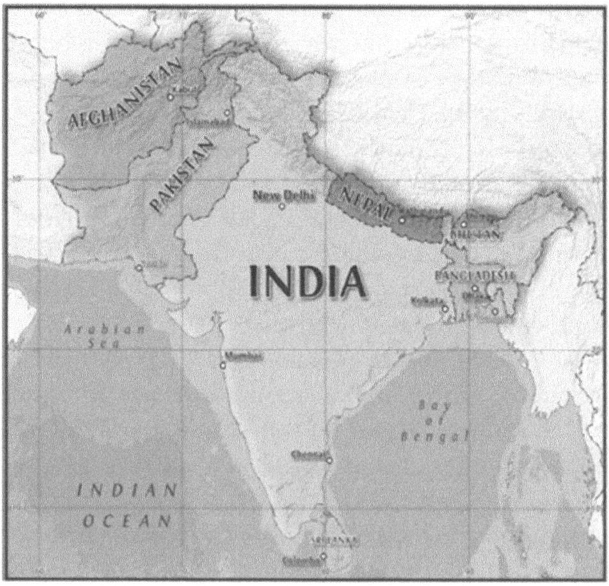

MAP-1: South Asia.
Source: Made from Open source and GIS data.

- The region is home to a wide diversity of geography. The region is home to a number of glaciers, rainforests, grasslands, deserts, and a large number of rivers. The region is surrounded by three water bodies: the Bay of Bengal, the Indian Ocean, and the Arabian Sea. These are known for having warm waters – all these features, along with varied climate zones, lend themselves well to power generation and exploitation of energy associated with these features.
- The total area of South Asia is about 5.2 million km^2, which is 11.71% of the total area of the Asian continent or 3.5% of the world's land surface area (https://en.wikipedia.org/wiki/South_Asia#cite_note-unreports.studycenters-8). The population of South Asia is about 1.891 billion or about one-fourth of the world's population. This makes the region one of the most populous regions in the world with a very high population density in the world (South Asia Regional Overview) – more population means more energy needs.
- The human development index (HDI) of the countries in the region is tabulated in Table 1.1.

TABLE 1.1 State of HDI of Countries in the South Asia

SL. No	Country	HDI	World Rank
1.	Afghanistan	0.511	169
2.	Bangladesh	0.632	133
3.	Bhutan	0.654	129
4.	India	0.645	131
5.	Nepal	0.602	142
6.	Pakistan	0.557	154
7.	Maldives	0.740	95
8.	Sri Lanka	0.782	72

Source: Data from the UNDP Report of Human Development Report, 2020.

It needs to be noted that except for Sri Lanka and Maldives, all other countries of the region have medium to low HDI. Therefore, the energy needs of the region are likely to grow much more than what is there in the current state of energy consumption in the region.

- As per the World Bank, the average GDP per capita of all the countries in the world is estimated to be $18,381 (data.worldbank.org). The Correlation between the GDP and the energy consumption can be understood with the analysis of Table 1.2.

TABLE 1.2 Correlation between Energy Consumption and the GDP

Country	Energy Consumption in MTOE Ref: https://www.ourworldindata.org/energy/countr/	GDP Per Capita in US$ in 2019 Ref: World Eco Outlook-GDP current prices per capita IMF Oct. 2019	GDP Growth Rate (2017) Ref: World Economic Outlook (Apr. 2017) The Real GDP Growth IMF: 26 Aug. 2017
Afghanistan	85.98 (data as of 2016)	513	3
Nepal	146.17 (data as of 2018/19)	1,048	7.7
Bangladesh	257.95 (data as of 2018/19)	2,104	4.5
Pakistan	386.83 (data as of 2016)	1,568	2.1
Sri Lanka	492.92 (-do-)	3,947	3.0
India	601.89 (-do-)	2,172	7.0
Maldives	1203.78 (data as of 2018/19)	15,563	4.1
Bhutan	2665.52 (-do-)	3,423	5.9

Two observations from Table 1.2 are relevant. Firstly, the growth in GDP can be supported by higher energy consumption. Secondly, except for Bhutan and Maldives, the energy consumption per capita is low. However, the trend for GDP growth is there for all the countries, and that suggests that there would be a need for additional energy in times to come to improve the quality of life.

- A study of the geography of the region brings out that many of the natural resources of the region are monolithic in nature. A case in point is the Lay of the rivers in the region. Most of the rivers are channels that have a trans-border presence. Also, the region is endowed with extensive forest cover, and as such, the scope for exploitation of available biomass is substantial. Similarly, scope exists for the exploitation of other renewables too. Cooperation and collaboration among the countries of the region to exploit the common resources and related research and development (R&D) can substantially reduce the dependence of countries of the region on energy resources from outside the region.

- However, it is also true that there is a wide variation in commercial energy resource endowments and commercial energy demand among the South Asian Countries (SACs). India, Pakistan, and Bangladesh have natural gas and coal. Bhutan and Nepal have large hydropower resources. Afghanistan is endowed with hydro resources (63.7%) and a limited quantity of fossil fuels (35.3%) (https://energypedia.info/wiki/Afghanistan_Energy_Situation#toc). Primary sources of energy for Sri Lanka are Coal, fuel oil, hydro, and renewables (solar and wind) (www.export.gov). Maldives, though, has scope for renewables (wind and solar) but essentially depends on the import of diesel. All the countries have vast renewable energy (RE) potential, and the sharing of these resources and related technology will lead to optimal utilization of resources.

1.4 RESOURCE APPRAISAL OF THE SOUTH ASIAN COUNTRIES (SAC)

1.4.1 AFGHANISTAN

1. In 2016, there was a wide gap between the production and consumption of electricity. While the production was 1.21 billion kWh, the demand was 5.526 billion kWh. However, this demand-supply gap was bridged by resorting to importing from Central Asia, and as such, 99% of the population could get access to electricity by 2018. Likewise, in the case of petroleum products, the import dependence is total because, against the nil, indigenous production demand in 2016 was 35,000 bbl/day. Luckily, natural gas is not imported, though its consumption was reported to be 164.2 million cum in 2017 (https://www.cia.gov/the-world-factbook/countries/afghanistan).

2. **Total Installed Generation Capacity:** 525 MW, including 254 MW from hydropower, 200 MW (39%) from thermal sources (furnace oil, diesel, and gas), and 65 MW (12%) from distributed generators. Import constitutes nearly 80% of the total power balance (https://www.adb.org/sites/default/files/linked-documents/47282-001-ssa.pdf).

3. Potential for energy generation in Afghanistan (Table 1.3).

TABLE 1.3 Potential for Energy Generation in Afghanistan

SL. No.	Renewable Resource	Potential	Status	References
1.	Renewable energy (https://energypedia.info/wiki/Afghanistan_Energy_Situation#toc) -Hydro	23,000 MW	• A number of rivers are trans-national, but there is no treaty with neighboring countries for equitable sharing. • Minimally exploited	https://energypedia.info/wiki/Afghanistan_Energy_Situation#toc
2.	Renewable energy (https://energypedia.info/wiki/Afghanistan_Energy_Situation#toc) -Wind	67,000 MW	Minimally exploited	https://energypedia.info/wiki/Afghanistan_Energy_Situation#toc
3.	Renewable energy -Solar	2,22,000 MW	Minimally exploited	https://energypedia.info/wiki/Afghanistan_Energy_Situation#toc
4.	Coal	250 MW	Located in Bamyan Province but not exploited	https://energypedia.info/wiki/Afghanistan_Energy_Situation#toc
5.	Gas	• 75 bcm • 444 bcm	• Proven reserve • Technically recoverable reserve	https://energypedia.info/wiki/Afghanistan_Energy_Situation#toc

4. **Regional Energy Corridors:** Afghanistan is located next to Central Asian Republics (CARs), which are rich in energy-bearing resources, and that is why there is a potential for getting access to energy/power from there, and such connectivity will also benefit countries like Pakistan and India. Some of the likely programs are as follows:

   i. **Turkmenistan–Afghanistan–Pakistan–India Gas Pipeline (TAPI):** It is an 1814 km pipeline starting from the Galkynysh gas field of Turkmenistan. Thereafter, it transits through Afghanistan and Pakistan and finally ends in India. Once completed, it will transport 33 bcm of gas for 30 years, with a distribution of 42% each to Pakistan and 16% to Afghanistan. Afghanistan will also receive $400 million/year as the transit fee. Work on this pipeline is in progress. However, the scheduled date of completion is 2022 but is likely to get delayed (https://

www.hydrocarbons-technology.com/projects/turkmenistan-afghanistan-pakistan-india-tapi-gas-pipeline-project/).

MAP-2: Turkmenistan–Afghanistan–Pakistan–India (TAPI) pipeline project proposed map
Source: Reprinted with permission from Danish et al., 2020. https://creativecommons.org/licenses/by/4.0/

ii. **Turkmenistan-Uzbekistan-Tajikistan-Pakistan (TUTAP) Grid:** It is an implementation of a regional interconnection concept through an Asian Development Bank (ADB) funded project for the development of a high-voltage direct-current (HVDC) power grid which will connect energy-rich CARs with energy-poor South Asia[1] which transits through Afghanistan (https://sites.google.com/site/iceafghanistan/regional-projects-and-masterplanning---overview/tutap). It is anticipated that this project will help to generate electrical trade between the initiators Kyrgyzstan and Uzbekistan and the beneficiaries namely Afghanistan and Pakistan to the tune of 1,500 MW (https://energypedia.info/wiki/Afghanistan_Energy_Situation#toc).

[1]Under the provisions of a framework agreement for the proposed Turkmenistan–Afghanistan–Pakistan (TAP) power interconnection project was signed in Feb. 2018. It was basically to support Afghanistan's energy needs and enable power trade between the three countries. The project is expected to be completed by 2022. Once completed, it will have a max of 4,000 MW getting transferred from Turkmenistan to Afghanistan and Pakistan.

MAP-3: Map made with data from USAID South Asia Regional Initiative for Energy Integration

Source: Adapted from USAID South Asia Regional Initiative for Energy Integration repository.

iii. **Central Asia-South Asia (CASA) Electrical Grid:** This electricity link of 1,200 km having a capacity of 1,300 MW is known as CASA-1000. It will connect Kyrgyzstan, Tajikistan, Afghanistan, and Pakistan. The objective of this link is to transfer surplus electricity from Central Asia to energy-deficient Afghanistan and Pakistan in South Asia. The share of Tajikistan in the supply of power is 75%, while that of Kyrgyzstan is 25%. Of the total power received, Afghanistan will receive 300 MW of electricity, and Pakistan will receive the remaining 1,000 MW (Zahid, 2016).

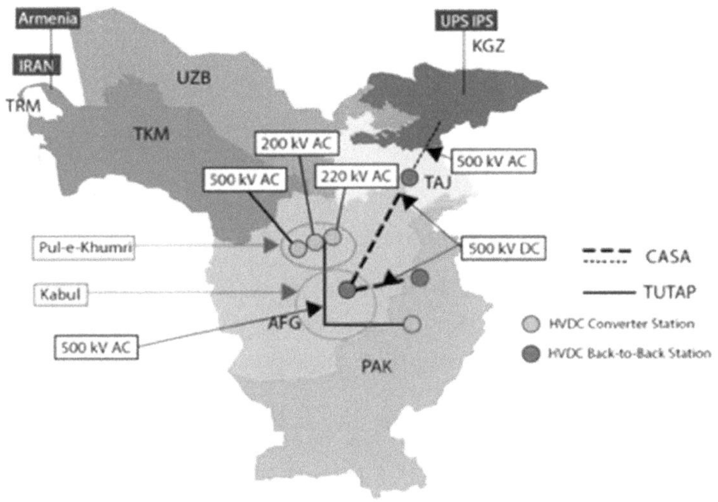

MAP-4: CASA-1000.

Source: USAID South Asia Regional Initiative for Energy Integration repository. https://sari-energy.org/program-activities/cross-border-electricity-trade/country-energy-data/

1.4.2 PAKISTAN

- **Indicators of Energy Usage:** Pakistan is a thickly populated country, having 2.56% of the world's population. The gap between demand and supply of electricity in 2009–2010 was 26.82%, which had gone up to 50% by 2012 (Farooqui, 2014). The current data with respect to electricity is as follows (https://www.cia.gov/the-world-factbook/countries/Pakistan/#energy).
  - **Installed Electricity Generating Capacity:** 26.9 million kW. There is a yearly increase in demand to the tune of 8–10%. Source-wise contribution is as follows:
    - **Fossil Fuels:** 62% (2016 estimate).
    - **Nuclear Fuel:** 5% (2017 estimate).
    - **Hydroelectrical:** 27% (2017 estimate).
    - **Other Renewables:** 7%.
    - **Electricity Production:** 109.6 billion kWh (2016 estimate).
    - **Electricity Consumption:** 92.33 billion kWh (2016 estimate).
    - **Electricity Import:** 490 million kWh (2016 estimate).
    - **Electricity Access:** Despite the above data, it is limited to only 79% of the population (2019 estimate).

Other indigenous sources of energy, their availability, and their status are as follows:

- **Crude Oil:**
  - o **Proven Reserve:** 332.2 million bbl (38%) (as estimated in Jan 2018).
  - o **Daily Production:** 90,000 bbl (estimate of 2018).
  - o **Import:** 1,68,200 bbl/day.
- **Refined Petroleum Products:** Production/Consumption in terms of bbl/day – 2,91,200/5,57,000 bbl/day (https://www.hydrocarbons-technology.com/projects/turkmenistan-afghanistan-pakistan-india-tapi-gas-pipeline-project/).
- **Natural Gas:**
  - o **Proven Reserve:** 588.8 bcm (27%).
  - o **Production/Requirement in Terms of bcm:** 39.05/45.05 (2017 estimate) (ibid. p. xxvii).
- **Hydro:** 32% (Existing capacity). Pakistan has a substantial potential for hydroenergy. The total hydropower generation capacity possible in the case of Pakistan in 2010 was 6,720 MW (ibid. p. xxvi), which went up to 7,320 MW in 2017, and by 2019, it had further improved to 9,827 MW (Country profile of Pakistan by IHA). A limiting factor in enhancing the exploitation of hydroenergy is the storage capacity of Pakistan, which currently is 18,502,227,829.8 m^3. This is sufficient to store only 13% of the annual river flow (Quereshi & Akintug, 2014). One of the issues with the storage capacity is the silting of reservoirs. Therefore, a technology-based desilting system needs to be part of storage management.
- **Coal:** Pakistan has a coal reserve of 184 billion tons, which works out to be 16.1% of the world. It is concentrated in Sindh, Balochistan, and Punjab. Till 2015, it contributed to the energy mix of Pakistan only 7%. However, it is now planned that by 2030, it will contribute to as much as 30% of the energy mix of the country (Farid & Engr, 2015).
- **Solar:** The total landmass of Pakistan is 7,96,096 km^2. The Latitude and longitude are latitudes 24° and 37° North and longitudes 62° and 75° East, respectively (Farooq & Kumar, 2013). This location helps Pakistan to get about 300 days of sunshine (Asif, 2009). As per the assessment of the Alternative Energy Development Board (AEDB) of Pakistan, Pakistan technically has a potential of about 1,600,000

MW of solar energy. Some of the areas inhibiting the full exploitation of this capacity are social prejudices, inadequate government policies to exploit this abundant resource, inadequate capacity of the local R&D organizations to develop the necessary technology, and finally, the limitation of the indigenous industry (Alauddin, 2012).

- **Biomass:** AEDB of Pakistan and the National Renewable Energy Laboratory (NREL) of the USA estimate that a potential of 1,800 MW exists from Sugarcane Bagasse (Mirza et al., n.d.). Pakistan also has a potential of 4,00,000 tons of ethanol/year and, as such, has adequate potential for biodiesel (Farooq & Hameed, 2012).
- **Waste to Energy:** Pakistan generates about 13–15 million cubic meters/day of municipal waste, which can be exploited to produce biogas using 'Waste to Energy' technology.

1.4.3 COOPERATION WITH IRAN AND CENTRAL ASIA

As explained in the section on Afghanistan, work is in progress on energy cooperation with Central Asia for TAPI, TUTAP, and CASA.

1. **Prospects of Iran-Pak Energy Cooperation:** Iran has an abundance of oil and gas, which can easily be transported to Pakistan due to geographical contiguity. The most significant area of cooperation is the supply of natural gas. In this connection, an agreement, commonly known as the Iran-Pakistan Gas pipeline, was conceptualized in 1994 with a view to transporting 750 million cubic feet of gas per day. The cost of the project was $7.5 billion. At some point in time, even India was interested in the project. However, she withdrew later, citing security concerns and the cost. The agreement was finally signed in 2013 but has still not been implemented, probably due to high cost. A ray of hope has emerged after the signing of an MoU between Iran and China, which stipulates that China will invest $400 billion in the next 25 years in the energy sector of Iran (Muhammad, n.d.).
2. **Inhibitors to the Growth of the Indigenous Energy Industry:** There are a number of issues that inhibit the growth of the indigenous industry in Pakistan, both in terms of GDP and energy consumption. The government of Pakistan wants to enhance the share of RE to 5% by 2030 (Khan & Pervaiz, 2013), which in real terms will be 10,000 MW and another 50,000 MW through the exploitation of wind energy through the wind-based energy systems (Pakistan Meteorological

Department (PMD), 2009). However, the indigenous industry is still not well developed to support this vision and may need substantial foreign support to realize this vision.

- *Note:*
  1. Indigenous sources Pakistan is quite capable of enhancing its power generation capacity substantially. However, that may need substantial foreign funding.
  2. It also needs to be noted that Pakistan can act as a conduit for the import of energy from energy-rich Iran, provided there is mutual trust between India and Pakistan about the construction of the Iran–Pakistan–India Pipeline, which is currently in a state of suspended animation without India's participation[2].

1.4.4 INDIA

1. **Indicators of Power Usage:**
   i. **Electricity (https://www.cia.gov/the-world-factbook/countries/india/)**
      a. **Electricity Production/Consumption:** 1.386/1.137 trillion kWh (2016 estimate).
      b. **Electricity Access:** 99% of the total population (as of 2019).
      c. **Electricity Export/Import:** 5.15/5.617 billion kWh (2015 estimate). Beneficiaries of exports are Nepal, Bangladesh, and Myanmar. Import to India is mainly from Bhutan (https://pib.gov.in/PressReleasePage.aspx?PRID=1607177).

   ii. **Source of Electricity:**
      a. **Fossil Fuel:** 71% (2016 Estimate).
      b. **Nuclear Fuel:** 2% (2017 estimate).
      c. **Hydroelectric:** 12% (2017 estimate).
      d. **Renewables:** 16% (2017 estimate).

[2]Iran–Pakistan–India Gas pipeline – This $7.5 billion was proposed in 1994 to bring gas from the South Pars Gas field in Iran to India via Pakistan. India left the project in 2009, and Iran and Pakistan signed the deal to go ahead. However, the project is not yet complete and unlikely, too, because of the financial problems of Pakistan. In 2017, an Indian Parliamentary panel recommended reviving it. The biggest impediment to its progress is the mutual distrust between India and Pakistan and probably the US sanctions against Iran. It needs to be noted that it will be a win-win proposal for all three countries. Iran will find buyers for her gas. Pakistan will not only get much-required gas but will also earn revenue to improve its economic condition, and India will get an alternative option to get gas, which is badly needed to spur economic activity in India.

2. **Indigenous Sources:**

   i. **Crude Oil: Production/Consumption/Proven Reserve/Import in Million Tons (2018 Estimates) (https://en.wikipedia.org/wiki/Oil_and_gas_industry_in_India):** 35.2/204.9/594.4/228.6 – these figures flag inadequacy of the indigenous production to meet the demand and therefore either alternative solutions will have to be found out to reduce import dependence or ways and means to be found to enhance the indigenous production. In this connection, it needs to be noted that the entire Landmass, as well as the offshore area of India comprising 3.36 million km^2, is spread over 26 sedimentary basins. Assessment of the entire area has now been completed, and it is encouraging to note that the assessment with respect to the availability of hydrocarbons has improved by 49.1% (https://online.dghindia.org/oalp/files/pdf/OALP5_MUMBAI_overview_of_Prospectivity_of_Indian_Sedimentary_Basins.pdf).

   ii. **Refined Petroleum Products: Production/Consumption in Terms of Million Metric Ton (MMT) for the Years 2017–2018–254.40/204.92 (Indian Petroleum & Natural Gas Statistics 2017–2018):** The surplus of refined products is meant to be exported to Singapore, United Arab Emirates, Netherland, Malaysia, United States, Israel, and Nepal (Bilal, 2019). India is planning to further strengthen its position as a leading refiner in the world. As per plans, by 2030, the refining capacity is likely to increase to 355 MMT, and by 2040, it will become 472 MMT (A PTI Report, 2020).

   iii. **Natural Gas: Recoverable Gas/Indigenous Production/Internal Consumption in Terms of bcm for 2017–2018 Period–1339.57/32.65/52.83 (Pakistan Meteorological Department (PMD), 2009):** This stat brings out that for its internal use, India depends on import currently, but with extra investment in infrastructure to exploit indigenous resources in medium to long term, it would be possible to reduce import dependence.

   iv. **Hydro Power:** India's economically exploitable hydroelectric potential is estimated to be 1,48,701 MW (https://cea.nic.in/reports/monthly/hydro/2016/hydro_potential_region-02.pdf). In addition to this potential, an additional exploitable potential through small hydro plants (SHP) of 6,780 MW (Shreya, 2016) exists. However, India currently has an installed capacity of only 46,000 MW (12.3% of the total power generation capacity of the country) (ET Energy

World Report, 2020). Although it is the fifth largest in the world, much more can be done to exploit the existing potential. It may be noted that an additional capacity of 4,883 MW has also been built, based on SHP, which can further be augmented based on the existing potential (A Central Electric Authority Report, 2020).

v. **Coal:** These projects have a share of 50% of the present installed capacity of power generation (An IANS Report, 2021). As per the 19th Electric Power Survey, it is estimated that the total installed capacity for power generation by 2026–2027 will be 6,19,066 MW. Out of this capacity, coal will have a share of around 2,38,150 MW (40% of the installed capacity). In this connection, it is relevant to note that the Indian coal is of poor quality, having as high as 30–50% ash content. In addition, over a period of time, the calorific value of coal in India has deteriorated due to better quality of coal reserves having been depleted. The calorific value of Indian coal currently ranges between 3,345 kcal to 4,703 kcal. This poses significant challenges (Craig & Mar, 2007). To an extent, it is handled by washing (A Report by NITI Aayog of India, 2020) or alternatively through import substitution. In the financial year 2021–2022, India imported close to 180–190 million tons of coal, which is likely to go up to 250 million tons in FY 2022–2023 (Shobha, 2022).

vi. **Renewables (An Envecologic Report, 2017):** India has an estimated RE potential of about 900 GW from sources like Wind – 102 GW, Bio-energy – 25 GW, Small Hydro – 20 GW, and Solar power – 750 GW. As of March 2017, the total installed capacity of RE was 57,260 MW, with a growth rate of 20% in the last five years. This works out to be 15.90% of the total installed capacity. Presently, wind energy dominates India's RE industry, accounting for 29151.29 MW by March 2017. In the domain of Wind, The Ministry of New and Renewable Energy (MNRE) has been authorized to explore and promote the deployment of offshore wind farms in the exclusive economic zone (EEZ) of the country. The National Institute of Wind Energy (NIWE) has been authorized as the Nodal Agency for the development of offshore wind energy in the country and to carry out the allocation of offshore wind energy blocks, coordination, and allied functions with related ministries and agencies. It is planned to set up the first offshore wind power project off the Gujarat coast soon. To spur the growth of solar energy, the Govt of India has established a National Solar Mission. The objectives of this mission entail the reduction of the cost of solar power generation in the country through long-term policy, large-scale deployment, aggressive R&D and the domestic production of critical components, and encouragement of foreign investment through the aegis of a dedicated financial institution – the Indian Renewable Energy Development Agency (IREDA). The mission also works on the skill development of the workforce. The Govt of India has also set up the Solar Energy Corporation with a view to facilitate the implementation of the National Solar Mission and the achievement of targets set therein. A policy for grid-connected Solar Roof-top Projects is being aggressively implemented in a number of states/UTs. It is heartening to see that the government power surveys are indicating that a steady rise in the RE capacity in the country is set to become the largest energy provider in the country over the next five years. It is set to increase from current levels of just about 1,00,000 MW to 175 GW by 2022 and over 2,75,000 MW by 2026–2027. This would be much more than coal capacity. Finally, it is planned to reach 450 GW (An Energy Policy Review by IEA, India, 2020). Increasingly, RE has started becoming cost-competitive as compared to fossil fuel. It, however, needs to be noted that in India, the growth areas for different forms of renewables are different, depending on the geography of the area and weather conditions. Having done a detailed analysis of the ways and means to enhance the renewable capacity, the Government of India is targeting to achieve a capacity of 450 GW by 2030 (Anmar, 2021).

vii. **Nuclear Energy:** Nuclear power is the fifth largest source of electricity in India, after coal, gas, hydro, and wind. As of November 2020, India had 22 nuclear reactors in operation in seven nuclear power plants, with a total installed capacity of 6,780 MW (A CEA Report, 2018). Seven more nuclear power reactors are under construction with a combined generation capacity of 4,300 MW. A plan was drawn in 2010 for India to reach a nuclear power capacity of 63 GW by 2032 (A PTI Report, 2010). However, the Fukushima nuclear disaster[3]Japan in 2011 gave a setback to the

[3]Fukushima Daiichi Nuclear Disaster – The Fukushima Daiichi nuclear disaster was a 2011 nuclear

growth of nuclear power world over (Sidharth, 2011). One of the problems of nuclear power in India is the 'low-capacity factor' of Indian Plants. Although it has shown improvement in recent years, it is consistently 74.4% these days, which is much lower than the global average of 82.5% in 2019 (Energy Availability Factor). One of the main reasons for the lower capacity factor (CF) is a shortage of nuclear fuel. Uranium deposits are known to exist in the Singhbhum and Hazaribagh districts of Jharkhand, the Gaya district of Bihar, and in the sedimentary rocks in the Saharanpur district of Uttar Pradesh. Some quantity of Uranium is also found in the copper mines near Udaipur in Rajasthan. A major source of Uranium is Monazite sands, which occur on the East and West Coast and some places in Bihar. The largest concentration, however, is found on the Kerala Coast. Around 15,200 tons of Uranium is estimated to be contained within Monazite sand in India. The total reserve of uranium is estimated to be 30,480 tons, which works out to be 2% of the World. Monazite sand is known to have a substantial quantity of Thorium. The Thorium reserves of India are estimated to be in the range of 4,57,000 and 5,08,000 tons (Ur & Th Distribution Across India and the World). The main locations where it exists are Kerala, Jharkhand, Bihar, Tamil Nadu, and Rajasthan. Beryllium is another element whose oxide is necessary for use as a moderator in atomic power generation. India has sufficient stock of beryllium. India has been making advances in the field of thorium-based fuels, working to design and develop a prototype for an atomic reactor using Thorium and low-enriched uranium, a key part of India's three-stage nuclear power program (Lisa, 2009). Prototype fast breeder reactor for producing 500 MWe being constructed at Madras Atomic Power Station Kalpakkam. There are plans to make six more such reactors after the completion of this reactor. Although it was supposed to have attained criticality in Dec. 2021, but it has still not gone critical (Ramachandran, 2020). It has now been mentioned in the Parliament by the GoI that it was presently in the Integrated Commissioning Stage (Lok Sabha Unstarred Question No.: 330).

---

accident at the Fukushima Daiichi Nuclear Power Plant in Okuma, Fukushima Prefecture, Japan. The accident was caused by the 2011 Tohoku earthquake and tsunami. It was the most severe nuclear accident since the Chornobyl disaster in 1986. It was classified as Level-7 on the International Nuclear Event Scale (INES). A total of 1,54,000 people were evacuated.

Despite the handicap of fuel supply and international geopolitical resistance, India's nuclear program is quite well developed, and India can technically help other countries of South Asia to go for nuclear power.

3. **Indian Assets Abroad:** Oil and Natural Gas Corporation (ONGC) has a vertical being called ONGC Videsh (OVL). It manages its assets located abroad. These are 14 in number, spread over a number of countries, and contribute substantially to the national kitty. In 2018–2019, its contribution was 12.8 MMTOE during 2018–2019 (Sanjeev, 2020). In addition to OVL, other state-run indigenous-producing agencies like OIL, IOC, and BPCL have also been contributing to the national kitty. Although, in recent times, there has been a reduction in output (OVL itself experienced a reduction of 11% in its output as against the output of 2017–2018), it is mainly a function of the current geopolitics of oil. What needs to be noted is that substantial potential exists, and as the financial muscle of OVL improves, more assets will be owned by OVL, and it will graduate to be a major player in the domain of Oil and Gas in the world.
4. **International Cooperation for the Exploitation of Hydro Resources:** Enough scope exists for the cooperation to exploit the common hydro resources with Bangladesh, Bhutan, and Nepal, respectively. There is also a possibility of transfer of hydropower through an HVDC connection from Tajikistan. In this connection, it is relevant to note that Tajikistan has invited India to explore the huge hydropower potential in that country as part of efforts to intensify bilateral economic cooperation and explore new areas of partnership between the two sides (PTI Report).
5. **Cooperation with Myanmar:** India has been a key stakeholder in the Myanmar–China gas pipeline. The potential exists for cooperation with Myanmar. In recent years, Myanmar has structured its energy sector and has identified gas as a major component of its energy strategy. It is believed that she is calling for fresh bids for the development of oil and gas blocks. Based on its experience working in the energy sector of Myanmar, India is well-positioned to bid for these blocks. It needs to be noted that this cooperation will also auger well for energy cooperation with other countries of South Asia, especially Bangladesh. It has been

opined by Dr. Mirza Sadaqat Huda, an energy expert based in Singapore in 2013, that it would be of great benefit to Myanmar, Bangladesh, and India if they join hands to build a trilateral pipeline to meet their energy needs (Saket, 2019). It is, therefore, of mutual interest and in the interest of the region that India should engage proactively with the Myanmar government and present a commercially viable case for a greater share in Myanmar's energy development. It would be pertinent to note that cooperation in the energy sector between India and Myanmar will also open new vistas for cooperation in the domain of building infrastructure and improving connectivity.

6. **State of Future Energy Mix of India:** Total Installed Capacity of the country as of 30.09.2019 was 363.4 GW. The contributors to this capacity are hydro: 45.4 GW; thermal: 228.6 GW; renewables: 82.6 GW; and nuclear: 6.8 GW (A Report on Optimal Generation Capacity Mix for 2029–2030). As per the recent study carried out by the Central Electricity Authority (CEA), the total installed capacity by 2029–2030 is likely to be 8,17,254 MW, and the contributors would be coal: 2,66,911 MW, gas: 25,080 MW, hydro: 71,128 MW, renewables: 4,35,155 MW and nuclear: 18,980 MW. The breakdown is tabulated in Table 1.4.

TABLE 1.4 Likely Installed Capacity by the End of 2029–2030

Fuel Type	Capacity (MW) in 2029–Percentage Mix 30 (%)	
Hydro *	60,977	7.46%
Pumped storage plant	10,151	1.24%
Small hydro	5,000	0.61%
Coal + Lignite	2,66,911	32.66%
Gas	25,080	3.07%
Nuclear	18,980	2.32%
Solar	2,80,155	34.28%
Wind	1,40,000	17.13%
Biomass	10,000	1.22%
Total	8,17,254	
Battery energy storage#	27,000 MW/1,08,000 MWh	

*Including hydro imports of 5,856 MW # Active Battery Storage.
Source: Adapted from Central Electricity Authority, 2020.

7. **Hydrogen: Fuel for the Future:** India has already started the work on hydrogen fuel cells. Presently, the bulk of hydrogen generation is based on fossil fuels, which is referred to as blue hydrogen. As per The International Energy Agency (IEA), hydrogen produced by this route adds roughly 830 MMT per year of carbon dioxide (CO_2). On the contrary, hydrogen produced from renewable sources, known as 'Green Hydrogen,' is currently far less in quantity and is more expensive. A recent report by the New Delhi-based, 'The Energy and Resources Institute' (TERI)' highlighted that by 2050, nearly 80% of India's hydrogen would be 'green.' In a speech in November 2020, the Indian announced a 'Hydrogen Energy Mission.' Accordingly, during the budget presentation of the FY 2021–2022, the FM of India announced the formation of this mission.

➢ ***Note:*** India essentially has a problem which can be summed up as TQQ Syndrome[4]. Thus, India, though, is short of energy to support its growth needs but, with her advances in technology, is fairly well placed to support the R&D for the energy cooperation.

1.4.5 BANGLADESH

1. **Demonstrated Energy Data (https://www.cia.gov/the-world-factbook/countries/bangladesh/#energy):**
   i. **Electricity:**
      a. **Installed Generating Capacity:** 11.9 million kW (2016 est.).
      b. **Production/Consumption:** In billion kWh – 60.51 (2016 est.)/53.65 (2016 est.).
      c. **Sources to Produce Electricity:**
         - Fossil fuels – 97% of total installed capacity (2016 est.).
         - Hydroelectric plants – 2% of total installed capacity (2017 est.).

[4]TQQ Syndrome: India does not have either quantitatively or qualitatively resources for which indigenous technology is available. For example, India has a full cycle of technology for energy conversion to exploit petroleum, but resources are inadequate. Similarly, India has an adequate quantity of coal for exploitation, but qualitatively, it is inferior. India has very high solar radiation but suffers from low efficiency of conversion. India has been able to master the full cycle of nuclear conversion based on the pressurized heavy water-based conversion of U-235, but the availability of Uranium locally is limited. On the other hand, India has an adequate quantity of Thorium, but even the first proto-type fast breeder reactor has still not reached criticality. Hydro conversion suffers from space, environmental, and silt management issues.

- Other renewable sources – 2% of total installed capacity (2017 est.).

d. **Population covered by Electrification** – 83% (2019).

ii. **Crude Oil:**

a. **Proved Reserves:** 28 million bbl (1 January 2018 est.).

b. **Production:** 3,000 bbl/day (2018 est.).

c. **Import:** 21,860 bbl/day (2015 est.).

iii. **Refined Petroleum Products**:

a. **Production/Consumption in bbl/Day:** 26,280 (2015 est.)/1,06,000 (2016 est.).

b. **Export/Import in bbl/Day:** 901 (2015 est.)/81,570 (2015 est.).

iv. **Natural Gas:**

a. **Production/Consumption in bcm:** 29.53 (2017 est.)/29.53 (2017 est.).

b. **Proved Reserves:** There have been 27 gas fields in Bangladesh till now. In 1993, the initial estimated recoverable gas reserve was around 12.43 trillion cubic feet (TCF). The amount became around 26.84 TCF by 2011. It increased slightly to 27.12 TCF at the end of 2017. From this amount, around 15.22 TCF gas has already been extracted. So, the remaining 12 TCF gas was only available for future use. As of 01 January 2018, it was established that Bangladesh has a reserve of only 185.8 bcm. Thus, it is projected that the country will be able to fulfill the growing demand for natural gas for the next 10–12 years only with the remaining gas reserves. There is, therefore, an urgent need to enhance the indigenous capacity by exploring new gas fields or go for import of gas, which could either be through an international pipeline or through a liquefied natural gas (LNG) route (Hassan & Moklesur, 2019).

2. **Additional Potential:**

i. **Hydro Power (Moirul et al., 2020):**

a. **Present State:** Although Bangladesh is endowed with plenty of rivers in most cases, it is lower riparian. As such, hydropower generation is not up to the mark. Kaptai Power Plant, built on Karnaphuli River in Rangamati District, having a capacity of 230 MW, is the only major operational plant. Even the Small-Scale Hydro Technology (SSHT) based generation contributes only 2.34% of the electricity generation. Two of the major issues are the geopolitics of water in South Asia and Bangladesh being the lower riparian suffers on account of this, and the second is the need for technology to exploit the existing hydroenergy sources.

b. **Potential:** Excessive rain offers a big opportunity to harness the power of the rainwater. Bangladesh Power Development Board (BPDB) is planning to establish a plant to generate 100 MW. River Brahmaputra, which is known as Jamuna in Bangladesh, has a potential of 1,400 MW for large-scale power generation. Plenty of potential for micro-hydro power exists. In Chittagong district alone, a potential of 135.6 MW exists. In addition, the Sangku and Matamuhari rivers have a potential of 300 GW/year and 200 GW/year, respectively. Plenty of scope exists to enhance power generation capacity in SSHT, provided technology is leveraged to optimize the output.

ii. **Renewable Energy (RE) (Viktor, 2021):**

a. **Potential:** Global Climatescope ranks Bangladesh 33 globally and claims that it has substantial unexplored potential. It is claimed by the experts that if the country utilizes only 1% of the Kaptai dam's surface area to install a solar floating system, it will result in generating 500 MW of solar energy. Similar unexplored potential is lying on the riverbanks and islands in the Meghna estuary. Similarly, industrial rooftop installation of solar panels adds to the potential another 5 GW of capacity. It has been estimated by the Sustainable and Renewable Energy Development Authority (SREDA) that Bangladesh has the potential to deploy solar systems to generate 30 GW of energy by 2041. NREL adds to this estimation by suggesting that 20,000 km^2 of land having wind speed in the range of 5.75–7.75 m/s can be utilized to generate another 30,000 MW by harnessing wind. As per a recent report investigating the RE, it has been forecasted that a technically feasible capacity of 156 GW of power by exploiting the incidence of solar radiation over a suitable area measuring 6,250 km^2 is feasible. Also, by harnessing prevailing wind conditions, another 150 GW of power can be generated.

b. **Current Status:** At the end of 2018, the total installed capacity of RE was 579 MW, which has gone up to 649 MW by July 2021. Although the country had set a target of accretion in the production of electricity through RE systems by 10% by the end of 2021, only 3% of electricity generation could be achieved by July 2021.

c. **Challenges:** The technology available for the exploitation of RE is quite limited, and full exploitation of the potential is possible only if technology up-gradation can be achieved.

3. **Cooperation within South Asia:** PM Modi of India visited Bangladesh on 25–26 Mar. 2021. Among other issues, cooperation in the energy sector was part of the agenda. The portion of the joint statement dealing with cooperation in the energy sector puts into perspective the agenda for cooperation. Salient aspects of the joint statement are as follows (Soumoy, 2021):
   i. Cooperation in the energy and power sector, including participation of the private sector on either side.
   ii. Indo-Bangladesh Friendship Pipeline.
   iii. Unit-I of Maitree Super Thermal Project.
   iv. Peaceful use of Nuclear Technology.

➢ *Note:*
   - This data suggests that there is only a limited scope for the export of gas. The exploitation of hydro resources is subject to tie up with India being a lower riparian. However, the exploitation of hydro resources in conjunction with India and Nepal presents a better picture and needs to be progressed.
   - There is definitely a scope for the exploitation of renewables to improve the power situation as well as earn revenue by exporting the generated surplus power.

1.4.6 NEPAL

1. **State of Electricity (https://www.cia.gov/the-world-factbook/countries/nepal/#energy):**
   i. **Installed Generating Capacity:** 9,43,100 kW (2016 estimate).
   ii. **Production/Consumption in Billion kWh (2016 Estimate):** 4.244/4.983.

iii. **Access:** 93% of the population (2019). This shows that the indigenous production is not adequate to meet the consumption. As such, there is a need to either enhance production based on indigenous resources or go for import.

iv. **Export/Import in Million kWh:** 2.69 (2017 estimate/2.175 (2016 estimate). Nepal imports electricity from India (Rastriya, 2020) and also exports to India.

v. **Sources for Energy Generation:**

   a. **Fossil Fuels:** 5% of the installed capacity (2016 estimate).

   b. **Hydroelectric Plants:** 92% of the installed capacity (2017 estimate). Here, it needs to be noted that the International Finance Corporation, World Bank Group, in its report, has mentioned that Nepal has the potential to generate upward of 50,000 MW, but the current capacity is merely 800 MW from 20 major hydropower plants and a number of small and micro-hydro power plants (A Report by International Finance Corporation). Thus, there is a need to support Nepal to exploit its full potential, which will be of great help to neighboring countries.

   c. **Other Renewable Sources:** 3% of installed capacity (2017 estimate). It substantially includes biomass and biogas. Harnessing biomass is not recommended as it causes health issues.

2. **Oil and Gas:** Neither produced nor any indigenous source exists.
3. **Refined Petroleum Products (PTI Report, 2009):** Although nothing is indigenously produced, consumption is 27,000 bbl/day (2016 estimate). The import is 26,120 bbl/day (2015 estimate). India is the sole source of supply to Nepal through tankers. Recently, a cross-border petroleum pipeline from Motihari in Bihar to Amlekh-gunj in Nepal was built, and it supplies two MMT/year of petroleum (Gopal, 2019).
4. **Potential for Export of Hydro Power (Deepak, 2019):** Nepal has the Potential to export hydropower from its fast-flowing rivers to the tune of a minimum of 40,000 MW. The exploitation of this capacity for exporting it to energy-deficient India has the potential to transform its economy. By an estimate, even if only 15,000 MW of this potential in the next 10 years is achieved, it will provide an additional revenue of $2.18 billion per annum to Nepal.

- *Note:* The above data suggests that Nepal has essentially only hydropower, but that itself has a great potential to earn revenue for It, and in this connection, the requirement of India and the potential of Nepal to supply complement each other and, as such, needs to be progressed.

1.4.7 BHUTAN

1. **Electricity (https://www.cia.gov/the-world-factbook/countries/bhutan/#energy):**
   i. **Production/Consumption in Billion kWh in 2016:** 7.883/2.184.
   ii. **Access:** 100% population (2020).
   iii. **Export/Import in kWh in 2016:** 5.763 billion/84 million.
   iv. **Installed Generation Capacity:** 1.632 million kW (2016 est.).
   v. **Source of Fuels:**
      a. **Fossil Fuels:** 1% of total installed capacity (2016 est.).
      b. **Hydroelectric Plants:** 99% of total installed capacity (2017 est.).
2. **Refined Petroleum Products:**
   i. **Production/Consumption:** 0/3000 bbl/day (2016 est.).
   ii. **Imports:** 3,120 bbl/day (2015 est.).
3. **Hydro Power Generation and Export:** Hydropower potential of Bhutan is 30,000 MW. Out of this, 23,760 MW has already been identified, and it is confirmed that it would be technically feasible to exploit it (Sonam & Bharat, n.d.). Efforts of Bhutan are bearing fruits, and the generation has recorded a growth of 14%, which is well beyond their needs. The country exports about 70% of the energy generated to India (A SASEC Report).

- *Note:* Bhutan is dependent on its needs for petroleum products but has a great potential for the generation of hydroelectrical power, which she is using effectively to generate revenue by exporting it to India. This is a very successful model and worthy of emulation by other countries in South Asia. Such a model has the potential of being a 'Win-Win' model, as the natural resources of the region are monolithic (Author's note).

1.4.8 SRI LANKA

1. **Electricity (https://www.cia.gov/the-world-factbook/countries/sri-lanka/#energy)**
   i. **Production/Consumption in Billion kWh (2016 Estimate):** 13.66/12.67-energy surplus country.
   ii. **Access:** 100% population (2019).
   iii. **Installed Generating Capacity:** 3.998 million kW (2016 estimate).
   iv. **Source of Generation (Of the Total Installed Capacity):**
      a. **Fossil Fuels:** 52% (2016 estimate).
      b. **Hydroelectric Plants:** 42% (2017 estimate).
      c. **Other Renewable Sources:** 6% (2017 estimate).

2. **Crude Oil Imports (PTI Report, 2009):** 33,540 bbl/day (2015 estimate).

3. **Refined Petroleum Products (PTI Report, 2009):**
   i. **Production/Consumption in bbl/Day:** 34,210 (2017 estimate)/1,16,000 (2016 estimate) – therefore, either refining capacity is increased or agreement with an energy-rich country will have to be made to import to bridge the gap between demand and supply.

   ii. **Export/Import in bbl/Day:** 3,871 (2015 estimate)/66,280 (2015 estimate).

4. **Energy Assessment of Sri Lanka: Availability and Potential (Aiming, 2019):** Petroleum, imported as crude oil and finished products, provides the highest share (43% in 2017) of energy to the national economy, followed by biomass (37%), coal (11%), hydro (6%), and new RE (3%). The details of contribution as well as the future potential of each form of resource are as follows:

   i. **Petroleum:** 43% of Sri Lanka's energy supply continues to depend on imports. About 30% of these imports are in the form of crude oil, which is processed at the Ceylon Petroleum Corporation (CPC) refinery. The balance requirement is directly imported as refined products. In 2011, the discovery of a gas field in the offshore region of the Mannar basin in the North-west of the country was reported. By 2015, two more positive gas discoveries were reported in the same region (http://www.prds-srilanka.com/exploration/mannarBasin.faces). The seismic study of the area pegs the findings to be holding a gas reserve of about 1 billion barrels of oil equivalent.

ii. **Biomass:** The Sri Lanka RE Master Plan estimates that the country has a potential for 2,400 MW of biomass-based generation capacity.
iii. **Hydropower:** Most of the major hydropower schemes are associated with Sri Lanka's two main rivers—Mahaweli and Kelani. Hydropower is primarily used for electricity generation. Almost all the potential has already been exploited by Ceylon Electricity Board (CEB) through large-scale power plants. In quantitative terms, it amounts to a total of 1,370 MW of large hydro and 20.5 MW of small hydro.
iv. **Wind:** The country has good wind energy resources that are concentrated in the northwestern coastal area and in the central highlands. In 2003, the Wind Resource Map of the Country was developed by the NREL. By the end of 2017, Sri Lanka had a total installed generation capacity of 131 MW. It is estimated that Sri Lanka can further exploit the available wind resources to the tune of 5,600 MW (Young & Vilhauer, 2003).
v. **Solar:** There is a good scope for the exploitation of Solar energy power by Sri Lanka as it lies within the equatorial belt. Annual average global horizontal irradiance (GHI) varies from 4.5 kilowatt-hours per square meter per day (kWh/m^2/day) to 6.0 kWh/m^2/day across the country. It is estimated that technically, a potential of 6,000 MW of electricity generation through the exploitation of solar energy exists.

➢ ***Note:*** There is a scope for developing biomass, solar, and wind resources. Also, there is a ray of hope for gas in the Mannar Basin. These are the areas where India can extend technical assistance to Sri Lanka to exploit these resources under the aegis of the South Asia cooperation program. The surplus capacity thus generated will have a ready market in India. Such a collaboration will help Sri Lanka to improve its revenue generation substantially.

1.4.9 MALDIVES

1. **Electricity (https://www.cia.gov/the-world-factbook/countries/maldives/#energy)**
   i. **Access:** 100% population (2020).
   ii. **Production/Consumption in million kWh (2016 estimate):** 402/373.9.

iii. **Installed Generating capacity:** 2,78,000 kW (2016 estimate).
iv. **Source of Generation:**
  a. **Fossil Fuels:** 96% of the installed capacity (2016 estimate).
  b. **Other Renewable Sources:** 4% of installed capacity (2017 estimate).

2. **Refined Petroleum Products (https://www.cia.gov/the-world-factbook/countries/sri-lanka/#energy):**
   i. **Consumption:** 11,000 bbl/day (2016 estimate).
   ii. **Imports:** 10,840 bbl/day (2015 estimate).

➢ *Note:* Maldives has no indigenous resources of energy, and as such, she totally depends on energy import in the form of finished products.

1.5 CONCEPT OF GRID CONNECTIVITY WITHIN SOUTH ASIA (EXEC SUMMARY, 2018)

The concept of cross-border power system connectivity is increasingly becoming relevant as it results in cost optimization and is more sustainable due to multi-directional diversified supply. A unique feature of South Asia is that it is well endowed with energy-bearing resources that are renewable in nature. The exploitation of these resources, many of which have trans-border footprints, makes them very useful in reducing carbon dependence. Such a shift in the approach towards energy management is in line with the desired mitigation strategy to cope with the dangers of climate change. The geography of South Asia, with a power surplus from Central Asia to its Northwest, is such that it is suitable for designing a power grid that can be stretched across the entire region. India, with its huge requirement, can be at the center of the power exchange/transit, wherein other countries of the region can either draw or feed the power into the grid. Also, Afghanistan and Pakistan can play an important role in getting power from the power surplus area of Central Asia and further sharing it with other countries of the region. It is encouraging to note that certain progress has already been made by bilateral interconnections and power trade between Bangladesh-India, Bhutan-India–India and India–Nepal. There is also an India–Myanmar link, which is in operation, and discussions are on to upgrade it (Interconnections with the Neighboring Countries). The proposed interconnection between India and Sri Lanka is still at the feasibility stage. Unfortunately, the Indo-Pak connection, though highly desirable, has unfortunately been a non-starter so far due to political differences between the two countries. The seamless power grid stretching across the entire South Asia is desirable as it can address the problems of energy affordability and energy sustainability and, with these two parameters, can help the region, as a whole, achieve energy security. Grid interconnection in South Asia between hydro power-rich countries like Bhutan, Nepal, and India can generate direct economic benefits for the former two. In fact, from these countries, power can be cycled to Bangladesh through India. The cost of generation can also be reduced by establishing a generation base in hydro power-rich countries, and transfer of power can be done through this trans-border grid. However, an efficient and flexible infrastructure and related organization with a certain statutory authority is a prerequisite for optimal utilization of resources and cross-border electricity trade (CBET). This authority can be created under the aegis of the SAARC Charter. By an estimate, the direct benefit of a comprehensive power grid connectivity in monetary terms may mean a saving of almost $9 billion, besides reducing Green House gases (GHG)/ emissions by more than 9% per annum compared with the current model of generation/transmission/distribution practiced by countries of the region. However, the impediments to its establishment are based on prevailing political differences, the absence of an institutional framework, and certain technical challenges. To overcome these impediments, an initiative by the SAARC will have to be taken. Which then can be supported by agencies like the World Bank, the ADB, the United States Agency for International Development (USAID), and other similar agencies. Certain progress in this regard has already been made. The most important initiative has been the establishment of the SAARC Energy Center in Islamabad, Pakistan, with the objective of driving subregional cooperation. Towards that end, the SAARC Framework Agreement for Energy Cooperation (Electricity) was signed in 2014. Consequent to that, India issued guidelines with provisions for its CBET in 2016. It is a pity that the progress has not been as planned. However, the impact of climate change and the requirement to progress international agreements on sustainable development clearly flag the need to address it on priority. Here, it is worth considering that the exploitation of the RE with a view to lower the emissions and thus achieve the sustainable development goals (SDGs), wherein the energy has a centrality across all the STGs, will be optimized through the South Asian Grid.

1.5.1 TRADE WITHIN THE REGION (SARI/EI)

The existing intraregional energy trade among the SACs is limited to electricity trade between India-Bhutan and India-Nepal (presently, within the framework of the South Asian Sub-Regional Economic Cooperation.[5] (SASEC), A total of 16 projects at the cost of $2.92 billion are in progress in the domain of the energy sector (https://www.sasec.asia/index.php?page=strategy).) The trade in petroleum products exists only between India–Nepal, India–Bhutan, India–Bangladesh, and India–Sri Lanka. The enhanced electricity trade will be based on the expansion of power transfer links between Bhutan and India and India and Nepal. The establishment of ongoing and proposed new power transfer links between Bangladesh and India, India and Sri Lanka would further strengthen the regional power trade. Presently, no system of power trade exists between India and Pakistan and Pakistan and Afghanistan, though through TUTAP and CASA-1000, certain projects are under execution to transfer power from Turkmenistan to Afghanistan and Pakistan (News Release by Asian Development Bank, 2018). It is hoped that notwithstanding their differences, both India and Pakistan will realize the need for the power trade for their mutual economic benefit and try to come to a common ground and start the power trade. Key issues that need to be addressed to improve the power trade within the region are the need to develop a framework and related infrastructure for a regional power market wherein energy supply availability is assured and the need to develop a common legal and regulatory framework to ensure a rule-based trade.

1.6 ENERGY COOPERATION WITHIN BAY OF BENGAL INITIATIVE FOR MULTI-SECTORAL TECHNICAL AND ECONOMIC COOPERATION (BIMSTEC) (ANASUA, 2020)

This grouping consists of Bangladesh, Bhutan, India, Myanmar, Nepal, Sri Lanka, and Thailand. Although, as can be seen, this grouping has nations that are not part of South Asia, a number of nations of South Asia are part of this grouping. Therefore, energy cooperation in this

---

[5]South Asia Sub Regional Economic Cooperation program is meant to bring together Bangladesh, Bhutan, India, Maldives, Myanmar, Nepal, and Sri Lanka in a project-based partnership with the objectives of promoting regional prosperity, improving economic opportunities, and building a better quality of life for the people of the subregion. Among its various verticals, energy is an important vertical. Some of the important elements of the energy strategy of this grouping entail improving cross-border electrical transmission, improving power trade between the countries of the grouping, improving cooperation in energy efficiency, clean power development, and capacity building.

grouping has a bearing on South Asia. A rising trend in the primary energy consumption per capita has been noted in Bangladesh, Bhutan, India, Nepal, Sri Lanka, and other members of the grouping. As per the 'BIMSTEC Energy Outlook 2030,' the total primary energy supply in the BIMSTEC region in 2008 was 772 million tons of oil equivalent (Mtoe), which is estimated to increase to 1,758 Mtoe by 2030. During the same period, the total primary energy consumption was 539 Mtoe in 2008 and is estimated to increase to 1,210 Mtoe in 2030. Currently, most of the countries in the group are importing energy assets to meet their primary energy needs. In this connection, it is important to note that while India's import dependency was 41.5% in 2018, Bangladesh was 16.8% in 2014, Nepal was 16.7% in 2014, and Sri Lanka was 57.5% in 2017, the region itself is endowed with abundant natural resources comprising of coal at 324 billion tons, oil at 664 million tons, natural gas at 99 TCF, biomass at 11 billion tons, hydropower (large) at 328 GW, and RE potential of more than 1,000 GW. Therefore, if the BIMSTEC countries cooperate among themselves, not only will their energy needs substantially be addressed, but they will also be able to trade the surplus energy with other countries of South Asia.

1.7 CONCLUSION

In view of the above discussion, it is clear that the countries of South Asia may be short of desired energy resources individually, but when considered as a whole, their resources complement each other, and an audit of the collective resources of the region brings out that the resources are adequate to support the economic growth of the region. To further add to the resources, spillover from Central Asia and Iran in the west and Northwest and the resources of Myanmar and Thailand in the East can add to the kitty. However, more sagacity is needed on the part of the leadership of these nations to share the resources. Yet another area that can further optimize the exploitation of the available resources is the need to invest in R&D in the domain of Renewables, nuclear, hydro, and even clean Coal technologies. India, with its economic muscle and lead in the technological domain, can contribute substantially to developing area-specific technologies to leverage them for the optimal exploitation of the available resources.

KEYWORDS

- **BIMSTEC**
- **centrality**
- **cooperation to exploit energy**
- **energy cooperation**
- **geopolitical equations**
- **grid connectivity**
- **research**

REFERENCES

A CEA Report, (2018). *India Installed Capacity*. Available on https://cea.nic.in/reports/monthly/installedcapacity/2018/installed_capacity-03.pdf (accessed on 12 August 2023).

A Central Electric Authority Report, (2020). *All India Installed Capacity of Power Stations*. Available online: https://cea.nic.in/reports/monthly/installedcapacity/2020/installed_capacity_03.pdf (accessed on 12 August 2023).

A Government of India, Ministry of Power, Central Electricity Authority, "Report on Optimal Generation capacity mix for 2029 –2030," p. 9, dated Mar. 2020, available online: https://cea.nic.in/old/reports/others/planning/irp/Optimal_mix_report_2029– 30_FINal.pdf.

A PIB Report from Ministry of Power, GoI. (2020). *Trade in Electricity*. https://pib.gov.in/PressReleasePage.aspx?PRID=1607177 (accessed on 12 August 2023).

A PTI Report, (2010). *India Eyeing 63000 MW Nuclear Power Capacity by 2032: NPCIL*. Pub by The Economic Times. Available on https://economictimes.indiatimes.com/industry/energy/power/india-eyeing-63000-mw-nuclear-power-capacity-by-2032-npcil/articleshow/6730724.cms (accessed on 12 August 2023).

A PTI Report, (2020). *India Plans to Nearly Double Oil Refining Capacity by 2030: Pradhan*. Pub by ETEnergyWorld.com.

A Report by International Finance Corporation, World Bank Group and Uploaded on: https://www.ifc.org/wps/wcm/connect/news_ext_content/ifc_external_corporate_site/news+and+events/news/nepal+hydropower (accessed on 12 August 2023).

A Report by NITI Aayog of India, (2020). *Economic & Environmental Impact of Coal Washing in India–2020*. Available on https://www.niti.gov.in (accessed on 12 August 2023).

A Report on Optimal Generation Capacity Mix for 2029–2030, (2020). p. 5 Available on https://cea.nic.in/wp-content/uploads/irp/2020/12/Optimal_mix_report_2029-30_FINAL.pdf (accessed on 12 August 2023).

A SASEC Report. (2020). *Bhutan's Electricity Export to India Continues to Increase*. Available on https://www.sasec.asia/index.php?page=news&nid=1178&url=bhu-hydro-export-aug2020 (accessed on 12 August 2023).

Aiming, Z., (2019). *Sri Lanka Energy Sector Assessment, Strategy, and Road Map*. Prepared by the Asian Development Bank. Available on https://www.adb.org/sites/default/files/

institutional-document/547381/sri-lanka-energy-assessment-strategy-road-map.pdf (accessed on 12 August 2023).

Alauddin, A., (2012). *500 Mw Will be Added to National Grid Soon: Alternative Energy Development Board Pakistan*. Pub in The Nation Pakistan, Lahore, Pakistan.

An Energy Policy Review by IEA, India, (2020). Available on: https://www.iea.org/reports/india-2020 (accessed on 12 August 2023).

An Envecologic Report, (2017). *Renewable Energy in India: Potential, Growth and Policies*. Available on: https://envecologic.com/renewable-energy-in-india-potential-growth-and-policies/ (accessed on 12 August 2023).

An IANS Report, (2021). *Polluting Coal to Remain Predominant Fuel in India's Energy Basket*. Pub in Business Standard. Available on https://www.business-standard.com/article/economy-policy/polluting-coal-to-remain-predominant-fuel-in-india-s-energy-basket-121040300312_1.html (accessed on 12 August 2023).

An internet upload: https://www.volkswagenag.com/en/news/stories/2020/03/lithium-mining-what-you-should-know-about-the-contentious-issue.html# (accessed on 12 August 2023).

An Overview by the Ministry of New and Renewable Energy. Government of India and uploaded on: https://mnre.gov.in/waste-to-energy/current-status (accessed on 12 August 2023).

Anasua, B. R. C., (2020). *Manifestation of SDG Goals and Scope for Cross-Border Energy Cooperation in BIMSTEC*. Published by ORF. Available on https://www.orfonline.org/expert-speak/manifestation-sdg-goals-scope-cross-border-energy-cooperation-bimstec/ (accessed on 12 August 2023).

Anmar, F., (2021). *India is Turning to 'Green Hydrogen' in a Bid to Decarbonize its Economy*. Reported by CNBC. Available on https://www.cnbc.com/2021/03/10/india-turns-to-green-hydrogen-in-a-bid-to-decarbonize-its-economy.html (accessed on 12 August 2023).

Asif, M., (2009). Sustainable energy option for Pakistan. *Renew Sust. Energy Rev., 13*, 903–909.

Bilal, A., (2019). *India's Petroleum Products Exports to Hit 8-Year Low in 2019*. Pub in ET Energyworld.com. Available on https://energy.economictimes.indiatimes.com/news/oil-and-gas/indias-petroleum-products-exports-to-hit-8-year-low-in-2019/68406477 (accessed on 12 August 2023).

BP Energy Outlook, (2020). pp. 3, 65, 77, 92, 104, 117, 124.

Country profile of Pakistan by IHA. Uploaded on: https://www.hydropower.org/country-profiles/pakistan (accessed on 12 August 2023).

Craig, D. Z., & Mar, A. S., (2007). *A Case for Enhanced Use of Clean Coal in India: An Essential Step Towards Energy Security and Environmental Protection*. Presented at: Workshop on Coal Beneficiation and Utilization of Rejects Ranchi, India. Available on https://fossil.energy.gov/international/Publications/Coal_Beneficiation_Workshop/coal_beneficiation_paper_zamuda.pdf (accessed on 12 August 2023).

Danish, Mir Sayed Shah & Senjyu, Tomonobu & Zaheb, Hameedullah & Sabory, Najib Rahman & Ahmadi, Mikaeel & Ibrahimi, Abdul Matin & Nazari, Zahra & Ahadi, Mohammad. (2020). Afghanistan as an emerging regional energy hub. 1. 10-14. 10.37357/1068/jso.1.1.02.

Deepak, A., (2019). *Nepal Power Export Plans in Doubt as India Reviews Options*. Pub by Nikkei Asia. Available on https://asia.nikkei.com/Politics/International-relations/Nepal-power-export-plans-in-doubt-as-India-reviews-options (accessed on 12 August 2023).

Electric Vehicle Market by Vehicle (Passenger Cars & Commercial Vehicles), Vehicle Class (Mid-priced & Luxury), Propulsion (BEV, PHEV & FCEV), EV Sales (OEMs/Models) Charging Station (Normal & Super) & Region – Global Forecast to 2030 and Uploaded on: https://www.marketsandmarkets.com/Market-Reports/electric-vehicle-market-209371461.html (accessed on 12 August 2023).

Energy Availability Factor. Pub by power reactor information system of IAEA available on https://pris.iaea.org/PRIS/WorldStatistics/ThreeYrsEnergyAvailabilityFactor.aspx (accessed on 12 August 2023).

ET Energy World Report, (2020). *India Overtakes Japan with Fifth Largest Hydropower Capacity in the World*. Pub by ET Energyworld.com. Available on https://energy.economictimes.indiatimes.com/news/power/india-overtakes-japan-with-fifth-largest-hydropower-capacity-in-the-world/76095023 (accessed on 12 August 2023).

Exec Summary, (2018). *Integrating South Asia's Power Grid for a Sustainable and Low Carbon Future*. Report compiled by The Economic and Social Commission for Asia and the Pacific (ESCAP).

Farid, A. M., & Engr, A. A., (2015). *Use of Coal in the Energy Mix of Pakistan*. Presented during 2015 International Pittsburgh Coal Conference Pittsburgh, PA, USA, Available on: https://www.engineering.pitt.edu/uploadedFiles/_Content/Sub_Sites/Conferences/PCC/Documents/Plenary/2015/Use%20of%20Coal%20in%20the%20Energy%20Mix%20of%20Pakistan.pdf (accessed on 12 August 2023).

Farooq, M. K., & Kumar, S., (2013). An assessment of renewable energy potential for electricity generation in Pakistan. Pub in *Renew Sust. Energy Rev., 20*, 240–254.

Farooq, N., & Hameed, (2012). Effective use of technology to convert waste into renewable energy sources. *J. Life Sci., 9*, 654–661.

Farooqui, S. Z., (2014). Prospects of renewables penetration in the energy mix of Pakistan. *Renew Sust. Energy Rev., 29*, 693–700.

Gopal, S., (2019). *India and Nepal Open South Asia's First Cross-Border Oil Pipeline*. Published by REUTERS. Available online: https://www.reuters.com/article/us-nepal-india-idUSKCN1VV135 (accessed on 12 August 2023).

Hassan, S. M., & Moklesur, R. M., (2019). Present status of Bangladesh gas fields and future development: A review. *Journal of Natural Gas Geoscience, 4*(6), 347–354. Available on https://www.sciencedirect.com/journal/journal-of-natural-gas-geoscience/vol/4/issue/6 (accessed on 12 August 2023).

http://www.prds-srilanka.com/exploration/mannarBasin.faces (accessed on 12 August 2023).

https://en.wikipedia.org/wiki/Oil_and_gas_industry_in_India (accessed on 12 August 2023).

https://en.wikipedia.org/wiki/South_Asia#cite_note-unreports.studycenters-8 (accessed on 12 August 2023).

https://energypedia.info/wiki/Afghanistan_Energy_Situation#toc (accessed on 12 August 2023).

https://online.dghindia.org/oalp/files/pdf/OALP5_MUMBAI_overview_of_Prospectivity_of_Indian_Sedimentary_Basins.pdf (accessed on 12 August 2023).

https://sites.google.com/site/iceafghanistan/regional-projects-and-masterplanning---overview/tutap (accessed on 12 August 2023).

https://www.adb.org/sites/default/files/linked-documents/47282-001-ssa.pdf (accessed on 12 August 2023).

https://www.cia.gov/the-world-factbook/countries/afghanistan (accessed on 12 August 2023).

https://www.cia.gov/the-world-factbook/countries/bangladesh/#energy (accessed on 12 August 2023)

https://www.cia.gov/the-world-factbook/countries/bhutan/#energy (accessed on 12 August 2023).

https://www.cia.gov/the-world-factbook/countries/india/ (accessed on 12 August 2023).

https://www.cia.gov/the-world-factbook/countries/maldives/#energy (accessed on 12 August 2023).

https://www.cia.gov/the-world-factbook/countries/Pakistan/#energy (accessed on 12 August 2023).

https://www.cia.gov/the-world-factbook/countries/sri-lanka/#energy (accessed on 12 August 2023).

https://www.hydrocarbons-technology.com/projects/turkmenistan-afghanistan-pakistan-india-tapi-gas-pipeline-project/ (accessed on 12 August 2023).

Iain, M., & Enda, C., (2021). A Bloomberg report. *As US Exits Afghanistan, China Eyes $1 Trillion in Minerals.* Reported by Aljazeera.

Indian Petroleum and Natural Gas Statistics 2017–2018, (2018). Publication by the Govt. of India.

Interconnections with the Neighboring Countries. Uploaded on: https://powermin.gov.in/en/content/interconnection-neighbouring-countries (accessed on 12 August 2023).

Internet upload: https://cea.nic.in/reports/monthly/hydro/2016/hydro_potential_region-02.pdf (accessed on 12 August 2023).

Internet upload: https://www.cia.gov/the-world-factbook/countries/nepal/#energy (accessed on 12 August 2023).

Internet upload: https://www.sasec.asia/index.php?page=strategy (accessed on 12 August 2023).

Khan, H. A., & Pervaiz, S., (2013). Technological review on solar PV in Pakistan: Scope, practices, and recommendations for optimized system design. *Renew Sust. Energy Rev., 23*, 147–154.

Lisa, P., (2009). *Considering an Alternative Fuel for Nuclear Energy*. Pub in The New York Times. Available online on https://www.nytimes.com/2009/10/20/business/global/20renthorium.html?_r=1 (accessed on 12 August 2023).

Lok Sabha Unstarred Question No.: 330, Budget Session 2021. Dept. of Atomic Energy, GoI 03 Feb. 2021.

Mirza, U. K., et al. (2008). An overview of biomass energy utilization in Pakistan. *Renewable and Sustainable Energy Reviews, 12*(7), 1988-1996

Moirul, I. M., et al., (2020). An overview of the hydropower production potential in Bangladesh to meet the energy requirement. *Environmental Engineering Research During.* https://doi.org/10.4491/eer.2020.514.

News Release by Asian Development Bank, (2018). *Power Interconnection Project to Strengthen Power Trade Between Afghanistan, Turkmenistan, and Pakistan.* Uploaded on: https://www.adb.org/news/power-interconnection-project-strengthen-power-trade-between-afghanistan-turkmenistan-pakistan (accessed on 12 August 2023).

Pakistan Meteorological Department (PMD), (2009). *Feasibility Report of the Establishment of Commercial Wind Power Plant of 18 MW at Gharo, Pakistan.*

PPP (current international dollars). www.data.worldbank.org. World Bank.

PTI Report, (2009). *Tajikistan Invites India to Invest in Hydro-Power*. Pub by The Hindu. Available on https://www.thehindu.com/news/national/

Tajikistan-invites-India-to-invest-in-hydro-power/article16879915.ece (accessed on 12 August 2023).
Quereshi, F. U., & Akintug, B., (2014). *Hydropower Potential in Pakistan*. Presented during 11th International Congress on Advances in Civil Engineering dated 21-25 Oct 2014 held at Istanbul/ Turkey and hosted by ITU.
Ramachandran, R., (2020). *India's First Prototype Fast Breeder Reactor has a New Deadline.* Should we trust it? Pub by The Wire.
Rastriya, S. S., (2020). *NEA to Import 250 MW of Electricity from India for Winter*. Pub by The Himalayan. Available on https://thehimalayantimes.com/nepal/nea-to-import-250-mw-of-electricity-from-india-for-winter (accessed on 12 August 2023).
Saket, S. K., (2019). *Opportunities for India in Myanmar's Energy Sector*. Pub by Observer Research Foundation. Available on: https://www.orfonline.org/expert-speak/opportunities-for-india-in-myanmars-energy-sector-47543/ (accessed on 12 August 2023).
Sanjeev, C. (2020). *Indian Companies' Oil and Gas Output from Overseas Fields Drops 11%.* Pub by The Economic Times. Available online: https://m.economictimes.com/industry/energy/oil-gas/indian-companies-oil-gas-output-from-overseas-fields-drops-11/articleshow/79782136.cms (accessed on 12 August 2023).
Shobha, R., (2022). *India's Coal Imports Likely to be Lower by 35-40 Million Tons in 2021–22.* Pub by Business line of the Hindu.
Shreya, J., (2016). *Govt Plans Push for Hydro Power*. Pub in Business Standard. Available on Line https://www.business-standard.com/article/economy-policy/govt-plans-push-for-hydro-power-116041100037_1.html (accessed on 12 August 2023).
Sidharth, S., (2011). *India's Rising Nuclear Safety Concerns*. Pub in Asia Sentinel. Available on https://www.asiasentinel.com/p/indias-rising-nuclear-safety-concerns (accessed on 12 August 2023).
Sonam, T., & Bharat, T. (2004). United Nations Symposium on Hydropower and Sustainable development held from 27–29 Oct 2004 at Beijing, China and uploaded on https://www.un.org/esa/sustdev/sdissues/energy/op/hydro_tsheringbhutan.pdf (accessed on 12 August 2023).
Soumoy, T., (2021). *PM Modi's Visit Shows India and Bangladesh Are Ready for Transformative Leap in Bilateral Ties Despite Some Sticky Challenges*. First Post. Available on: https://in.news.yahoo.com/pm-moddi's-visit-shows-india-130755574.html (accessed on 12 August 2023).
South Asia Regional Initiative for Energy Integration-Regional Market for CBET. Pub by SARI/ EI also Available on: file:///Users/majgenajay/Desktop/Regional%20Power%20Market%20For%20CBET%20%7C%20USAID%20SARI:Energy%20Integration%C2%A0.html (accessed on 12 August 2023).
South Asia Regional Overview. Uploaded on: https://web.archive.org/web/20081121043924/http://www.sardeg.org/marketana.asp (accessed on 12 August 2023).
The Flagship Report, *Global Energy Review 2019*. Pub by IEA in April 2020 and Uploaded on: https://www.iea.org/reports/global-energy-review-2019 (accessed on 12 August 2023).
UNDP Report of Human Development Report, 2020. A Government of India, Ministry of Power, Central Electricity Authority, "Report on Optimal Generation capacity mix for 2029 –2030," p. 9, dated Mar. 2020, available online: https://cea.nic.in/old/reports/others/planning/irp/Optimal_mix_report_2029 –30_FINal.pdf.
Ur and Th Distribution Across India and World. Pub by PMF IAS and Uploaded on: https://www.pmfias.com (accessed on 12 August 2023).

USAID South Asia Regional Initiative for Energy Integration repository. Accessed on 23 February 2021, URL: https://sari-energy.org/program-activities/cross-border-electricity-trade/country-energy-data/

Viktor, T., (2021). *The Renewable Energy Potential of Bangladesh.* Pub in Energy Tracker Asia. Uploaded on: https://energytraker.asia/the-renewable-energy-potential-of-bangladesh/ (accessed on 12 August 2023).

www.export.gov (accessed on 12 August 2023).

Young, M., & Vilhauer, R., (2003). *Sri Lanka Wind Farm Analysis and Site Selection Assistance*. National Renewable Energy Laboratory. Global Energy Concepts, LLC Kirkland, Washington. https://www.nrel.gov/docs/fy03osti/34646.pdf (accessed on 12 August 2023).

Zahid, K., (2016). *CASA-1000: Options for Energy Corridor, Trade Projects, and Regional Stability.* Pub in Foreign Policy News. https://foreignpolicynews.org/2016/05/28/casa-1000-options-energy-corridor-trade-projects-regional-stability/ (accessed on 12 August 2023).

CHAPTER 2

Towards a Regional Energy Security Model: A Cooperative Approach

A. K. CHATURVEDI

AVSM, VSM – Freelance Contributor (Retired)

2.1 ENERGY SECURITY

The 'Energy Security,' defined by the International Energy Agency (IEA), is the sustained energy availability at an affordable price. 'Energy security' needs to be considered both in the long and short term. Long-term energy security entails necessary infrastructure with a view to ensuring continuous supply even if a *force majeure* occurs on account of an adverse geopolitical event or a natural calamity. Two of the enabling conditions to achieve such a state are timely investments and steady economic growth, which go hand in hand with the growth in energy needs. Simultaneously, planners need to take into account the environmental impact of the development with a view to guard against its adverse effect on the sustainability of the system thus created. For short-term energy security, the focus will have to be on the ability of the energy system to react promptly to the sudden changes in the 'Demand-Supply' balance. Thus, an appraisal of the energy security flags the need for the sustained availability of energy-bearing resources in desired quantities, with desired quality to guard against adverse environmental impact, necessary conversion technology which lends itself for optimal utilization of resources, and finally, adequacy of the energy storage infrastructure to cater for a *force majeure.*

The South Asian Energy Corridor: Toward Sustainability and Security. Amritanshu Shukla, PhD, Atul Sharma, PhD & Saurabh Mishra, PhD (Eds.)

2.1.1 NECESSITY OF RESILIENCE OF THE ENERGY SYSTEMS FOR ENERGY SECURITY

Disruptions in the timely availability of energy has the potential to cause major setback to economic and societal development. As such, modern energy systems should be capable of bearing shocks from a wide range of sources, including natural disasters, geopolitical conflicts, and the new and emerging threats related to the ongoing digitalization of energy systems. Resilience is necessary to withstand the impacts of climate change within the framework of the technical viability of the energy sector and its ability to remain cost-effective to meet the increase in demand on account of greater automation, to cater to rising economic needs and, above all rising energy needs to meet the basic needs of the growing population. In this connection, it needs to be noted that the interdependence between energy and water is set to increase in the future, as water is essential for all phases of energy generation. Therefore, it is important that the energy-water linkage is managed in such a way that the systemic optimization is achieved without disturbing environmental balance. As automation of the energy systems the World over is increasing by the day, so is its vulnerability to cyber-attacks. Thus, a system for the regular safety and security audit of the concerned energy system has to become an integral part of the operational plan.

2.1.2 THREATS TO ENERGY SECURITY

Energy is a coveted input for the economic growth of a country, and the energy-bearing resources are not enough to meet the rising demand. This 'demand-supply' gap results in 'too many people/agencies competing for too few resources.' Such a state leads to threats to the rights, including the sovereignty of weaker and vulnerable countries. Therefore, there is a need to do an appraisal of the likely threats. Some of the threats to energy security are the political instability of the countries that hold the energy resources, manipulations by the organizations that control the energy supplies, the competition among the seekers over the control of the energy sources, attacks on supply infrastructure by the forces inimical to national security, as well as accidents, natural disasters, terrorism, and excessive reliance on foreign countries for the supply of oil, gas, and other energy bearing resources with its attended issues. Avoidance of threats to energy security is essential for its sustainability. Hence, a rule-based energy-sharing system is the need of the hour to meet the aspirations of all countries.

2.1.3 INDIVIDUAL VS. COLLECTIVE ENERGY SECURITY

Energy Security is essentially a sum total of self-reliance in terms of indigenous resources, interdependence with other energy-rich countries, the capability of the concerned country to leverage it to its advantage, the economic strength of the country to buy the resources it lacks, and the availability of technologies to harness its own resources. It needs to be appreciated that while for a big country, it may be possible to manage her energy needs individually, for smaller nations, it may be a tall order. In such a situation, a collective response is always a way ahead. More so, in case the region itself has adequate resources collectively, then a cooperative model can be worked out, which would entail building necessary infrastructure, regulatory mechanisms, and energy-sharing agreements. However, for such a system to work, an organization will have to come up to oversee the seamless integration and smooth implementation. South Asia is one such region where the inadequacy of individual countries as far as energy-bearing resources are concerned and the lack of capacities of individual nations in terms of available technologies to harness their respective resources and financial muscle can be complemented to achieve a win-win situation. An attempt is being made in this chapter to examine the capacity and capability of South Asia to achieve collective energy security.

2.2 SOUTH ASIA

2.2.1 GEOGRAPHY

South Asia consists of Afghanistan, Pakistan, India, Nepal, Bhutan, Bangladesh, Maldives, and Sri Lanka. Its boundary in the South is the Indian Ocean; in the North, it is bound by the Himalayas, Karakoram, and Pamir. The Amu Darya forms the Northwestern extremity, and the Eastern Land Boundary is with Myanmar. The maritime boundary in the West is with Iran, and in the East, it is with Myanmar. Thailand and Indonesia form the Southeastern maritime boundary. The total area over which South Asia is spread is 5.2 million square km, which is 11.71% of the Asian Continent or 3.5% of the World's surface area. Area-wise, India is the biggest country, and it shares boundaries, either land-based or maritime-based, with every country in the region. Even with Afghanistan, she shares a land boundary (in Wakhan Corridor), though that area is presently under the illegal occupation of Pakistan.

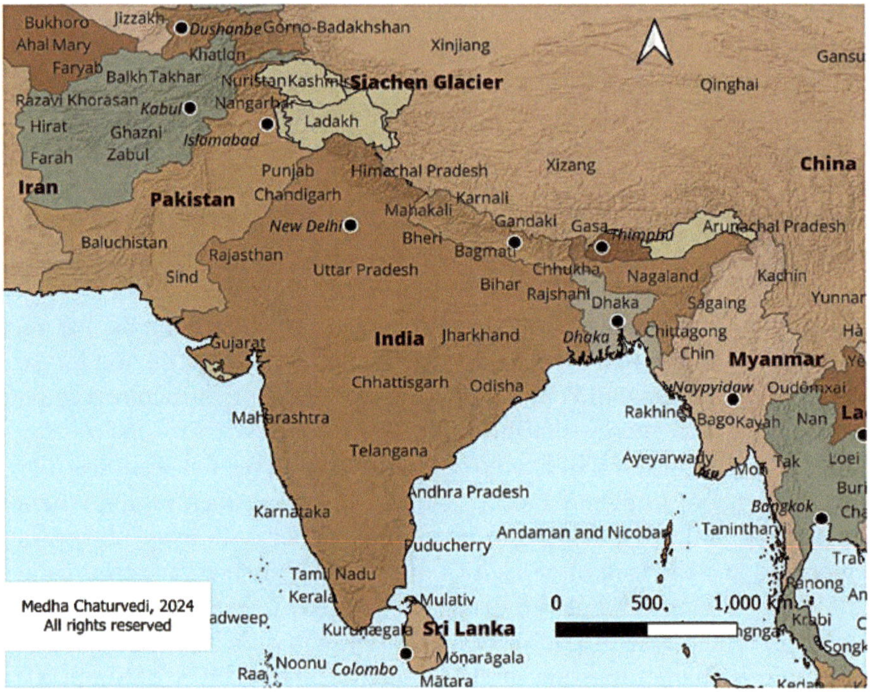

MAP-1: South Asia.

2.2.2 POPULATION

The population of the region as a whole is 1.89 billion, which accounts for 39.41% of the population of Asia and 24% of the World's population. This makes the region the most populous region of the World and also the most densely populated region of the World. This state wherein a larger population is accommodated in the smaller areas has its own problems, and one of the major issues is a shortage of energy per capita, resulting in violation of two basic elements that define energy security, and these are erratic energy supply which impacts its sustainability of the timely energy supply and the scarcity which leads to a higher cost of available energy.

2.2.3 ENERGY RESOURCES

The region is endowed with diverse, rich natural energy resources like coal, wind, thermal, water, gas, solar, etc. (Abas et al., 2017). The details are estimated coal reserves of 1,33,237 million tons, hydropower potential of

2,96,431 MW, natural gas reserves of 85 trillion cubic feet (TCF), potential for solar energy of 3,65,639 MW; wind 3,78,594 MW (Ibid. p. ii; Gulagi et al., 2017). Biomass is 223 million tons, and for oil, it is 5906 million barrels (Ankush, 2018).

2.2.4 APPRAISAL OF AVAILABLE RESOURCES

Interestingly, it needs to be noted that the region of South Asia, as a whole, is actually not short of energy resources but suffers from a lack of technology to exploit the available resources. This results in a number of SACs experiencing a severe shortage of electrical power with numerous blackouts (Rehman & Rashid, 2017) despite the availability of an abundance of energy resources in the region.

2.2.5 CHANGING CONSUMPTION PATTERN

The primary energy consumption pattern within SACs is undergoing subtle changes in terms of type, volume, and mix, depending on the availability of indigenous resources and ease of access to energy sources from abroad, the economic viability of identified energy-bearing sources, both of indigenous origin or of foreign origin, the technical feasibility of the available technologies to exploit the available resources and finally on the international tie-ups, wherever required. In this connection, a case study of Bangladesh is interesting. Bangladesh, despite having indigenous sources, has one of the lower per capita energy consumption (Islam & Khan, 2017). Bangladesh has a vision of achieving 100% electricity access by 2021 by integrating more solar PV and Biomass in its energy mix as the country is rich in these sources (vi: Islam et al., n.d.). Likewise, despite coal being the significant fuel in the energy mix of India (India has adequate coal reserves), only 6% of the World's primary energy is utilized in India (Sahoo, 2016; Manju & Sagar, n.d.), probably because of ill effects of coal on the environment. Thus, to sustain its current pace of growth, India needs to look for alternative energy sources. By 2022, The Jawaharlal Nehru National Solar Mission (JNNSM) 2010 had predicted an accretion of 22 GW of grid-tied solar power (Kaur, 2016). According to India's Nationally Determined Contributions (INDC) to fight climate change, it has been reported that India has plans to develop 100 GW of solar power capacity in the next few years (Yu et al., 2018; NITI Aayog, 2017). In fact, India has set a target of 450 GW by 2030 (A PTI Report, 2020). Similarly, the Islamic Republic of Pakistan also has plans and energy policies for the exploitation of the indigenously available renewable energy (RE) sources with the objective of almost 20% capacity accretion by the year 2025 and at least 30% by 2030 (Alternative and Renewable Energy, 2020).

2.2.6 GROWTH PATTERN IN COMPARISON TO REST OF THE WORLD

According to the US Energy Information Administration (US EIA), the region's primary energy consumption has increased by 58% between 1991 and 2000, and it is expected to increase by another 40% in the next three decades. However, South Asia's per capita primary energy consumption of 0.45 tons of oil equivalent (toe) is much lower than the world average of 1.68 tons. Also, in terms of per capita electrical energy consumption, the SACs lag far behind the world average. Surprisingly, the region's energy intensity of total energy use per unit of gross domestic product (GDP) is 0.65 toe/US $1,000 as compared to the world average of 0.29 toe/US $1,000 (Ritu, 2006). The gross electrical power consumption and the power consumption per capita are tabulated in Table 2.1. Although the data is a bit dated, it indeed gives a fairly good picture of the trend of growth vs. power consumption. This data flags the need for better collaboration among SACs to achieve optimization of indigenous resources to give impetus to economic growth.

TABLE 2.1 Gross Power Production and Consumption

Country	Gross Electrical Power/Year (2014 Data) Unit: GW-h/Year	Power Consumption/Capita (2013 Data) Unit: kgOE/Capita
Afghanistan	552.6 (2016 data)@	85.98%
Pakistan	90,000#	474.9$
India	1,547,000#	606.1$
Bangladesh	70594#	215.5$
Nepal	6562#	369.7$
Bhutan	2386#	2461.32^
Sri Lanka	13438#	487.5$
Maldives	565#	605.44&

Source:

@: https://www.cia.gov/the-world-factbook/countries/afghanistan/#energy.

#: International Energy Consumption-EIA dated 19 Nov 2014.

$: Developmental Indicators data. Worldbank.org dated 17 Sep. 2016.

%: https://www.adb.org/sites/default/files/linked-documents/47282–001-ssa.pdf.

^: https://www.adb.org/sites/default/files/linked-documents/47282–001-ssa.pdf.

&: https://www.worlddata.info/asia/maldives/energy-consumption.php.

2.2.7 RESOURCE ANALYSIS

Resource analysis shows that coal dominates the energy sector in India. Pakistan and Bangladesh, besides fossil fuels, are also substantially dependent on natural gas. Nepal and Bhutan depend on their indigenous resources of hydro resources to an extent but substantially depend on the import of either energy-bearing resources or the finished products (particularly oil and gas) of energy generated. Sri Lanka, Maldives, and Afghanistan substantially depend on energy/power imports as their indigenous energy-bearing resources/energy generation capacity is quite limited.

2.2.8 AN APPRAISAL OF ENERGY RESOURCES WITHIN SACS

South Asia is substantially monolithic in terms of hydro resources. A substantial number of major water channels in the northern portion of India and the adjacent region are flowing across the political boundaries of the nations of the region. Thus, there is a complementarity of hydro resources. Such a complementarity facilitates interdependence and can be used for mutual good. For example, water-rich Nepal and Bhutan can share their riches with lower riparian India. Similarly, Bangladesh and India can work out arrangements not only to share the water but also the hydro energy generated by the common rivers. It may be noted that there are 54 Rivers that flow into Bangladesh from India (Sudha, 2006). The Indus Water Treaty of 1960 and the Ganges Water Treaty of 1996 are fine examples of collective utilization of hydro resources for the common good between India–Pakistan and India–Bangladesh, respectively. It is heartening to note that these two treaties have withstood the test of time despite tensions bedeviling the relationships at times. This clearly underlines the relevance of cooperative collectivism, wherein the common resources have been exploited by the two countries in a 'win-win' situation. As against these examples, there is a dispute simmering between Pakistan and Afghanistan about the sharing of the Kabul River, for which the two countries have not been able to come to a common understanding. An early understanding between Pakistan and Afghanistan will benefit people in the Peshawar region of Pakistan and also people in the Kabul valley of Afghanistan. A resource that is practically not yet utilized is the warm water of oceans/seas surrounding SACs. India, Pakistan, Maldives, Sri Lanka, and Bangladesh, among the SACs, are the littoral countries of the Indian Ocean, Arabian Sea, and the Bay of Bengal.

Ocean thermal energy conversion (OTEC) has immense potential to provide much-needed energy to the littoral countries. Although this type of conversion is still in the domain of frontier technology, collective Research and Development (R&D) could help the littoral countries to exploit this common resource for energy generation. The third energy domain is RE, other than hydro energy, where collective R&D effort can also help the region develop necessary technologies to optimally utilize this resource, which is available in plenty. Some of the sub-domains of renewables that have the potential for commercial exploitation immediately or with a bit of R&D support are solar, wind, geothermal, and waste (Municipal solid waste and agriculture waste). In recent times, Hydrogen has also been added to this list.

The SACs are part of developing countries whose growth rate is quite impressive. Also, they boast to be centers for a rising middle-class population. This stimulates major growth in the prime energy utilization sector due to improving economic conditions and an enhancement in per capita revenue. However, it is significant to note that in 2017, the South Asian Association of Regional Cooperation (SAARC) collectively accounted for only 4% of the World's collective GDP, with approximately 24% of the World's residents (Ul-Haq et al., 2020). This shows that the region as a whole still has a long way to go on the path of economic well-being to catch up with the more affluent West. Based on the current trends, it is predicted that over the next few decades, the energy demand for SACs will increase enormously due to rapid urbanization, the surge in population, and continuous economic growth.

2.2.9 SCOPE FOR COOPERATION WITH AREAS LOCATED IN THE NEIGHBORHOOD

An aspect of the geography of South Asia is that the energy-rich West Asia, Central Asia, and Myanmar in the East have surplus energy resources that can be leveraged to enhance the energy availability in South Asia. However, access to these resources is a function of the favorable geopolitical climate, relations between the concerned countries, and, above all, a will to share the resources. There are a number of examples where the sagacity of the leadership of the concerned countries has helped each one of them to come out with a solution that is 'win-win' for both countries. The elephant in the room is 'Big Power rivalry,' which can put impediments to such cooperation. Cases in point are Chinese attempts to appropriate resources of Indus and its tributaries in the West in Pakistan, her efforts to exploit Karnali and other rivers common between Nepal and Tibet in the Central Sector, and her recently reported efforts to divert water from Yarlung Tsangpo (Brahmaputra) in the East.

2.2.10 SOURCES/SCOPE OF INTER/INTRA REGIONAL COOPERATION FROM CENTRAL ASIA

The Central Asian region has the potential to produce a power surplus to its requirement and, as such, is likely to emerge as a hub for the energy trade with South Asia, especially with Afghanistan and Pakistan (The chapter on this cooperation has already been alluded to). The capacity of Central Asia in power generation is given in Figure 2.1.

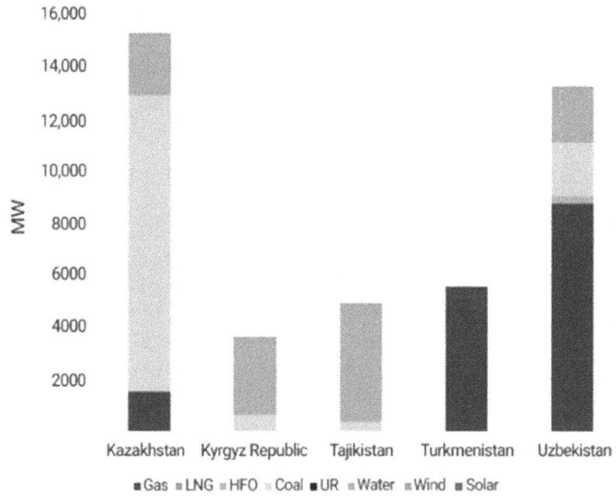

FIGURE 2.1 Power generation mix in Central Asia–2018.
Source: Adapted from The World Bank, 2021.

Figure 2.1 clearly shows that gas is available in plenty in Turkmenistan and Uzbekistan and to a limited extent in Kazakhstan, water is available in plenty in Tajikistan and Kyrgyz Republic, and coal is available in plenty in Kazakhstan and to a limited extent in Uzbekistan. Some more resources are available but in a very limited quantity. However, an assessment of these resources is adequate to confirm that there is a scope for sharing them with energy-starved South Asia, especially with Afghanistan.

MAP-2: CASA-1000.
Source: World Bank, 2020. https://creativecommons.org/licenses/by/4.0/

It is felt that the development of both Central Asia–South Asia (CASA)-1000 and Turkmenistan–Uzbekistan–Tajikistan–Afghanistan–Pakistan (TUTAP) with cross-regional power interconnection projects, connecting Central and South Asian countries are positive steps in the direction of electrical trade between Central Asia and South Asia.

MAP-3: Map made with data from USAID South Asia Regional Initiative for Energy Integration
Source: Adapted from USAID, 2021

There is a possibility of hydro energy-based cooperation between Tajikistan and India using HVDC supply, subject to working out the technical feasibility.

2.2.11 POTENTIAL FOR COOPERATION WITH IRAN

Iran, located in the Northwest of South Asia, is another energy surplus country. In quantitative terms, Iran has the fourth-largest reserve of oil and the largest reserve of natural gas (CIA-The World Fact Book). Iran also gets about 300 days of sunshine over most of its land mass (Solar Power Plants to Replace Fossil Fuel, 2010). The abundance of these three energy resources throws up another possibility for energy trade/collaboration with South Asia through Afghanistan and Pakistan/directly through Pakistan or through the maritime domain to India. In a reality check, it emerges that presently, a system of energy trade exists only between Iran and Pakistan, but even that is quite insignificant in terms of volume. The Iran–Pakistan–India gas pipeline, which held promise at one point in time, is a non-starter due to India's decision to withdraw from the project. As against this pipeline, India at one point in time showed interest in an undersea pipeline from Iran, avoiding Pakistan because of advancement in the technology (A PTI Report, 2017), but even that project is currently in limbo due to Iran being under US sanctions under the provisions of US law, "Countering America's Adversaries Through Sanctions Act (CAATSA)."

2.2.12 COOPERATION WITH MYANMAR

There is a potential for yet another inters regional cooperation existing in the East. A collaboration between India and Myanmar is taking shape in the form of a Hydroelectric project, and a pipeline is being planned to transport natural gas from Myanmar to India via Bangladesh (An ADB Report). In this connection, it needs to be noted that energy cooperation with Myanmar has the potential to grow to 60,000 MW by the year 2045, subject to strengthening of the intra-/inter-regional power cooperation between the countries of the South Asian region, particularly between India, Bhutan, Bangladesh, Nepal, Sri Lanka, Pakistan, and Myanmar.

2.2.13 CROSS-BORDER ELECTRICITY TRADE (CBET)

According to Integrated Research and Action for Development (IRADe), a Delhi-based independent research organization, till 2018, a mere 3,000 MW of power was being traded between SACs, which was expected to double by 2020 (Ankush, 2018). Generally, it has been noticed that across the SACs, there is an overarching dominance of a single fuel in the energy mix (petroleum) and limited exploitation of indigenously available RE due to related technology deficits. This kind of poor management of energy has resulted in increasing energy deficits and, accordingly, rising import dependence with consequent impact. That is why there is a need to go for an energy management model, which entails higher cross-border energy trade within SACs and shortfall, if any, to be made up by having trade with countries of Central Asia, Iran in the West, and Myanmar in the East. At present, the status of energy exchange between the SACs and collaboration with countries of Central Asia/West Asia/within the region is as tabulated in Table 2.2.

TABLE 2.2 Domestic Supply/Import of Power (FY-18/CY-17)

Countries	Electricity (GWh)		Surplus/ Deficit	Deficit/ Surplus in Percent Terms	Major Importing/ Exporting Countries
	Gross Demand	Domestic Supply			
Afghanistan	4,981	1,076	Deficit	(–)78	Iran, Tajikistan, Uzbekistan, Turkmenistan/–
Bangladesh	65,124	58,300	Deficit	(–)10	India/–
Bhutan	2,186	7,603	Surplus	249	–/India
India	1,205,266	1,202,099	Deficit	0	Bhutan/Bangladesh, Nepal
Maldives	1,405	1405	Self-sufficient	0	–
Nepal	5,557	4,476	Deficit	(–)19	India/–
Pakistan	1,20,392	1,18,916	Deficit	(–)1	Iran/–
Sri Lanka	15,763	15,763	Self-sufficient	0	–

Source: Country Power Reports, CRISIL Research.

Table 2.2 brings out that generally, there is an energy deficit scenario in most of the countries of the region, but there is a complementarity of demand and resources within the countries of the region, which are united by geography but are divided by history. Cooperation in the R&D of technologies to exploit the indigenous energy-bearing resources and collaboration in the utilization of common resources within the region can substantially address the existing energy deficit. This fact has finally been realized by the SACs, and as such, certain steps are being planned/taken collectively to address the deficiency of energy. In 2004, at the 12th South Asia Association for Regional Cooperation (SAARC) summit, a concept of —the Energy Ring, after due deliberations, was included in the declaration with a view to promoting regional cooperation. The discussions continued, and in 2014, during the 18th summit at Kathmandu, The SAARC Framework Agreement for Energy Cooperation was signed (SAARC Energy Outlook).

2.2.14 SAARC ENERGY RING

As explained above, most of the member countries of SAARC lack the requisite energy infrastructure for the resources that they have, and thus, there is a mismatch between the indigenously available technology and the indigenous resources, which brings out a peculiar problem wherein their own resources remain under-utilized, and they remain perpetually depended on import of energy bearing resources/energy products. However, this problem can be addressed only when the SAARC countries complement their resources and technical know-how. The SAARC Energy Ring was envisioned to evolve a model that facilitates regional cooperation in energy management with the twin objectives of optimizing the exploitation of the available indigenous resources and building the necessary infrastructure with the view to ensuring sustained availability of energy to all member countries at an affordable cost. It entailed the creation of a ring for possible cross-border energy transfers. Development of the SAARC Market for Electricity (SAME) is one of the key components of the SAARC Energy Ring. To establish the SAARC power grid, the focus will be on creating a Bhutan–Bangladesh–India–Nepal (BBIN) link in the East, an India–Sri Lanka link in the south, and an India–Pakistan link in the West. These linkages will help in expanding the electricity trade; with resource-rich Central Asian countries (Tajikistan, Kazakhstan, Uzbekistan, Turkmenistan), through Afghanistan to Pakistan in the West, and with Myanmar in the East. It was envisaged that initially, energy trade would progress in two clusters, namely, East and West, with the ultimate goal of merging the two clusters into a greater South Asian Energy Network, which Sri Lanka could join later. However, there is always a gap between plan and execution. It is worth noting that while in the East, certain progress has been made between Bhutan–Bangladesh–India–Nepal (BBIN) region, in the West, due to existing cross-border tension between India and Pakistan, the desired progress is still a distant dream. One of the implications is that the energy surplus of Central Asia is getting stopped at Pakistan only and is not getting further passed on to India.

2.2.15 CURRENT COLLABORATIONS

India's power trade with its neighbors is largely one-sided. The country annually imports around 1,200 mw power from Bhutan, exports 1,200 MW to Bangladesh, exports 450 MW to Nepal, and 3 MW to Myanmar.

2.2.16 POWER GRID

To enhance the energy trading within the region, greater integration of respective national power grids is required. Such an integration will help in achieving optimal use of resources by seamless access to energy resources that are distributed unevenly across the region. This will facilitate smooth energy management within the region as a whole. Regrettably, there are no regional grids in operation as yet, to help transfer power across long distances and reap the benefits of energy trade. However, things are not that bad. A beginning has already been made with the power grid of India getting interlinked with those of Bhutan, Bangladesh, and Nepal through a number of cross-border interconnections, such as India—Bangladesh, India-Nepal, and India–Bhutan (BBIN). Five more interconnections—one each with Bhutan, Bangladesh, and Sri Lanka, and two with Nepal—are either underway or in the planning stage. The 400 kV Punatsangchu-I hydroelectric project (Bhutan)–Alipur Duar (India) and 400 kV Dhalkebar (Nepal)–Muzaffarpur (India) lines have already been completed in 2015. The 400 kV Tripura (India)–Comilla (Bangladesh) line was accorded approval in early 2015. In addition, a 400 kV line is also being discussed with Pakistan, and a subsea link project to connect India and Sri Lanka has also been in planning for a long time. The scope of the project is to develop an undersea cable for trade with Sri Lanka (https://www.saarcenergy.org/wp-content/uploads/2019/05/SAARC-Energy-Outlook-2030-Final-Report-Draft.pdf). In this connection, it is flagged that India recently submitted the draft report to the Sri Lankan government for this project. In fact, the Prime minister of India gave a vision of 'One Sun-One World-One Grid' in October 2018 during the First Assembly of the International Solar Alliance, which he reiterated during the

UN Climate Change Conference in Glasgow in November 2021. Indeed, it would be a game changer in years to come in the domain of power sharing.

2.2.17 PROJECTS IN BBIN REGION

The progress made in developing the infrastructure for the energy trade in this region is a ray of hope for complete South Asia as a number of projects of cooperation/collaboration among the member nations are progressing. One of the major reasons is that the borders are stable, and there is no major dispute among the member nations. Important projects of mutual cooperation in the energy exchange domain are as follows:

Source: Base map - Wikimedia Commons;

MAP-4[1]: BBIN region-hydro projects and cross-border transmission lines.
Source: Base map – Wikimedia Commons; https://upload.wikimedia.org/wikipedia/commons/3/34/Ganges-Brahmaputra-Meghna_basins.jpg; Local of dam projects – Data from the Grand Dams Database, compiled by International Rivers and uploaded by The Third Pole in 2014; http://data.thethirdpole.net/layers/136; Location of cross-border transmission lines – Vaidya et al. (2019).

1. **India-Bhutan Hydro Energy Cooperation:** Bhutan has the potential to generate 30,000 MW of hydro energy, of which 23760 MW is considered economically exploitable. The cooperation on harnessing

[1]This map is courtesy of Udisha Saklani et al. "Hydro-energy cooperation in South Asia: Prospects for transboundary energy and water security," Pub in Environmental Science and Policy 114 (2020), page 23 and available on: www.elsevier.com/locate/envsci.

the hydroenergy potential of Bhutan with India commenced in 1955. Initially, it was based on a project-to-project basis, but in 2002, Bhutan came out with its National Water Policy, which was updated in 2007. This policy guides the approach of Bhutan in making use of its water resources. Now, Indo-Bhutan water cooperation relations are governed by this policy. The Indo-Bhutan energy relations are governed by three agreements: the Agreement on Cooperation of 2006 for Hydroelectric Power; the Protocol Agreement of 2008 to increase electricity transfer from Bhutan to India from 5000 MWh to 10000 MWh by 2020 (the deadline has already been postponed), and Inter-Governmental Order of 2014 for developing joint venture hydropower through public sector undertakings of the two governments (An Hindustan Times Report, 2021). Currently, four hydroelectric projects of bilateral cooperation (the 336 MW Chukha Hydroelectric Project, the 60 MW Kurichhu Hydro Electric Project, the 1,020 MW Tala Hydro Electric Project, and the 720 MW Mangdechhu Hydro Electric Project) totaling over 2,100 MW are operational in Bhutan (Udisha et al., 2020). In 2020, India and Bhutan signed a pact for the construction of the 600 MW Kholongchhu joint venture hydroelectric project in Bhutan that is to be completed in five years (Elizabeth, 2020). The Kholongchhu project is one of four additional projects agreed to in 2008 as a part of India's commitment to help Bhutan create a total of 10,000 MW of installed capacity by 2020. The other three projects are 180 MW Bunakha, 570 MW Wangchu, and 770 MW Chankarchu. Projects in the pipeline are 1200 MW Punatsangchu-I and 1020 MW Punatsangchu-II (Ibid. p. xxvi).

2. **Indo-Bangladesh Collaborations:** Indo-Bangladesh energy cooperation is another good example of regional cooperation. The first interconnection came up in 2013. It was based on high voltage direct current (HVDC) to supply 1160 MWh from India to Bangladesh. A new connection is under implementation to transmit 1540 MWh. Yet another agreement has been signed for providing a transmission line to transfer more than 3600 MWh from India to Bangladesh. India is also setting up a coal-fired thermal power station of 1329 MW capacity at Khulna in Bangladesh, located Southwest of Dhaka on the Banks of Rupsha and Bhairab rivers. In the spirit of collective growth of the region, India has agreed in principle to allow Nepal to export 500 MWh of energy to Bangladesh using the Indian Transmission line. Besides power transmission, India and Bangladesh are

also cooperating in the domain of petroleum products. A 130 km India–Bangladesh Friendship diesel pipeline project with Indian grant and assistance is coming up. Further talks are on to identify the possibility of a 'Power Corridor' that would allow India to use the Bangladesh territory to connect to India's Northeast States located in the East of Bangladesh (Suhasini, 2020).

3. **Indo-Nepal Collaborations:** Indo-Nepal cooperation commenced as early as 1954 when the first project was built by India in Nepal. It was the 'Koshi Barrage Project.' Actual construction happened during the period 1958–1962. This project had twin objectives. Firstly, to control floods downstream and secondly; to generate 44 MW of power (Udisha et al., 2020). The project laid the foundation for the future Indo-Nepal energy exchange. An umbrella agreement for the 'Electric Power Trade Agreement' (EPTA) was signed in 1996. In 2014, a 'Cross border transmission interconnection and grid connectivity' agreement was signed between the two countries. Currently, there are 21 power exchange points that exist between the two countries, which can transmit power through 11 kV, 33 kV, and 132 kV power lines (Koshi Barrage Over the Largest River of Nepal, 2018). In 2018, a 140 km long, 400 kV, Muzaffarpur (India) Dhalkebar (Nepal) transmission line was completed (Udisha et al., 2020). It is meant for transferring a high volume of electricity. A plan to construct yet another 140 km long, 400 kV transmission line between Gorakhpur (India)-New Butwal (Nepal) has been finalized (SB Pun). As far as hydropower is concerned, Nepal's Hydropower policy prioritizes, firstly, meeting the internal demand and only then considering export if surplus power is generated. It is ironic that a country Like Nepal, which has a technically feasible potential upward of 50,000 MW, was in 2018–2019 a net importer of power (37.25% of her requirement). As of date, there are 209 hydropower plants under construction with a total capacity of 7952 MW. Once these are completed, Nepal will become a net exporter state with destinations in India and Bangladesh. There are two more projects under execution. The first one of these is the 679 MW Arun-3 hydro-electric power project, from which the power generated less than 21% would be exported to India by a dedicated transmission line. 21% of power will be allocated to Nepal (A New Business Age Report). The second project is the 900 MW Upper Karnali power project. A part of the power produced would be exported to Bangladesh using an

Indian transmission line (An All-India Radio News Report, 2021). India-Nepal energy cooperation is not limited to Hydropower alone. However, much more is needed to be done to exploit the hydropower potential of Nepal. That will not only help Nepal to earn much-needed revenue from the export of power generated to India and through India to Bangladesh. India Nepal is also collaborating in the transportation of petroleum products from India to Nepal. A 69 km long pipeline to transport petroleum products from Motihari in Bihar state of India to Amlekhgunj in Nepal has also been constructed with a capacity of 2 million metric tons/annum (A NS Energy Report).

4. Nepal Bangladesh Collaboration:

i. In August 2018, for Bangladesh – Nepal bilateral energy cooperation, an energy cooperation agreement was signed to oversee investment, development, and trade in hydroelectricity between the two countries. As per the provisions of this agreement, Bangladesh will import 9000 MW of hydropower from Nepal by 2040. The Upper Karnali Project is the first concrete step towards getting 9000 MWh of power from Nepal in the next 10 years.

ii. Consequent to the agreement, a Joint Working Group (JWG) was set up to facilitate the cooperation, and a Joint Steering Committee (JSC) was formed to review the progress of the JWG.

iii. It needs to be appreciated that the progress of this deal is likely to reach fruition only once the desired tripartite agreement between Bangladesh, Nepal, and India is signed. Unfortunately, this agreement has still not been signed (Report by a Special Correspondent, 2019).

iv. Finalization of a Power Tariff Agreement for the import of 500 MWh of hydropower out of 900 MW of power generated from the Upper Karnali project has been completed.

v. Cross-border interconnection – the power from Nepal to Bangladesh will move on the Indian power grid. That is why, in this connection, Indian policy is of importance, which is based on two relevant documents, namely, 'Guidelines for Import/Export of Electricity–2018' and 'Cross Border Trade of Electricity Regulation–2019.'

vi. During the first meeting of the JSC in 2018, among other things, the scaling up of the cooperation in the field of RE, particularly in the field of 'Solar Home Systems,' was also discussed (Muhammad & Umesh, 2020).

vii. Cooperation in the field of capacity building, conservation, reforms, and energy efficiency were also discussed during the meeting of the JWG.

These collaborations confirm that the current state of intra-regional energy trade within the South Asia Region is far from optimal. In terms of the volume of total power and POL traded among the SACs, it is grossly insignificant. Following aspects, in this connection, merit considerations:

- Energy trade is limited to only POL and electricity. Broadly, it entails:
- Bhutan selling electricity to India; India selling electricity to Bangladesh and Nepal.
- India exporting refined petroleum products to Nepal, Bhutan, Sri Lanka, and Bangladesh (only diesel).
- There are no existing intra-regional gas pipelines. Also, a regional power grid stretching across all SACs is still not there.
- The region is heavily dependent on crude and POL imports from the Middle East. Earlier, certain dependence was on Iran too, but due to US sanctions against Iran, it has gone down substantially.
- There is no energy trade between India and Pakistan, though a World Bank-financed power trade deal for 1200 MW was discussed in 2014, and it did make some progress, but due to political differences between the two countries, which has got accentuated post-suspension of Article 370 of the Indian Constitution, it is lying in a state of suspended animation (Udisha et al., 2020).
- The geography and the resource base of the SACs suggest that there is immense scope for expansion of intra-regional energy trade through India's open-access power transmission system, power interconnection between adjacent countries, and gas and oil transmission through pipelines connecting two or more SACs. However, pipelines call for more deliberation and funding, for which not only a consensus among the countries but also consent and support of a funding agency from outside will be needed.

2.3 ISSUES RELATED TO CROSS-BORDER ENERGY TRADE WITHIN SACS

Infrastructure builds up in terms of interconnections between the national grids of the SACs is essential for the cross-border power trade to flourish, though ideally, a regional grid in the ultimate analysis will be the answer for a seamless transfer of power. Such interconnections in short to medium terms will help power surplus countries like Bhutan to sell the power to neighboring India or to Bangladesh through India. It is projected that Bhutan is likely to achieve a power surplus to the tune of 600% by 2024 (A PTI Report, 2014). However, for this purpose, dedicated transmission lines or through transit rights on the existing lines will have to be planned. This flags for improvement to necessary infrastructure and building up of necessary dedicated organization with a view to establishing a 'cross border power exchange (CBPE)' by the interconnected nations within the region. This CBPE may eventually become part of the regional grid as and when that comes up. The scope of work of this CBPE would entail the facilitation of the generators/consumers of one country seamlessly selling/buying power on a short- or long-term basis.

South Asia is likely to become self-sufficient in petrol and diesel (refined petroleum products) by 2030 (A CRISIL Report, 2018). However, the dependence on inter/intra-regional trade for liquefied petroleum gas (LPG) and crude oil shall continue. By setting up product pipelines, petroleum oil lubricants (POL) can be seamlessly supplied between any two countries. The proposed west coast refinery of 60-million-ton capacity at a cost of $ 3 trillion in the Ratnagiri district of Maharashtra, India, will be able to serve as one of the possible options to supply POL products to other SACs and promote intra-regional trade (Ibid. p. xl).

To increase energy cooperation, some of the important steps which can be taken are as follows:

1. **System of Identification of the Technically and Economically Feasible Cross-Border Interconnections:** This establishment of an independent South Asia Energy authority under the aegis of SAARC would be a way ahead. May the time be ripe to revive the SAARC Energy Ring and Power Exchange Market for Electricity trading conceptualized during the 12th SAARC Summit in Islamabad, Pakistan, in 2004 to achieve cooperation within South Asia in the domain of energy, particularly in the areas of electricity, oil, and gas and RE (Ibid. p. xl). The concept has already got a concrete shape when through the Dhaka Declaration in 2005, a special purpose vehicle (SPV) was established in the form of the SAARC Energy Center at Islamabad in 2006 (A South Asia Sub Regional Economic Cooperation (SASEC), 2018).
2. **Development of a Structure for a Regional Power Exchange:** It would entail synchronizing the power system structures of SACs and evolving a mechanism with necessary independent regulatory authority and a common commercial procedure for cross-border energy trade.

3. **Investment in the Generation and Transmission Systems:** It is vital, particularly in underdeveloped but energy-rich Nepal and Bhutan. For this concept to further progress, the support of international agencies and private entrepreneurs would be useful.
4. **Harmonizing Legal and Regulatory Framework:** As the trade grows, a new set of rules/laws will have to be evolved in such a way that the legal dispute arising between the suppliers and consumers can be addressed. In this regard, an 'Energy Charter Treaty' will be required to be put in place to achieve greater security for cross-border energy transfer-related investments and transactions.

2.4 COLLABORATIONS WITH COUNTRIES OUTSIDE THE REGION

1. **Collaboration between Afghanistan and Pakistan of South Asia and Turkmenistan of Central Asia:** A framework agreement for the proposed Turkmenistan–Afghanistan–Pakistan (TAP) Power Interconnection Project was signed in Feb 2018. TAP is being phased into two concurrent stages. The project is under execution. The first phase is planned to be completed by 2021, the TAP countries will utilize the existing infrastructure under the ADB-financed Turkmenistan-Uzbekistan-Tajikistan-Afghanistan-Pakistan (TUTAP) power interconnection project, which aims to export power from Turkmenistan to Afghanistan and thereafter to Pakistan. During the second phase, which is planned to be completed by 2022, power will be transferred from Turkmenistan through the border of Serhetabad (in Turkmenistan) and Torghundy into Herat, Kandahar, and Spin Boldak in Afghanistan and will be exported to Chaman and Quetta in Pakistan. The power trade will utilize existing power infrastructure as well as promote collaboration through new investments to improve the transmission of power in the three countries. Options will also be explored to especially optimize and use any excess power or transmission capacity during the winter months—from October to April, through a planned integration with the Central Asia-South Asia (CASA) system. The project will include the construction of around 500 km of 500 kV transmission line between Turkmenistan, Afghanistan, and Pakistan. Once completed, the project will be able to transfer up to 4,000 MW of power from Turkmenistan into Afghanistan and Pakistan (https://www.saarcenergy.org). It will meet Afghanistan's energy needs substantially and also facilitate power trade among the three countries.

2. **Iran–Pakistan Energy Collaboration:** Currently, Pakistan is importing a meager 35 MW from Iran basically for Gwadar, and the work is in progress to upscale it to 100 MW (News Release by the ADB, 2018). Iran has the second-largest proven gas reserves. Iran also holds a huge amount of crude oil. Further collaboration between energy-deficient Pakistan and energy surplus Iran will be beneficial for both countries. However, this can happen only once sanctions imposed on Iran are lifted.
3. **Cooperation with BIMSTEC:**
   i. The Bay of Bengal Initiative for Multi-Sectoral Technical and Economic Cooperation (BIMSTEC) was formed in June 1997. Subsequently, in July 2004, it was renamed as the Bangladesh–India–Myanmar–Sri Lanka–Thailand Economic Co-operation (BIMSTEC).
   ii. The area is reasonably rich and highly diverse in energy resources. India is not only the biggest in size but has maximum resources, particularly coal, and has a great potential for renewables. The resources of the Group countries are tabulated in Table 2.3.

TABLE 2.3 BIMSTEC Resource Potential

Country	Coal	Oil	Gas	Biomass#	Hydro	Renewables*
	Million Tons	Million Tons	Trillion Cubic Feet	Metric Tons	GW	GW
Bangladesh	3,300	–	5.7	218	–	3.67
Bhutan	1	–	–	62.5	23.8	12.76
India	3,19,020	600	45.5	4,150	145	1,000
Myanmar	120	64.3	41.3	3,303	100	60.7
Nepal	<1	–	–	1,056	42	4.829
Sri Lanka	–	–	–	155.5	2	11.6
Thailand	1,063	–	6.6	1,838	15.2	23.0
Total	3,23,504	664	99	10,783	328	1,117

–: *Denotes: Either the particular resource is nil, or its value is less than 0.5.*
*: *Renewables: include; Solar; wind and small hydro.*
#: *Biomass: includes Forest and other wooded land.*
Source: BP Statistical Review 2019; Sectoral and National Plans of individual countries, IRENA, UN, ADB, Government Statistic Report, Government Portals, European Journal of Sustainable Development Research dated 27/02/2020.

➢ ***Note:*** Table 2.3 shows that there is a scope for collaboration between Myanmar and SACs, especially in the domain of gas trading and also hydro-based energy trade.

iii. During the sixth Ministerial meeting in 2004, Nepal and Bhutan also joined the forum. One of the many areas that were identified for cooperation among the member nations was energy. The First BIMSTEC Energy Ministers Conference was held in 2005. It was decided during the conference that a BIMSTEC Energy Center would be established in India to share experiences in reforms, restructuring, regulation, and best practices in the energy sector. Other decisions taken during the meeting were as follows:

a. Establishment of a Trans BIMSTECH gas pipeline.
b. Establishment of the BIMSTECH Trans Power Exchange.
c. Development of new and RE sources; small hydro projects (less than 50 MW Capacity); R&D in the field of rice husk cogeneration (Ibid. p. xlv).

2.4.1 *GRID INTERCONNECTIONS*

A MoU was signed in 2018 to create a broad framework for the countries to cooperate towards the implementation of grid interconnections for the trade in electricity. Some of these interconnections have already been created. Bhutan and Nepal's interconnections with India offer scope for electricity trade with other BIMSTECH countries. India also has grid interconnections with Bangladesh, Myanmar, and Sri Lanka. These networks can facilitate short-, medium- and long-term trading arrangements for the import as well as export of electricity with neighboring countries. During the early part of 2020, it was revealed that trilateral cooperation between India-Myanmar-Thailand was underway to set up a 3000 km long power grid (https://web.archive.org/web/20171206152918/http://bimstech.org/sectors/energy/).

2.4.2 *ASIAN ENERGY HIGHWAY (AEH)*

It is significant to note that a need for a regional grid has been recognized by most Asian countries to create a pan-Asian integrated energy grid or an Asian Energy Highway (AEH) (Mihir, 2020). Several sub-regional initiatives, which are also either being planned or are underway to enable greater cooperation in the domain of energy, can be co-opted in this initiative to optimize its efficacy. The SAARC Energy Ring discussed in the earlier part of the chapter and its related Market Mechanism for Electricity trade will be its part. Similarly, CASA-1000 and CAS Regional Electricity Market (CASAREM) will also be part of the contributory network of AEH and will augment supply to SACs.

2.4.3 GAS PIPELINES

South Asia does not have much oil or gas, but it is located next to Central Asia, West Asia, and Myanmar, which are quite rich in these two energy resources, and still, for some time, these two will continue to remain the main sources of energy conversion. South Asia will, therefore, continue to look for transportation of these resources from North, West, and, to a limited extent, from East. There are two significant pipelines, namely, the Turkmenistan–Afghanistan–Pakistan–India (TAPI) pipeline and the Myanmar–Bangladesh–India (MBI) pipeline. In addition to these two pipelines, there are two more important pipelines whose future is not very certain, but when we talk of energy cooperation, these also have a very profound bearing on the energy sharing between Iran–India, and Iran–Pakistan. The details of all these pipelines are as follows:

1. **Turkmenistan–Afghanistan–Pakistan–India Gas Pipeline (TAPI):** It is an 1814 km natural gas pipeline originating at the Galkynysh gas field of Turkmenistan (Concept Note – Asian Energy Highway). Once completed, it will transport 33 bcm of gas for 30 years with a distribution of 14 bcm (42%) each to Pakistan and India and 5 bcm (16%) to Afghanistan (Abdujalil, 2015). Afghanistan will also receive a sum of $ 400 million/year as the transit fee. Work on the Turkmenistan portion commenced on 13 Dec. 2015 (Michael, 2015) and was completed in mid-2019 (A Reuters Report, 2015). Work on the Afghan side commenced on 24 Feb. 2018 (Imaduddin, 2019). The work on the Pak side was planned to commence in Oct 2019, and it was planned to be completed by 2020 (Ibid. p. xlx). Work on this pipeline is in progress. The scheduled date of completion is 2022, but it is likely to get delayed due to a very unstable political and security situation.

MAP-5: Turkmenistan–Afghanistan–Pakistan–India (TAPI) pipeline project proposed map

Source: Reprinted with permission from Danish et al., 2020. https://creativecommons.org/licenses/by/4.0/

2. **Iran–India Gas Pipeline:** For the proposed pipeline, a feasibility Study has been done by the South Asia Gas Enterprise Pvt Ltd. (SAGE). It will be a 1300 km pipeline with the start point at Chabahar in Iran. It will be an undersea pipeline, coming to Porbandar in Gujarat, India, via the Gulf of Oman and skirting the EEZ of Pakistan. On a future date, a pipeline from Oman also may join to augment the supply. In India, it will be connected to the National Gas Grid. The cost of construction will be $ 4 billion, and it will take two years to build it from the time the necessary sanction is given and the signing of the Gas Sale Purchase Agreement (GSPA). Once completed, it will transport 31.5 million standard cubic meters of gas/day. The cost of gas will be cheaper by $ 2 as compared to the imported LNG. To further bring the cost down, it has been planned by the company that the pipeline would be available for rent for those who want to buy gas from Iran and transport it back (Ibid. p. xlivii). However, the limiting factor is that Iran is currently under US Sanctions[2]. Therefore, it would be in Fitness of Things that India tries to get a waiver from the US on the invocation of CAATSA.[3] Or try to get the ban

[2]On 08 May 2018, the US announced its withdrawal from the Iran Nuclear Deal. Consequently, Sanctions were reinstated in November 2018.

[3]CAATSA- It is a US Federal Law that imposes sanctions on Iran, North Korea, and Russia.

on Iran Lifted. Maybe the involvement of Oman would be able to circumvent the CAATSA provisions.

MAP-6: Iran–Pakistan (IP) pipeline project proposed map
Source: Reprinted with permission from Danish et al., 2020. https://creativecommons.org/licenses/by/4.0/

3. **Iran–Pakistan Pipeline**[4]Initially conceptualized as the Iran–Pakistan–India pipeline, but it was reduced to be only the Iran-Pakistan pipeline when, in 2009, India withdrew from the project on the issue of security of the pipeline and the pricing. Be as it may, this pipeline starts from Asalouyeh using gas from the South Pars gas field and stretches 1172 km through Iran (A PTI Report, 2017). In Pakistan, the length is 785 km (A Mehr News.com Report, 2010). In Pakistan, it will pass through Baluchistan and Sindh and finally terminate at Multan. The pipeline will carry 8.7 bcm of gas/year and 40 bcm as the maximum capacity. The cost of pipelines is estimated to be $3–4 billion (Shardul, 2013). It will have the potential to get stretched to Delhi should India change its policy at a later date (Daily Times Report, 2010). The portion of the pipeline on the Iranian side has been completed. On 11 Mar 2013, the construction on Pakistan's side was inaugurated jointly by the then Presidents of Pakistan and Iran

[4]It was conceptualized as an I-P-I Pipeline, but as of date, it is only an I-P Pipeline, as India has withdrawn from the project.

(Sohaib, n.d.). It was proposed to be built in 22 months in conjunction with Iran. However, it is far from getting ready due to sanctions against Iran since Feb 2014. As of early 2019, the project remains substantially delayed (Pakistan–Iran Gas Pipeline Defies US, 2013). (*The project is unlikely to reach fruition unless either the US changes its policy with respect to sanctions against Iran or Iran on its own comes out clean on the nuclear proliferation Issue*).

Myanmar - Bangladesh - India Proposed Natural Gas Pipeline

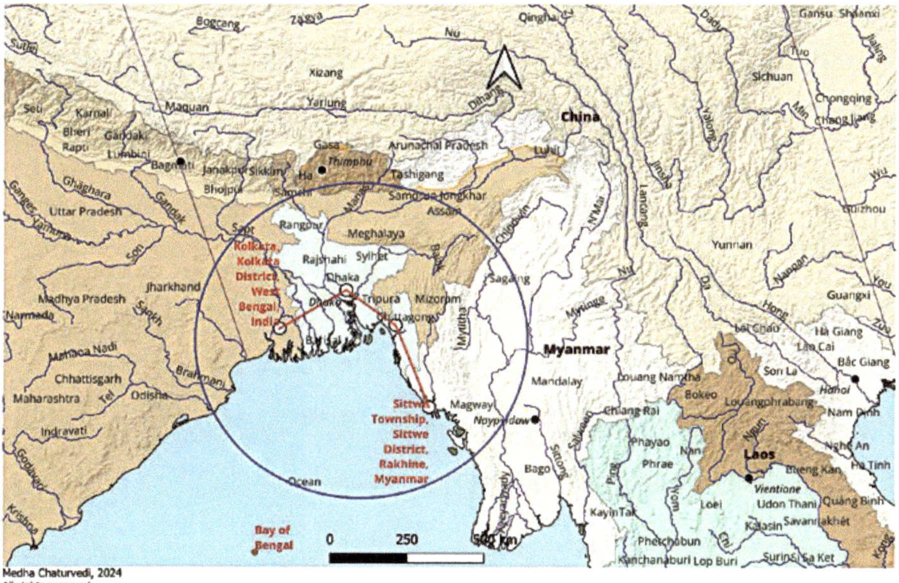

Medha Chaturvedi, 2024

MAP-7: Myanmar-Bangladesh-India proposed natural gas pipeline

4. **Myanmar–Bangladesh–India Pipeline:** India, ever hungry for gas, had proposed a 6900 km pipeline from Myanmar for the gas available in the offshore waters of Sittwe in early 2000, but the project could not be progressed because of the attitude of the then Govt of Bangladesh, who had put very difficult conditions, which India did not accept and the project became a non-starter (Khalid, 2019). To make matters worse, in Dec 2008, the Chinese National Petroleum Corporation signed a 30-year hydrocarbon purchase and sale agreement with Daewoo International to draw gas from the Shwe gas project located offshore (off Sittwe) in the Bay of Bengal. China also

agreed to build a pipeline to transport the gas from this gas field. The pipeline finally became operational in July 2013. It was a strange coincidence that the blocks A-1 and A-3 of the Shwe Gas Project were meant to be exploited by a consortium led by Daewoo International of South Korea. Other members of the consortium included two Indian companies, namely ONGC Videsh Limited (OVL) and Gas Authority of India Limited (GAIL), one South Korean company; a Korean Gas Corporation, and a Myanmar company; Myanmar Oil and Gas Enterprise companies. The project entailed an offshore production platform, an underwater pipeline, and an onshore gas terminal in Kyauk Phyu town on the Arakan Coast at a cost of $3.73 billion to develop. The anticipated production capacity was 500 million cubic feet of natural gas per day (India–Myanmar–Bangladesh Gas Pipeline, 2020). Thus, the failure on the part of Bangladesh and India to come to a common approach led to the failure of the 'Trilateral gas pipeline' (Shwe Gas and Pipelines Project Myanmar, 2016).

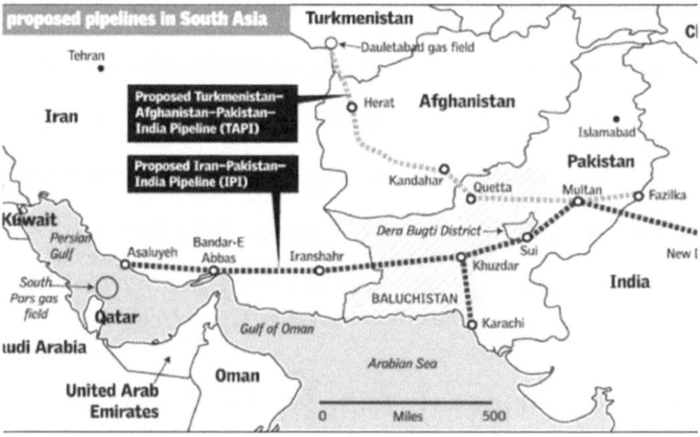

MAP-8: Proposed Myanmar–Bangladesh–India Pipeline.
Source: Irrawaddy online magazine open-source image.

It is believed that the exploitation of this gas field by the Chinese has left very little gas in the gas field (this fresh input came up during a private interaction with an official of the GAIL), as such, the pipeline project was almost scrapped. However, the issue figured once again during the visit of Indian Prime Minister Sri Narendra Modi to Bangladesh in June 2015. The PMs of India and Bangladesh directed their respective officials to review the issue and consider finalizing the terms for the establishment of a Joint Venture Company that will construct and operate the pipeline. This proposed pipeline will connect Sittwe in Myanmar with the Northeastern States of India via Chittagong in Bangladesh. Under the agreement, an LPG plant will come up at Chittagong, from where the gas will be piped to the Northeastern Region of India. In Dec 2018, India's minister of Petroleum and Natural Gas, Sri Dharmendra Pradhan, stated that India was continuing to explore ways to expand its gas network to Bangladesh through Myanmar (Mahendra, 2020). The project will include a Gas Pipeline to transport gas coming from Russia to make the Northeastern part of the country a Hydrocarbon hub. This hub will come up in the next 15 years and will cater to the demands of Bhutan, Myanmar, and other ASEAN countries. The project will also entail exploration of existing and new oil and gas blocks, setting up of the pipeline, which will have four new branches, five new petroleum, oil, and lubricants product pipelines, and three new LPG pipelines. The approximate cost of the project is likely to be Rs 1,30,000 Crores (Ibid. p. lxi).

At the policy level, the Indian Hydrocarbon Vision 2030 for Northeast India stipulates that the availability of any surplus supplies of Natural gas in Myanmar should be imported to India's Northeastern States. This conclusion is based on the understanding that the Northeastern gas grid of India and Myanmar's promise of energy reserves complement each other (as a user and the supplier, respectively), and there is a need to reconsider reviving the idea of a bilateral/trilateral natural gas pipeline. It also needs to be taken note of the fact that Myanmar is reportedly planning to open fresh bids for its oil and gas blocks to international companies. This offers an opportunity to go for these blocks as they will benefit SACs in general and India in particular (Prabin, 2016).

2.5 WAY AHEAD

2.5.1 METHODOLOGY TO ACHIEVE RESOURCE OPTIMIZATION

Aspects merit consideration for the effective exploitation of energy-bearing resources, both indigenous as well as imported ones, to achieve their optimal utilization. These are, firstly, the need to exploit available energy-bearing resources of the region to optimize the availability of energy; the need to find engineering solutions to integrate the available diverse sources to convert them all into a usable form, the need to create a management organization a regulatory mechanism and an organization dedicated to capacity building and fourthly; there is the need to build an infrastructure for achieving systemic optimization. It needs to be noted that the impediments to the smooth functioning of the desired system are bilateral disputes. The desired standard operating procedure should ensure that the disputes are separated from the economic developmental issues, and they should not be allowed to impact the economic development. It has been noticed that wherever disputes do not exist or they are manageable, energy cooperation becomes better, and that helps in improving the GDP. A case in point is the BBIN Region, where good progress in collectively exploiting energy-bearing resources has helped the region to grow economically.

2.5.2 RESOURCES WHICH NEED TECHNOLOGICAL UPGRADATION FOR OPTIMAL EXPLOITATION

Strategy, including necessary R&D policies, needs to be formulated to develop technologies to exploit the potential of resources that so far have not been exploited:

1. **Coal:** The quality of coal in South Asia is such that it needs technologies that can help its optimal utilization like; coal gasification, integrated gasification combined cycle, carbon capture and storage, and many others. Since coal is available in regions in sufficient quantity but qualitatively not the best, investing in those technologies that can be leveraged to exploit it optimally and ensure that the ill effects of coal that cause pollution are reduced substantially.
2. **Renewables:** These are those which are area and country-specific. However, in addition to those, common renewable resources are waste, biomass, and small hydro need to be addressed. In this connection, the formation of the International Solar Alliance in 2015 at the time of COP-21 is quite significant. However, systems related to the exploitation of renewables suffer from low efficiency. There is a need to go for research in terms of materials and designs to address this issue. Waste management in SACs is a major issue. However, a cost-effective technology can convert the menace of waste into a major source of energy.
3. **Ocean Thermal Energy Conversion (OTEC):** Four out of seven SACs, namely; Bangladesh, India, Sri Lanka, and Maldives, are maritime nations that have access to warm waters of the sea. Exploitation of energy conversion of this form will help substantially to enhance the availability of energy. However, again, this conversion

calls for R&D of materials and systems in addition to related equipment that can work efficiently under seawater as well as in saline conditions, which have an adverse impact on the moving parts of the equipment. India has had a head start in the domain of this technology, and as such, her experience can be of great value to the other maritime countries in the region.

4. **Nuclear Energy:** While Pressurized Heavy Water reactor-based technology using Uranium is fairly well established, the fast breeder reactor-based technology using Thorium and Thorium-based reactor technology is still not fully developed. It is essential that these technologies are speedily brought up for their commercial exploitation because India has almost 25% of the World's Thorium reserve, and such progress will help the region as a whole to achieve energy sufficiency. In this connection, it is relevant to realize that both India and Pakistan have domain knowledge, and thus, a collective effort will help to enhance the availability of energy that can be shared/traded with other countries of the region.
5. **Hydrogen:** It is the fuel of the future. Therefore, there is a need to work both on blue as well as green hydrogen energy-based systems, including fuel cells. In this connection, it is significant to note that during his 75th Independence Day Address, the PM of India announced the National Hydrogen Mission (Sanket, 2019). The fruition of this mission will benefit the entire South Asian Region in the future.
6. **Engineering Capability:** It needs to continuously evolve. While systems and equipment for the exploitation of traditional equipment and technologies are unlikely to diminish any time soon in the near future, the manufacturing capability to exploit new forms of energy has to be developed. Future equipment will substantially be smart; therefore, the capacity and capability to design and manufacture systems based on the internet of things (IoT), Artificial Intelligence (AI), robotics, and sensor-based systems will have to be built up.
7. **Capacity Building and Research and Development:** Necessary training institutions/research centers based within the region will have to be established, and existing ones will have to be further strengthened and upgraded. In this connection, the SAARC Energy Center at Islamabad is a good idea; however, it needs to be strengthened to do R&D in the frontier technologies for augmenting energy generation and efficient transmission. The

Research Center of BIMSTEC at Bengaluru and the Department of Oceanography at IIT Madras also need to be tapped in to get the necessary technologies developed. Country-specific Energy Centers should also come up so that technologies specific to the resources of that country can come up. There should be a system of seamless integration between all these centers so that a cooperative model for the entire region can also be developed. Besides South Asian universities, there is a need to establish a Skill Training Center for the SACs so that capacity building can be strengthened. Development of the Industry in the region by complementing the strengths of each other will help South Asia to develop indigenous manufacturing capability, which will help in cutting the cost of energy infrastructure build-up. R&D setup will also be required to develop technologies for resources like Carbon Hydrates and storage technologies.

8. **Management:** These resources for optimal utilization, furthering the cooperation of the individual countries with regions/countries outside South Asia, like cooperation and collaboration with the countries of Central Asia, Iran, and BIMSTEC, need to be encouraged. Such an approach will not only contribute towards the optimization of these additional resources but will also help in getting additional revenue to the transit countries. As a part of management, a common power/gas grid needs to be developed, which will help in seamless integration. In this connection converting into reality the concept of AEH would be useful. The first step towards creating an AEH has already been taken, with the various regional grid integration initiatives underway. Further, the AEH concept has been integrated into the Asia-Pacific regional cooperation framework for energy, with analytical work for the same being addressed. The next steps required to be worked out are the nuts and bolts of the AEH concept and the evaluation of the socio-economic and environmental benefits of creating the AEH. As and when established, the AEH is expected to improve energy efficiencies in power production and trade through greater regional integration. It will also help in optimizing resource allocation between supply and demand centers and open up trading opportunities and resource security alternatives for all Asian countries.

9. **Formation of an Independent Regulatory Mechanism:** It needs to be formed for the region under the aegis of SAARC basically to help regulate and optimize energy resources within the region. Although the region is rich in resources keeping in view the climate change issue in mind, slowly and gradually, SACs will also be required to make a transition to renewables. While micro-management of resources and infrastructural assets will continue to be the responsibility of the individual countries, the issues at the policy and macro-level will be addressed by this regulating authority. As the energy cooperation grows, so will the disputes based on individual perceptions and aspirations. This calls for the establishment of a dispute resolution mechanism under the aegis of this regulating authority.
10. **Building Up New Infrastructure:** Infrastructure needs to continuously transform to a changing form of energy conversion. It will have to cater to generation, transmission, utilization, and construction practices.

Finally, all this would be possible only if mutual trust exists between the countries of the region. In this connection, a case study of the failure of the earlier edition of the MBI pipeline is relevant. The pipeline became a non-starter because India and Bangladesh could not resolve their differences in time, and China managed to get the deal through with Myanmar. Even today, it appears to be a clear case of lost opportunity. On the contrary, despite having disputes of all sorts, the Indus Water Treaty between India and Pakistan has remained intact for more than 60 years, which shows that the countries, when sectionizing the problems between economic benefits and the political differences between them, can move ahead. A positive example in this connection is Indo-Bhutan energy cooperation which has been a 'win-win' case for both the countries; the prosperity of Bhutan has gone up, and the energy needs of the Eastern part of India are substantially addressed.

2.6 CONCLUSION

An analysis of resources and the impact of existing energy resources on the environment emerges that in the future, RE systems will play an important part in achieving a sustainable energy mix. However, such a change calls for regional interconnection, which presently is a captive of bilateral dynamics of the countries of the region. With the help of the identified challenges and the proposed strategies, all these South Asian countries can ensure a reliable energy mix by encouraging effective utilization of fossil fuels, embracing RE sources and effective energy policies, and promoting regional energy cooperation and energy market integration. Although the countries are attempting to implement their respective energy vision with the main emphasis on the environment, it is considered challenging for the developing nations to ensure the security of their increasing energy demand with a paradigm shift to RE development. In Bangladesh, Nepal, and Pakistan, prospects for growth and development are huge, but their current economic situation is not that sound to support such ambitious energy projects without foreign direct investment and funding from international monetary bodies. Importantly, energy access to remote and rural areas of the countries can be safeguarded by fostering dispersed power generation and developing mini grids in the region.

KEYWORDS

- **collective energy security**
- **consumption pattern**
- **energy resources**
- **geography**
- **intra-regional cooperation**
- **population**
- **resilience**

REFERENCES

A CRISIL Report Compiled Under the Aegis of SAARC Energy Center. *SAARC Energy Outlook*. Available on https://www.saarcenergy.org/wp-content/uploads/2019/05/SAARC-Energy-Outlook-2030-Final-Report-Draft.pdf.

A CRISIL Report, (2018). *SAARC Energy Outlook 2030*. Compiled under the aegis of SAARC Energy Center during.

A Mehr News.com Report, (2010). *Khatam-ol-Anbia to Build 2nd Phase of IPI Gas Pipe* Available on https://web.archive.org/web/20111120100109/http://www.mehrnews.com/en/NewsDetail.aspx?NewsID=1131606 (accessed on 12 August 2023).

A New Business Age Report. *New Butwal-Gorakhpur Transmission Line: Preparations for Joint Investment Agreement in Final Phase*. Available on https://www.newbusinessage.com/Articles/view/12172 (accessed on 12 August 2023).

A NS Energy Report. *Upper Karnali Hydro Power Project Nepal.* Available on https://www.nsenergybusiness.com/projects/upper-karnali-hydropower-project-nepal/# (accessed on 12 August 2023).

A PTI Report, (2014). *Pakistan to Import Electricity from India: Report.* pub by The Economic Times / Industry.

A PTI Report, (2017). *Under the Sea, the Iran-India Gas Pipeline can bring Cheaper LNG to India.* Pub in Live Mint. available on https://www.livemint.com/industry/wj8PfRvGGyAfh1p8aoPX9K/undersea-IranIndia-gas-pipeline-can-bring-cheaper-LNG-to-I.html (accessed on 12 August 2023).

A PTI Report, (2017). *Undersea Iran-India Gas Pipeline Can Bring Cheaper LNG to India.* Pub in Mint.

A PTI Report, (2020). *India to Have 450 GW of Renewable Energy by 2030: President.* Reported by The ETEnergyWorld.com. Available on https://energy.economictimes.indiatimes.com/news/renewable/india-to-have-450-gw-renewable-energy-by-2030-president/73802626 (accessed on 12 August 2023).

A Reuters Report, (2015). *Turkmenistan Starts Work on Gas Link to Afghanistan, Pakistan, and India.*

A South Asia Sub-Regional Economic Cooperation (SASEC) news release (2018). *SAARC Council of Experts of Energy Regulators (Electricity) Hold Third Meeting.* Available on: https://www.sasec.asia/index.php?page=news&nid=965&url=saarc-council-of-experts-energy-regulators-third-meeting (accessed on 12 August 2023).

Abas, N., Kalair, A., Khan, N., & Kalair, A. R., (2017). 'Review of GHG emissions in Pakistan compared to SAARC countries. *Renew. Sustain. Energy Rev.*, *80*, 990–1016.

Abdujalil, A., (2015). *Is Turkmenistan Gas Line a Pipe Dream*? Reported by the BBC News.

Alternative and Renewable Energy, (2020). ARE Policy 2019, Government of Pakistan. [Online]. Available: http://www.aedb.org/images/ARE_Policy_2019_AEDB.pdf (accessed on 12 August 2023).

An ADB Report. *Energy Trade in South Asia Opportunities and Challenges.* Available on https://www.adb.org/sites/default/files/publication/29703/energy-trade-south-asia.pdf (accessed on 12 August 2023).

An All-India Radio News Report, (2021). *Nepal Decides to Award Hydropower Project to India's Satluj Jal Vidyut Nigam.* Available on http://newsonair.com/News?title=Nepal-decides-to-award-hydropower-project-to-India%26%2339%3Bs-Satluj-Jal-Vidyut-Nigam&id=408960 (accessed on 12 August 2023).

An Economic Times Report, *India Independence Day: PM Modi Announces National Hydrogen Mission for Self-reliance in Energy.* Uploaded on: https://m.economictimes.com/news/india/india-independence-day-pm-modi-announces-national-hydrogen-mission-for-self-reliance-in-energy/videoshow/85344456.cms (accessed on 12 August 2023).

A Hindustan Times Report, (2021). *One Sun, One World, One Grid: PM Modi Calls for Global Solar Grid at COP-26.* Uploaded on: https://www.hindustantimes.com/india-news/one-sun-one-world-one-grid-pm-modi-calls-for-global-solar-grid-at-cop26/ (accessed on 12 August 2023).

Ankush, K., (2018). *Electricity Trade in South Asia Could Grow up to 60,000 Mw Through 2045.* Pub in ET Energyworld.com. Available on https://energy.economictimes.indiatimes.com/news/power/electricity-trade-in-south-asian-could-grow-up-to-60000-mw-through-2045/66477793 (accessed on 12 August 2023).

Concept Note – Asian Energy Highway. Available on https://www.unescap.org/sites/default/d8files/Session_3_concept_note_on_Asian_Energy_Highway.pdf (accessed on 12 August 2023).

Danish, Mir Sayed Shah & Senjyu, Tomonobu & Zaheb, Hameedullah & Sabory, Najib Rahman & Ahmadi, Mikaeel & Ibrahimi, Abdul Matin & Nazari, Zahra & Ahadi, Mohammad. (2020). Afghanistan as an emerging regional energy hub. 1. 10-14. 10.37357/1068/jso.1.1.02.

Elizabeth, R., (2020). *India, Bhutan Sign Pact for First Joint Venture Hydro Power Project.* Pub by Mint.

Gulagi, A., Choudhary, P., Bogdanov, D., & Breyer, C. (2017). 'Electricity system based on 100% renewable energy for India and SAARC *PLoS One*, *12*(7), Art. no. e0180611.

Ibid. p. ii.

Ibid. p. lxi.

Ibid. p. xl.

Ibid. p. xlivii.

Ibid. p. xlv.

Ibid. p. xlx.

Ibid. p. xx.

Ibid. p. xxvi.

Imaduddin, (2019). *Pakistan- Turkmenistan to Discuss TAPI Gas Pipeline Groundbreaking in Aug*. Pub by Business Recorder. Available on https://www.brecorder.com/news/508207/ (accessed on 12 August 2023).

India–Myanmar–Bangladesh Gas Pipeline, (2020). Pub by Global Energy monitor Wiki. Available on https://www.gem.wiki/India-Myanmar_Bangladesh_Gas_Pipeline#cite_note-ti-1 (accessed on 12 August 2023).

Internet upload: https://web.archive.org/web/20171206152918/http://bimstech.org/sectors/energy/ (accessed on 12 August 2023).

Internet upload: https://www.saarcenergy.org (accessed on 12 August 2023).

Islam, S., & Khan, M. Z. R., (2017). 'A review of the energy sector of Bangladesh. *Energy Procedia*, *110*, 611–618.

Kaur, R., (2016). Evaluation of solar photovoltaic programs under Jawahar Lal Nehru solar mission in India. *Dyn. Public Admin.*, *33*(2), 164–177.

Khalid, H., (2019). *IP Gas Project is in Limbo: Pakistan Wants Iran to Interpret Sanctions*. Pub in the DAWN. Available on https://www.dawn.com/news/1454952 (accessed on 12 August 2023).

Koshi Barrage Over the Largest River of Nepal, (2018). Boss Nepal.

Mahendra, P. L., (2020). *How India Lost a Gas Pipeline to China*. Pub in The Kathmandu Post.

Manju, S., & Sagar, N. (2017). *Progressing Towards the Development of Sustainable Energy: A Critical Review on the Current Status, Applications, Developmental Barriers, and Prospects of Solar Photovoltaic Systems.*

Michael, T., (2015). *A Fillip for the TAPI Pipeline*. Pub by The Diplomat. Available on https://thediplomat.com/2015/12/a-fillip-for-the-tapi-pipeline/ (accessed on 12 August 2023).

Mihir, B., (2020). *Prospects for Energy Co-operation Under BIMSTEC*. Pub by the ORF.

Muhammad, H., & Umesh, R. R. (2020). *Bangladesh- Nepal Energy Cooperation: The Horizon of New Possibilities.* Pub by The Himalayan. Available on https://thehimalayantimes.

com/business/bangladesh-nepal-energy-cooperation-the-horizon-of-new-possibilities/ (accessed on 12 August 2023).

News Release by the ADB, (2018). *Power Inter Connection Project to Strengthen Power Trade Between Afghanistan, Turkmenistan, and Pakistan*. Available on https://www.adb.org/news/power-interconnection-project-strengthen-power-trade-between-afghanistan-turkmenistan-pakistan (accessed on 12 August 2023).

NITI Aayog, (2017). *Draft National Energy Policy*. National Institution for Transforming India. Government of India, New Delhi, India. [Online]. Available: http://niti.gov.in/writereaddata/files/new_initiatives/NEP-ID_276 (accessed on 12 August 2023).

Pakistan–Iran Gas Pipeline Defies US, (2013). Reported by BBC News.

Prabin, K., (2016). *Center May Revive Myanmar- Bangladesh-India Gas Pipeline*. Pub in the Times of India. Available on https://timesofindia.indiatimes.com/city/guwahati/center-may-revive-myanmar-banladesh-india-gas-pipeline/articleshow/52904470.cms (accessed on 12 August 2023).

Rehman, M. U., & Rashid, M., (2017). 'Energy consumption to environmental degradation, the growth appetite in SAARC nations. *Renew. Energy*, *111*, 284–294.

Report by a Special Correspondent, (2019). *India Nepal Inaugurate Motihari- Amlekhganj Oil Pipeline*. Pub by The Hindu.

Ritu, G., (2006). *How is South Asia Faring on Energy?* Pub in Down to Earth. Available: on https://www.downtoearth.org.in/news/how-is-south-asia-faring-on-energy-8591 (accessed on 12 August 2023).

Sahoo, S. K., (2016). Renewable and sustainable energy reviews solar photo- voltaic energy progress in India: A review. *Renew. Sustain. Energy Rev.*, *59*, 927–939.

Sanket, S. K., (2019). *Opportunities for India in Myanmar's Energy Sector*. Pub by ORF.

SB Pun. *The 400 kV Dhalkebar-Muzaffarpur 140 km Transmission Line*. Available on 11273-Article Text-39487-1-10-20141017.pdf.

Shardul, (2013). *Pakistan Ratifies Iran- Pakistan Pipeline Agreement*. Pub by Natural Gas World. Available on https://www.naturalgasworld.com/pakistan-ratifies-iran-pakistan-pipeline-agreement (accessed on 12 August 2023).

Shwe Gas and Pipelines Project Myanmar, (2016). by BankTrack. Available on https://www.banktrack.org/project/shwe_gas_and_pipelines_projects/pdf (accessed on 12 August 2023).

Sohaib, S., (n.d.). *Iran-Pak-India Gas Pipeline: Implications and Prospects*. Pub in Business and Finance Review and Uploaded on: http://jang.com.pk/thenews/jan2007-weekly/busrev-15-01-2007/p8.htm (accessed on 12 August 2023).

Solar Power Plants to Replace Fossil Fuel, (2010). Pub in ZAWYA. Available on: https://web.archive.org/web/20110610071343/http://www.zawya.com/mobile/default.cfm/sidZAWYA20101220044922/actstory (accessed on 12 August 2023).

Sudha, R. C., (2006). *India, Bangladesh Fight Against the Current*. Pub by Asia Times. Available on http://atimes.com/atimes/South_Asia/HF08Df04.html (accessed on 12 August 2023).

Suhasini, H., (2020). *India, Bhutan Sign Pact for First Joint Venture Hydropower Project*. Pub in The Hindu. Available on https://www.thehindu.com/news/national/india-bhutan-sign-pact-for-first-joint-venture-hydropower-project/article31945878.ece (accessed on 12 August 2023).

The World Bank, *Central Asia Electricity Trade brings Economic Growth and Fosters Regional Cooperation,* Accessed on 8 February 2021 URL: https://www.worldbank.org/en/news/

feature/2020/10/20/central-asia-electricity-trade-brings-economic-growth-and-fosters-regional-cooperation

Udisha, S., et al., (2020). Hydro-energy cooperation in South Asia: Prospects for transboundary energy and water security. *Environmental Science and Policy, 114*, 22–34. Para 4.1. Available on www.elsevier.com/locate/envsci (accessed on 12 August 2023).

USAID South Asia Regional Initiative for Energy Integration repository. Accessed on 23 February 2021, URL: https://sari-energy.org/program-activities/cross-border-electricity-trade/country-energy-data/

Ul-Haq, A., et al., (2020). *Energy Scenario in South Asia: Analytical Assessment and Policy Implications.* Pub by IEEE Access. Available on file:///Users/majgenajay/Desktop/IEEE%20Xplore%20Full-Text%20PDF:.html (accessed on 12 August 2023).

Yu, S., Evans, M., Kyle, P., Vu, L., Tan, Q., Gupta, A., & Patel, P. (2018). 'Implementing nationally determined contributions: Building energy policies in India's mitigation strategy. *Environ. Res. Lett.*, *13*(3), Art. no. 034034.

CHAPTER 3

India's Role in Clean Energy Initiatives in South Asia

MANOJ KUMAR SRIVASTAVA

Department of Physics, Army Cadet College, Indian Military Academy, Dehradun, Uttarakhand, India

ABSTRACT

In ancient India, people were taught to maintain harmony with nature in all their deeds. Hence, they accepted this doctrine and practiced to worship nature and its creations. However, over the ages, we have plundered our natural resources without showing serious concerns about sustainability. We forgot that we are the only custodian of natural resources, and it is to be passed on without depleting them to our future generations.

Recently, India has accepted this global challenge as a leader in the South Asian region so that natural resources are harnessed as per the doctrine of sustainability. Referring to the Paris Accord on climate change, India paid due attention to harnessing renewable and clean energy sources, mitigating carbon emissions, creating carbon sinks by forestation, etc. India has started a widely spread solar energy harnessing program as the founder nation of the International Solar Alliance in the South Asian region. India is chasing its ambitious goal of 175 GW of Solar Energy power generation by 2022. These efforts will definitely bring down the undesired climate changes in this region. Nationally Determined Contributions (NDCs) of the Indian government have aimed to reduce the emissions intensity of GDP to 33–35% below what it was in 2005 by the year 2030 with multiple initiatives taken both in industrial and domestic sectors.

The South Asian Energy Corridor: Toward Sustainability and Security. Amritanshu Shukla, PhD, Atul Sharma, PhD & Saurabh Mishra, PhD (Eds.)

India has taken initiatives through NDCs to strengthen efforts in the years ahead. India is committed to reducing the share of fossil fuels in its energy basket with non-conventional renewable and clean energy sources. It has set a futuristic, ambitious RE capacity expansion program at a projected growth of 40% of its total power generation by 2030 to meet NDCs target. India has also set a goal to eliminate its fossil fuel-driven vehicles by 2030. Hence, it can be concluded that India's concerted efforts are on so that it is able to respect its commitments to Climate Change.

This proposed chapter will include the ancient age commitment of the Indian subcontinent towards climate change through its rich cultural heritage. The deviation of the modern developing generation from ancient culture will also be discussed, which is leading us to a climatically polluted and unsafe South Asian region. India's concerns about addressing its commitments as a responsible nation of the International Solar Alliance to harness clean energy sources, along with different initiatives taken, will be discussed in the chapter proposed.

3.1 INTRODUCTION

The word 'energy' was derived from the ancient Greek '*energeia*,' which means activity or operation. Energy is always conserved, meaning that it can neither be created nor be destroyed, but it is interring convertible into different forms. Energy can be obtained in many ways, but energy in the form readily available to humankind is limited. In the present scenario, energy consumption per capita in any country has become scale for the economic health of that country. It is the main component that directly commands the economic growth of a country. There is a proportional relation between the energy availability and growth of any nation. In the 19th century, coal was harnessed as the main component of energy, followed by oil in the 20th century, and now, sustained by natural gas. The world total primary energy consumption has been continuously increasing. It has escalated from 6,100 million tons of oil equivalents in 1973 to 13,699 million tons of oil equivalents in the year 2016. Presently, major fraction of energy consumed worldwide is derived from fossil fuel which is exhaustible in nature and bears continuous price hikes. This demanded alternative sources of energy worldwide. Also, environmental pollution by greenhouse gasses produced by the burning of fossil fuels has attracted global attention.

The global emission of CO_2 increased by 2.5% per year, resulting in a 2°C rise in the world's temperature during the last decade. In 2015, around 194 countries signed the Paris Climate Conference (popularly known as the Paris Agreement), which aims:

- to bring down the global temperature increase by 2°C;
- to mitigate the greenhouse gaseous emission; and
- to explore alternative energy sources.

To protect the environment and meet the future energy demand, it is imperative to look for renewable and eco-friendly clean energy alternatives. Accordingly, the exploration of alternative and economically feasible clean energy sources is being carried out worldwide.

As per Indian topography and concerns shown by the government of India towards the RE sector, it can be inferred that India has the potential to harness the clean energy initiatives not only for its own territory but in the whole South Asian region. India is the fourth most attractive RE market in the world. India is ranked fourth in wind power, fifth in solar power, and fifth in overall renewable power installed capacity as of 2018. In recent years, the Indian installed renewable power generation capacity has gained a reasonable pace, resulting in a combined annual growth rate (CAGR) of 17.33% between the years 2016 to 2020. With the support of the government and improved economic policies, the clean energy sector of India has attracted lot many foreign and domestic investors too. India targets to meet its energy requirements indigenously, which is expected to reach 15,820 TWh by 2040. Hence, clean and RE alternatives will play an important role in Indian perspectives. Indian government is committed to installing 175 gigawatts (GW) of RE capacity by 2022. India has also set a goal to install an RE capacity of 500 GW by the end of this decade.

3.2 IMPERISHABLE ENERGY FOR ALL

Environmental pollution, global warming, depleting biodiversity, different health hazards to living beings, etc., have drawn the serious attention of the global community. Consumption of fossil fuels in can be held responsible for all these dangerous global effects. Depleting fossil fuel resources and their hazardous environmental effects called the attention of the global community for serious thinking towards alternative imperishable and clean energy sources. As a result, global summits and conventions started in the

21st century. Paris agreement 2015 was agreed and signed by every nation of this planet with the commitment to save this earth through mitigating the carbon emissions with the use of alternative clean energy sources such as solar energy, wind energy, hydrogen energy, nuclear fusion energy, etc. India has started a comprehensive policy for clean energy uses and support for the same with its neighboring South Asian countries Bhutan, Afghanistan, Bangladesh, Nepal, Myanmar, Pakistan, Sri Lanka, and Maldives. The Hydropower sector has been hailed as the centerpiece of relations between both countries. The Hydropower sector serves as an excellent example of a mutually beneficial relationship between two partner countries. Like Bhutan, several other South Asian countries in carry immense potential for hydroelectricity-based trade with interregional energy transfer capacities. This will develop further with financing mechanisms and institutional and public support.

To accept the responsibility in the South Asian region, India has shown its serious concern in framing policies to attract the best global investments in the sector. India has installed an RE capacity of 87.26 GW till April 30, 2020, which includes solar and wind components as 34.81 GW and 37.74 GW, respectively. Biomass and small hydropower have comparatively less contribution, 9.86 GW and 4.68 GW, respectively. This indicates that India should achieve its target of 175 GW RE installations by 2022. By December 2019, 15.10 GW of wind power projects were sanctioned, of which 12.16 GW capacity projects have already been awarded. Power generation from RE sources in India reached 127.01 billion units (BU) in the year 2020. During the year 2019 only, India installed 7.3 GW of solar power, securing its position as the third-largest global solar market.

India has always supported its neighboring South Asian countries on all fronts, including environmental, political, and social issues, to ensure better living standards in the South Asian region. India's proactive clean energy initiatives and futuristic projects have motivated the South Asian countries to take the required support from India for their efforts in the clean energy sector.

3.3 CAPITAL PROGRESS IN CLEAN ENERGY INITIATIVES IN INDIA

Over the period, India has become a favorable investment destination for the RE sector. Varieties of global stakeholders are investing in RE projects in India, including institutions, banks, and industries. Indian Renewable Energy

Development Agency (IREDA) has been serving as a major source of funding for renewable. The investor-friendly foreign policy of the Government of India allowed global players to collaborate with Indian partners for financial and technical mutual support to set up RE power generation projects in India.

The international investment in India's clean energy sector has been gradually increasing, with US$ 283 million in 2016, US$ 532 million in 2017, and US$ 1.02 billion in 2018. Adani Group in India aims to become the world's largest solar power company by 2025 and the biggest RE entrepreneur by 2030. As of 2019, India was set to open the world's largest solar plant with a capacity of 2,255 MW Bh,adla Solar Park in Rajasthan. This plant is committed to serving the nation with cost-effective solar electricity. Around Rs 36,729.49 crore investment was made in the RE sector during April-December 2019 by private companies. Central public sector entities under the administrative control of the MNRE implement its schemes; the Solar Energy Corporation of India (SECI) is the main central sector entity dedicated to the solar energy sector. SECI has since realized its mandate broadened to include the entire RE domain. Along with the National Thermal Power Corporation (NTPC), a central public sector power generation utility, SECI conducts competitive auctions for the award of solar and wind energy generation capacity. An investment of about US $ 80 billion at today's prices is expected in renewable plants without transmission lines up to the year 2022. Thus, on an annual basis, investment opportunity for over US$ 30 billion per year is expected to come up in the coming years. All this investment data and interest of the private sector shows the commitment of India towards clean energy initiatives.

3.4 GOVERNMENT'S CLEAN ENERGY INITIATIVES

Different initiatives taken by central and state Governments of India to boost India's RE sector have shown India's focus on clean energy initiatives. These initiatives will not only help this country to contribute towards global initiatives to curb carbon footprints but also motivate the neighboring countries for the same. India has ever been anxious to help south Asian countries in all such initiatives. Some of the clean energy initiatives are as follows:

- The Ministry of New and Renewable Energy (MNRE) has decided to provide tax benefits to the solar energy sector to reduce the cost of solar power generation.

- India is committed to exploiting its coastal desert region for the installation of solar energy capacity on its western borders, such as Gujarat and Rajasthan.
- Indian Railways has also decided to gradually switch over to sustained energy-efficient measures. Railways are committed to cut down its emission level by 33% by 2030.
- A new decade for 2018–2028 hydropower policy is proposed for the growth of hydro projects in the country.
- The Government of India has also given out plans to implement a National Mission on advanced ultra-supercritical technologies for cleaner coal utilization.
- Several state governments, with their initiatives, exempted solar energy from electricity duty to motivate people to utilize solar power in their daily agricultural and household needs. Few state governments are also buying back the excess electricity produced from house solar rooftops.

3.4.1 ROAD AHEAD

The Indian Government is committed to clean energy sources and is undertaking various power projects to promote the same. The clean energy potential in India will not only create many employment opportunities at all levels, especially in rural areas but will take the country and its neighbors to a sustainable environment friendly development also. The Ministry of New and Renewable Energy (MNRE) has set an ambitious target to set up RE capacities to the tune of 175 GW by 2022, of which about 100 GW is planned for solar, 60 GW for wind, and the rest for hydro and bio, among others. India's efforts show that the target will certainly be attained. India's RE sector is expected to attract investment worth US$ 80 billion by the year 2023. According to Bloomberg New Energy Outlook 2018, India is expecting to produce and consume 49% of the total electricity by RE sources using more efficient batteries to store electrical energy. This will further cut the solar energy cost, which has been a great concern in the solar energy sector.

According to an estimation by the International Energy Agency (IEA), India will be the fastest-developing energy consumer and market by 2040. It is clearly perceptible that RE, with rapid reducing costs, will turn solar and wind energy into the major operators of growth in the energy sector. India has enthusiastically accepted its global responsibility to reduce carbon emissions and ensure a more sustainable future not for itself but for our planet as a whole, including neighboring countries. With an ambitious aim of achieving 175 GW by 2022, India has been making various efforts towards cleaner energy with considerable cost reduction. Progressing towards this goal, India has recently announced that RE projects have to surpass 200 GW of clean energy capacity by 2022. India is working hard to attract new investment in its RE infrastructure with the commitment to lead the clean energy initiatives in the South Asian region.

3.4.2 CHALLENGES AND INTEGRATED APPROACH

With various grid management challenges in power distribution, India has to evolve as per the need of RE production to manage future power distribution challenges. It needs to be technologically equipped to handle the creepy nature of power generated from RE sources. This includes advanced forecasting and frequent rebalancing to fill the gap created by erratic generation in renewable generation. Experts believe that digitization of the grid will help in prompt and smooth control of the power distribution. Experts also warn that it will be inevitable to have accurate weather forecasting with respect to solar and wind generation such that demand and supply balancing is achieved through advanced analytics. To reduce transmission and distribution losses, Internet of things (IoT) devices will be needed throughout the transmission and distribution network. Accordingly, complimentary assets and investments to manage huge volume of renewable power generation will be imperative for faithful transmission. Though some distribution companies are already addressing these ways of distribution and transmission, a lot of efforts are still to be done.

3.5 CLEAN ENERGY INITIATIVES AND EMPLOYMENT

India's clean energy initiatives will improve energy availability to its poorer communities. This will facilitate their children's self-studies at home. Energy access will also support greater productivity and income for their families, which will improve their overall living standard, including health. Because of this huge futuristic RE project, India can see a growth in demand for skilled and non-skilled manpower in this sector. India's clean energy initiatives can also help to address poverty by providing steady incomes through skilling the unskilled and semi-skilled workforce. Keeping in view the access of the RE sector to every part of the country, lot many demands for different types of workforces will be generated. This will certainly offer an ocean of opportunities to both the rural and urban populations of the country. In a similar way, India's leadership role in the RE sector will also inspire the neighboring South Asian countries.

3.5.1 THE ROAD AHEAD

Solar and wind energy sites are continuously progressing with the government's special thrust to this RE sector. This sector has huge potential for jobs in the services sector in India and abroad. The Indian government is trying hard to push the manufacturing sector of solar photovoltaic components with a lot of tax rebates for new startups from within the country and abroad. The government plans to seek global participation for RE generation capacity by companies ready to set up manufacturing facilities in India, with huge investment potential in days to come.

The solar energy sector demands a big challenge in its energy storage. Economic and durable energy storage options will certainly boost this sustainable energy sector worldwide. Particularly, the automobile sector will get a big boost with efficient and cost-effective energy storage options. The future is of e-vehicle only. As battery technologies improve, environment friendly electric vehicles will be replacing fossil fuel-based vehicles rapidly, not only in India but across the globe. India's Central Electricity Authority (CEA) reiterated the coupling of all solar and wind projects above a certain size with respect to energy storage pooling. Many global companies are investing heavily in the R&D of batteries as energy storage devices. Now, India should also investigate this area so that it develops self-reliance and does not depend on imports of technology. Research and production within the country will also generate job opportunities.

At present, India's major component of electrical energy requirement is borne by thermal power generation while renewables contribution is extremely poor. Addressing its global commitment, India has set an ambitious but achievable goal of 275 GW of renewable capacity installation by the year 2027. India is performing well in addressing the challenges in the field of the RE sector, but still a long way to go.

3.6 RENEWABLES AND ENERGY SECURITY

India planned to reduce its carbon emissions intensity by 33 to 35% between 2005 and 2030 with its commitment under Intended Nationally Determined Contribution (INDC). The central government notified a Renewable Power Obligation (RPO) to primarily power distribution companies to enhance their RE share to 21% by the year 2022. This resulted the share of grid integrated Renewable Energy to increase at a faster rate. With numerous benefits associated with RE sources, India is striving hard to reach its INDC. Though Wind and Solar generation are less predictable depending on natural weather conditions, India is still committed to switching over from conventional sources of power to renewables. Prices of RE have also fallen significantly in recent years. This decline is facilitated, in turn, by the large-scale expansion of production. As a result, the global oversupply situation also contributed to the reduction in solar PV module prices.

COVID-19's impact on Indian initiatives to become self-dependent Bharat is giving impetus to all sectors, including the RE sector. India is showing its commitment to reduce its imports. The present policies and announcements by the Indian government reflect India's initiatives to make India a producing hub and grab the export market along with satisfying domestic requirements. Initiatives like Make in India, offering export stimulus, and launching a phased manufacturing program (PMP) have helped India to become global destination for investment in the RE sector. Special economic zones (SEZ) are built to cater to the industrial potential of the nation. Significantly, in an address to the nation on 12th May 2020, the Prime minister of India asked the Indians: to 'Vocal for local' for self-reliant India, i.e., Atma Nirbhar Bharat. In the case of the RE sector, this will be possible by ensuring a conducive and industry-friendly environment for domestic manufacturers in RE production.

Merely clean energy production to mitigate carbon emissions will not help, but smart and efficient consumption of energy also contributes a lot. Indian government's Ujala scheme to distribute LED bulbs and lamps at a very subsidized rate really enabled significant energy savings with the concept of "energy saving means energy producing." All such initiatives and programs in India are motivating the neighboring South Asian countries to learn and earn.

Actually, the present menace of the global carbon footprint challenge can only be addressed with the use of cost-effective RE. Solar energy can be beautifully exploited to promote mandatory usage of ICT gadgets in far-flung areas having erratic and intermittent power supplies. A solar power-producing unit is independent and can be installed at the domestic level without any cumbersome infrastructure required, provided we have cost-effective, efficient energy storage devices to maintain the reliability and usefulness of digital services delivery in every part of our country.

India is in a leadership capacity in the South Asian region with its initiatives towards the RE sector. The year 2020 would have been India's year in the global scenario, but unfortunately, COVID-19 has brought many key activities to a standstill. Amidst these unprecedented times, encouraging news to the photovoltaics is that National Hydro Power Corporation (NHPC), a mini-Ratna company that has been developing hydropower projects across the country, has recently raised its funds through private bonds to build its solar regime. The creation of a dedicated ministry, such as the Ministry of New and Renewable Energy (MNRE), to address all concerns of RE stakeholders is evidence of India's concern towards its commitments. Additionally, the Prime Minister reiterated on several occasions that India is putting serious efforts in the field of RE and will very soon be installing 450 GW of RE capacity. This reveals the seriousness and commitment of India towards looking at its RE growth and inspiring its neighbor nations for the same.

Under hydropower obligation (HPO) and subsequent amendments in the electricity act, hydropower purchase is becoming obligatory for all stakeholders. To address all concerns of RE developers in India, hydropower production has come under the scope of the Ministry of New and Renewable Energy. Here, energy storage is really a big challenge to exploit the RE sector in an efficient and convenient manner. Hence, it would be wiser to bring all energy storage-related aspects also under the purview of MNRE. This will support the RE proliferations to achieve future targets of the country. India's global ranking in the RE sector has improved significantly over the last few years in the ease of doing business. Prime Minister of India rightly mentions that 'Minimum Government and Maximum Governance' is the key to a successful government of any democratic nation. To cut down the renewable tariff, MNRE opined that RE storage is to be made feasible for both users and producers. Hydrogen as a clean energy option should be explored with other existing clean energy options with respect to its cost-effectiveness and feasibility to use. Recently, India has shown its serious concern towards hydrogen as a clean alternative fuel. The Indian government has also set up a green energy corridor with an estimated investment of USD 5.8 billion to ensure the transmission of RE from generation points to the load centers by creating adequate and efficient transmission infrastructure. MNRE has provisioned liberal and attractive bank loans to stakeholders to set up large, medium, and small-scale RE startups.

3.7 CONCLUSION

Here, it is pertinent to see that renewables should be capable of replacing the retiring thermal power plants to meet the enhanced future energy demand to run the South Asian regional economy smoothly. Further, RE capacity should be coupled with storage systems to provide round-the-clock power to address the peak time demand. This will not only support the domestic and industrial energy requirements but also help to keep pace with the economy of our growing nation. This will be a win-win for economy, environment, and politics. India has also asked telecom service providers to collaborate with RE solution providers for uninterrupted services. Solar energy alternatives should be exploited in rural India for both domestic and small-scale commercial applications to address erratic grid electricity supply. As a result, there have been hundreds of rural branches of micro-finance institutions that run on solar power today. Hence, with feasibility testing, the requirement now is to build solar energy as a green alternative, an essential element of this global concern. This would also support India's digital economy initiatives towards globally demanded sustainable development goals (SDGs) with low-carbon footprints.

South Asian countries are striving to strengthen mutual integration and cooperation in trade, infrastructure development, and other socio-political affairs. In this connection, the integration of the power grids of the countries in South Asia will be a major component that can realize benefits across the social, environmental, and economic development of the whole South Asian region. South Asia offers a unique example of the application of a regional approach to electricity. At present, this sub-region faces electricity deficits due to a lack of coordination among the power grids in the whole South Asian region. There has been enormous unexploited power generation potential in the region. India has taken this challenge to be handled through the RE sector. After the concept of the SAARC Energy Ring was advanced in 2004, several factors have emerged to address the concerns over GHG emissions and the emergence of cost-effective RE power generation. With these Indian initiatives and collaborations with the South Asian countries, appropriate use of the RE sector will be a sustainable development gift for the whole region. These sincere efforts to mitigate carbon footprints and become self-reliant in power generation will not only save the interests of the region but also save our planet for future generations. Though India has significantly contributed to the RE capacity over the past few years, but existing and evolving challenges need to be addressed to the level of the ambitions of the South Asian region.

KEYWORDS

- **clean energy**
- **energy security**
- **NDC**
- **Paris accord**
- **power generation**
- **RE capacity**
- **renewable energy**
- **road ahead**

REFERENCES

APP [Asia Pacific Partnership on Clean Energy and Climate] (2011). APP project roster. Online: www.asiapacificpartnership.org/english/project_roster.aspx.

Agarwal, A., & Narain, S. (1991) *Global Warming in an Unequal World: A Case of Environmental Colonialism.* New Delhi: Centre for Science and Environment

Atteridge, A., Nilsson Axberg, Nitu, G., Kumar, A., Lazarus, M., Ostwald, M., Tollefsen, P., Torvanger, A., Upadhyaya, P., & Zetterberg, L. (2009) *Reducing Greenhouse Gas Emissions in India: Financial Mechanisms and Opportunities for EU-India Collaboration.* Stockholm: Stockholm Environment Institute.

Bairiganjan, S., Cheung, R., Delio, E., Fuente, D., Lall, S., & Singh, S. (2010). *Power to the people: investing in clean energy for the base of the pyramid in India.* Washington and Chennai: WRI and IFMR.

Bhattacharyya, S. C. (2010) .'Shaping a sustainable energy future for India: Management challenges.' *Energy Policy 38*(8), 4173–4185.

Cecilia Tortajada, & Udisha Saklani (2018). Hydropower-based collaboration in South Asia. *The Case of India and Bhutan Energy Policy 117,* 316–325.

Cutler, J., Cleveland A., Robert, K., Kaufmann, & David I. Stern. (2000). Aggregation and the role of energy in the economy. *Ecological Economics 32,* 301–317.

Dubash, N., (2011). From norm taker to norm maker? Indian energy governance in global context. *Global Policy 2*(Suppl. s1), 66–79.

Edgerton, R. H. (1982). In: *Available Energy and Environmental Economics.* USA: Lexington Books, 246–250.

Friedlingstein P., Andrew, R. M., et al. (2016). Persistent growth of CO_2 emissions and implications for reaching climate targets. Nature Geoscience. Available at http://hdl.handle.net/10871/20695.

Houghton, J. T. et al., (eds.). (2001). *Climate Change 2001: Scientific Basis* (Cambridge University Press, New York).

Kammen, D. (2021). Is the renewable energy target for India within reach? Online: http://blogs.worldbank.org/climatechange/renewable-energy-target-india-within-reach (accessed on 15 March 2021).

Liming, H. (2007). 'A study of China–India cooperation in renewable energy field.' *Renewable and Sustainable Energy Reviews 11*, 1739–1757.

MNRE (2010). Renewable Energy in India: Progress, Vision and Strategy. *Paper presented at the Delhi International Renewable Energy Conference (DIREC) 2010, New Delhi*, 27–29 October.

Dresselhaus, M. S., & Thomas, I. L. (2001). Alternative energy technologies. *Nature. 414.*

Mukul, J. (2011). *'Govt's Clean Energy Fund Stays Pristine.'* Business Standard, 7 March.

Newell, P. (2008). 'Civil society, corporate accountability and the politics of climate change,' *Global Environmental Politics, 8*(3), 124–155.

Sargsyan, G., Bhatia, M., Banerjee, S. G., Raghunathan, K., & Soni, R. (2010). *Unleashing the Potential of Renewable Energy in India.* World Bank South Asia Energy Unit, Sustainable Development Department.

Schneider, L. (2009). 'Assessing the additionality of CDM projects: practical experiences and lessons learned'. *Climate Policy 9*(3), 242–254.

Schultz, P. W. (2001). Assessing the structure of environmental concern: Concern for the self, other people, and the biosphere. *Journal of Environmental Psychology, 21,* 327–339.

TERI, "Bhutan: 25 MW of renewable energy by 2025," http://www.terienvis.nic.in/index1.aspx?lid=1763&linkid=1261&langid=1&mid=4, accessed on 23 Mar 2021.

Vihma, A. (2011). India and the global climate governance: between principles and pragmatism. *Journal of Environment & Development, 20*(1), 69–94.

CHAPTER 4

Indo-China Renewable Energy Hub in the Asian Energy Corridor

ANKIT SRIVASTAVA,[1,2] ABHISHEK ANAND,[1] SHAILENDRA SINGH,[1] AMRITANSHU SHUKLA,[3] and ATUL SHARMA[1]

[1]*Non-Conventional Energy Laboratory, Rajiv Gandhi Institute of Petroleum Technology, Jais, Amethi, Uttar Pradesh, India*

[2]*Feroze Gandhi Institute of Engineering and Technology, Ratapur, Raebareli, Uttar Pradesh, India*

[3]*Department of Physics, University of Lucknow, Lucknow, U.P., India*

ABSTRACT

Renewable energy (RE) is now one of the most significant aspects and the world's expectation of preserving a stable environment and earth's prosperity for subsequent generations. India and China have played a vital role on this front, showing the emerging nations of the world a path ahead for socio-economic development. India and China are very optimistic about their RE objectives. There is a huge possibility available for India and China in the RE sector. RE solutions produce marketable energy by the transformation of natural phenomena and resources into accessible energy. These resources provide an enormous energy market far beyond the resources of fossil fuels. The use of RE as a potential solution to fossil energy is a positive development. This chapter comprises RE possibilities in both countries and a SWOT analysis for the RE possibilities of India and China. Government policies regarding the growth of RE are also discussed as targets for implementing policies in the right direction. The Government of India is constantly working towards the 2030 goal of producing 450 GW of RE. China's ambitious goal

The South Asian Energy Corridor: Toward Sustainability and Security. Amritanshu Shukla, PhD, Atul Sharma, PhD & Saurabh Mishra, PhD (Eds.)

is to achieve carbon neutrality by the year 2060. It also aims to reduce the CO_2 intensity by more than 65% by 2030.

4.1 INTRODUCTION

As nations deal with the increasingly damaging effects of climate change, primarily caused by ecocentric development activities, the importance of renewable energy (RE) in power and electricity has become extremely important. Both developed and developing countries around the world have begun to give priority to rapidly increasing the percentage of RE in their economies' energy supply mix. RE has become one of the most significant considerations and the world's hope of maintaining a healthy climate and the wealth of the earth for future generations. India and China have led the world on this front, showing the world's developed nations a way forward for socio-economic growth. In its goals to encourage green energy, India and China both countries are very optimistic. Renewable technology has continued to play a major role in India and China in terms of enhancing grid capacity, supplying energy availability, reducing fossil fuel usage, and supporting India and China in pursuing a low-carbon growth path.

India submitted its Planned Nationally Defined Contribution (INDC) to the UNFCCC ahead of COP 21, outlining the post-2020 climate behavior of the country. By establishing a new aim to raise the nation's share of non-fossil-based installed electricity potential to 40% before 2030, India's INDC is increasing its target of building 175 GW of sustainable energy by 2022 (IEA, 2020).

In April 2020, China issued a proposed Energy Law that would enforce almost all of the country's existing energy transformation policies, emphasizing renewables but supporting coal, nuclear, and some fossil fuel technologies. We examine the frequency of different terms in the law in this section, highlighting its focus on planning and monitoring, as well as the strong emphasis on renewables (Sino-German Energy Transition Project, 2020).

4.2 EXISTING RENEWABLE ENERGY (RE) OF CHINA

Energy is an important need of people all over the world, including food and shelter. Especially in China, as a developing country with 1.25

billion people, Energy is extremely important for its rapid economic growth (average annual GNP growth rate between 8% and 9%). As the economy is experiencing rapid growth, the fair use of natural resources and environmental conservation will need to be ensured. This has been put forward in the Chinese Agenda 21 (Song, 2020). RE uses in China have a long history, including biomass, solar, geothermal, sea, and wind. RE technologies generate marketable energy through the transformation of natural phenomena and resources into a usable type of energy. These resources represent a huge energy opportunity far beyond the resources of fossil fuels. The use of RE options as a potential alternative to fossil energy is a promising prospect. However, the proportion of the other renewable energies used is considerably lower than traditional energy resources, except biomass energy. In recent decades, we will incorporate the different RE sources that have been used in China.

4.2.1 SOLAR ENERGY

China has a lot of solar power. In western China, there are over 3,000 hours of sunshine each year. Considering the huge areas and low population density, small electricity plants such as photovoltaic panels are more cost-effective than extending the existing power grid. Photovoltaic devices will also play a key role in the West's control of China's mainland.

In general, solar radiation in China is highest in the west and lowest in the east. The Inner Mongolia Plateau on the east, Taihang Mountain, Qinling Mountain, the Sichuan Basin on the west, and the Yunnan-Guizhou border are all part of the annual solar radiation average. Solar radiation is weaker in the east and south of the boundary, although it is stronger in the west and north. Tibet receives the most solar radiation in the southwest, at 7347.55 MJ/m^2. The higher provinces also include Qinghai, Xinjiang, and Gansu in the northwest and Hainan in central and southern China. Chongqing and Guizhou are provinces with poor solar radiation in the eastern portion of southwest China and Heilongjiang in northeast China. The spatial distribution map of China's average annual solar radiation is shown in Figure 4.1. By comparing the solar energy resources of each province, the higher altitudes of the southwest and northwest regions are better, whereas the lower altitudes of the eastern regions are poorer (Zhang et al., 2020).

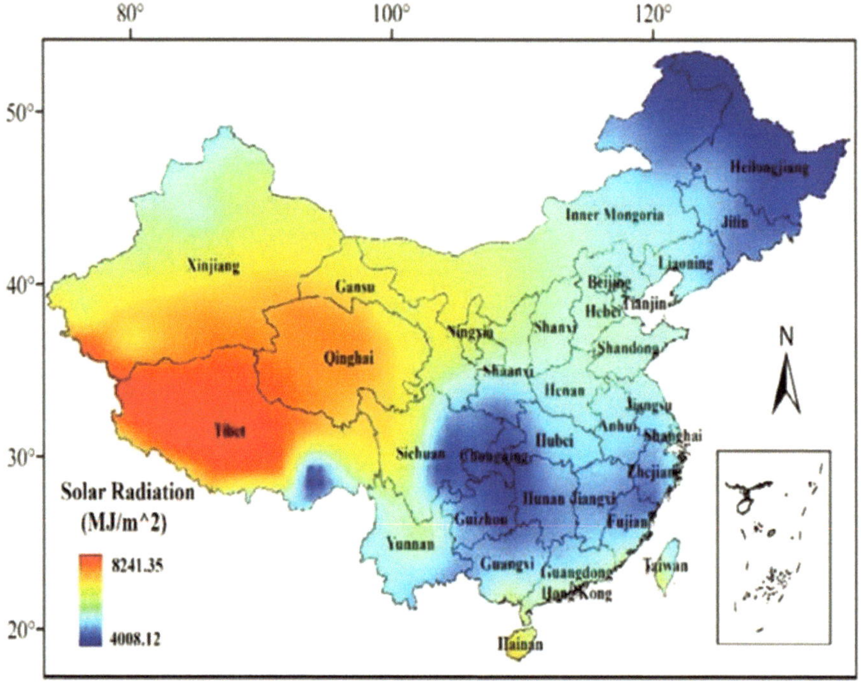

FIGURE 4.1 Spatial distribution map of China's average annual solar radiation (Zhang et al., 2020).

Source: Reprinted with permission from Zhang et al., 2020. Copyright 2020. Elsevier

4.2.2 WIND ENERGY

China has been harnessing wind energy for hundreds of years. Windmills were used to pump water and grind grain from ancient Holland to China, enabling the use of wind energy. A wind turbine, which is today's version of a windmill, uses wind energy to generate electricity. Wind turbines with advanced technology can generate electricity for homes, businesses, and even infrastructure. In 2016, the National Growth and Reform Commission (NDRC) announced the development plan for the 13th Five-Year Period (2016–2020) (shown in Figure 4.2). According to the project, offshore wind power capacity deployed in five provinces and two municipalities exceeds 5 GW, with capacity under construction reaching 10.05 GW before 2020 (Compilation and Translation Bureau, 2016).

Province	Installed Capacity (MW)	Under Construction (MW)
Liaoning	0	100.00
Hebei	0	500.00
Tianjin	100.00	200.00
Shandong	0	0
Jiangsu	3000.00	4500.00
Shanghai	300.00	400.00
Zhejiang	300.00	1000.00
Fujian	900.00	2000.00
Guangdong	300.00	1000.00
Guangxi	0	0
Hainan	100.00	350.00
Total	*5000.00*	*10050.00*

FIGURE 4.2 Global 13th five-year plan (2016–2020) for offshore wind power installed and under construction capacity (MW).

Source: Compilation and Translation Bureau, 2016

4.2.3 OCEAN ENERGY

Design of ocean energy conversion in China follows basic strategies in line with China's special conditions as follows: (i) location Installation should be carefully placed in areas where traditional energy is not available; (ii) To make the conversion of tidal or ocean energy feasible, the design of ocean energy conversion systems should be diversified to meet various needs and requirements, and economic considerations should be taken seriously; (iii) In the immediate future, it will not be possible to build large-scale energy conversion systems. Instead, the development of a small and medium-sized grid in the coastal area would be more feasible.

China's ocean energy supplies are very rich and well-distributed to the energy needs of the coastal regions of China. There are substantial reserves of tidal energy, wave energy, marine current energy, ocean thermal energy, and salinity gradient energy, with great growth potential. The theoretical power of marine current in the top nine provinces in China is shown in Figure 4.3. Ocean energy harvesting technology has reached maturity after years of growth. Some forms of ocean energy technologies are planned to be engineered and commercialized and then industrialized. From the point of view of technological sophistication, tidal power generation technology is the most advanced and could have a potential environmental effect than other technologies. The theoretical power of wave energy in the top 11 provinces in China is shown in Figure 4.4. Its production should therefore be sponsored based on careful decision-making. In the short term, China's ocean energy production is mainly related to the generation of tidal power, and a 10 MW tidal power plant is currently being constructed. Ocean RE is facing a strong growth opportunity, and China has provided a favorable climate, so there is much reason to believe that ocean energy will grow further in the future and contribute more to the national economy (Wang et al., 2011).

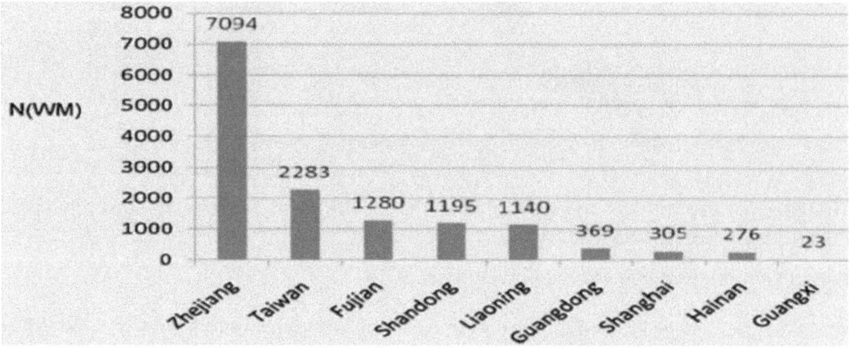

FIGURE 4.3 Theoretical power of marine current in the top nine provinces in China.
Source: Adapted from Zhang et al., 2014

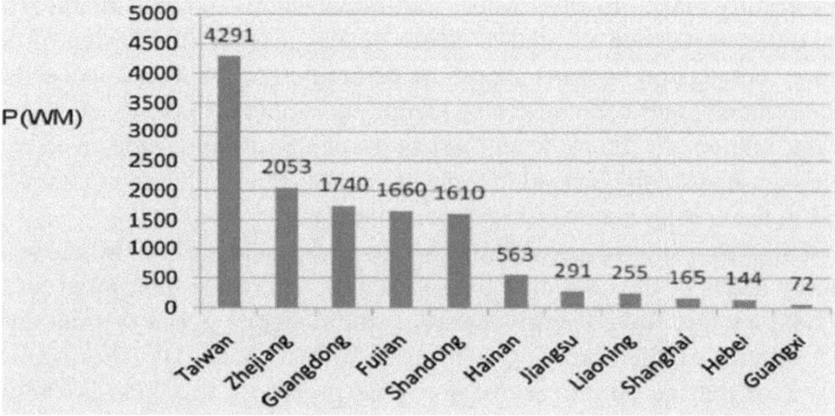

FIGURE 4.4 Theoretical power of wave energy in the top 11 provinces in China.
Source: Adapted from Zhang et al., 2014

4.2.4 GEOTHERMAL ENERGY

Geothermic energy mainly applies to the earth's inner heat and the process of radioactive substance decay in molten magma. Geothermal energy can be classified into three categories: shallow geothermal power, underground thermal water, and hot dry rock supplies (Donatini, 2018). Shallow geothermal energy, for example, applies to low-temperature heat in dirt, rocks, and groundwater at a depth of 200 meters below the earth's surface (Hähnlein et al., 2013). Underground thermal water sources are hot water and steam that are found at a depth of fewer than 4,000 meters and a temperature of more than 25°C. High-temperature rocks buried at depths ranging from 3 km to 10 km below the atmosphere, with a temperature of more than 150°C, with or without a small quantity of water and steam, and mainly consisting of a mixture of metamorphic or crystalline rocks, are referred to as hot dry rock sources. This section, in particular, describes the current storage, development, and consumption situation, as well as the rules, regulations, and patterns that apply (China Mineral Resources, 2015).

During the 12th Five-Year Plan, the Ministry of Land and Resources' China Geological Survey Office conducted many surveys. Surveys to determine the region's geothermal resource reserves for the first time. Around 336 cities above the prefecture-level and 31 provinces (districts, villages) have completed surveys on shallow geothermal resources and deep thermal water sources. Also, surveys of hot, dry rock deposits have begun. These surveys describe the existing conditions, allocation characteristics, advancement, and use of Chinese geothermal energy.

The China Geological Survey Office of the Land and Resources Ministry organized a series of surveys during the 12th Five-Year Plan. Studies initially to evaluate the region's geothermal reserves. Surveys were completed in 336 towns over prefecture-level, and 31 provinces (districts, towns) were subdivided into the subterranean thermal water supply. Surveys on hot and dry rock resource bases were also started. These surveys describe existing conditions, distributional characteristics, and Chinese geothermal energy production and exploitation. Geothermal energy distribution in China is shown in Figure 4.5 (Wang et al., 2020).

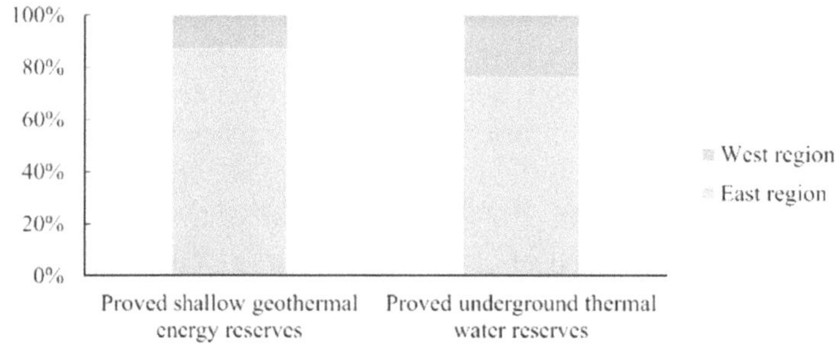

FIGURE 4.5 Geothermal energy distribution in China (Wang et al., 2020).

Source: Reprinted with permission from Wang et al. 2020. Copyright 2020. Elsevier

4.2.5 *BIOMASS ENERGY*

People have been using biomass energy for thousands of years, ever since humans began burning wood to cook food or keep it warm. Wood is still today our primary biomass energy resource. However, several other sources of biomass are currently being used, including trees, agricultural or forestry residues, and the organic portion of municipal and industrial waste, and the gases from the landfill can be used as a source of energy for biomass. In the biomass market, forecasts for plant investment and operating costs have been generated both from current data for existing generation facilities in China and from recent ventures in other countries. Table 4.1 presents the forecasts for new growth in biomass power plants. For straw and bagasse capital, the evaluation included cogeneration technology, direct combustion, co-fired with existing coal boilers, and large-scale advanced gasification. As stated in the table, the expected capacity for 2020 is much higher than the capacity that was built in 1998 but is also much lower than the predicted total potential (Jingjing et al., 2001).

TABLE 4.1 Projections of Added Biomass Power Capacity (Jingjing et al., 2001)

Resource	1998 Installed Capacity (MW)	Capacity in 2020 (MW)	
		Projected Installed	Potential Capacity
Landfill gas	0	600	4,650
MSW	15	960	23,330
Biogas	2	302	630
Bagasse	410	600	2,380
Straw	3	303	43,850

4.3 EXISTING RENEWABLE ENERGY (RE) OF INDIA

India is changing rapidly the RE market, while RE has become substantially more scarce and costly to import as traditional energy (e.g., coal) has begun to play an important role in the nation's energy security. India has also set ambitious goals for RE. In 2020, the National Action Policy on Climate Change (NAPCC) set a goal of obtaining 15% of electricity from renewable sources, as well as the response of India to climate change to address various issues such as the protection of energy and industrial competitiveness, with an initial target of 5% in 2010, up by 1% per year (Pandve, 2009). The government of India, under the Jawaharlal Nehru National Solar Mission (JNNSM), is also planning to produce 20,000 MW of solar energy by 2022 (MNRE, 2001). NITI Aayog, a non-statutory and advisory authority in India, has taken responsibility for creating an integrated and combined view of the complex socio-economic and substantive status of the country in India. It also assessed the progress made by India and its state in achieving sustainable development goals (SDGs). India's energy demand is rising dramatically, with India's overall energy demand more than doubling by 2030 and electricity demand nearly tripling today (Kumar & Majid, 2020). Also, existing conventional sources have both infinite potential and responsibility for the environment. Therefore, cleaner and limitless alternative forms of generation are necessary. The installed capacity of RE reflects, by July 2019, 22.5% of the overall installed capacity of India for power generation. India, a nation of abundant natural resources, has the enormous capacity to generate electricity through RE resources. Current technologies have made it possible to make more effective use of these renewable resources in producing electricity (MNRE, 2020). Existing renewable resource conditions in India are described here.

4.3.1 SOLAR ENERGY

In the Indian energy scenario, the visible effect of solar energy has been seen over the last few years. Decentralized and distributed applications focused on solar energy have helped millions of people in Indian villages by meeting their cooking, lighting, and other energy needs in an environmentally friendly way. Also, over the years, India's solar energy sector has emerged as a major player in grid-connected power generation potential. It supports the government's sustainable development agenda, though at the same time emerging as an integral part of the solution to meet the energy needs of the nation and as an important player in energy security. The National Solar Energy Institute has assessed the country's solar capacity of approximately 748 GW, predicting that 3% of the wasteland area will be covered by solar PV modules. Solar energy has played a major role in India's National Action Plan on Climate Change, with the National Solar Initiative being one of the major missions. The government started the National Solar Mission in January 2010. The NSM is a major initiative by the Government of India to promote sustainable, environmentally sensitive development, with active participation by the States, while addressing

India's energy-protection challenges. It will also be India's biggest contribution to the global initiative to resolve climate change challenges. The Mission's goal is to make India a global leader in solar energy by setting policy conditions that will enable solar power to spread globally as rapidly as possible. Under this mission, the main aim is to install 100 GW capacity solar power plants before 2022. This is in line with India's Expected Nationally Defined Contributions (INDCs) goal of achieving approximately 40% of the installed capacity of electricity from non-fossil fuel-based energy supplies before 2030 and reduced its GDP pollution intensity by 33% to 35% compared to 2005 levels. It would have produced India recently achieved its fifth global position in the deployment of solar power by surpassing Italy. Solar power capability has grown by greater than 11 times over the last five years, from 2.6 GW in March 2014 to 30 GW in July 2019. The top 10 States of India in Solar Installation are shown in Figure 4.6 till 2019. Solar tariffs in India are already very competitive, and grid parity has been reached (Ministry of New and Renewable Energy, 2021a).

FIGURE 4.6 Top 10 states of India in solar installation (capacity in MW as of 31-12-2019).

The capacity for solar power in the country has been assessed to be about 750 GWp based on the availability of land and solar radiation. State-wise details of estimated solar energy potential in the country and the cumulative installed capacity (as of 31-12-2019) are given in Tables 4.2 and 4.3, respectively, which are assessed by the National Institute of Solar Energy (MNRE, 2020).

TABLE 4.2 State-Wise Details of Estimated Solar Energy Potential in the Country (MNRE, 2020)

SL. No.	State/UT	Solar Potential (GWp)
1.	Andhra Pradesh	38.44
2.	Arunachal Pradesh	8.65
3.	Assam	13.76
4.	Bihar	11.2
5.	Chhattisgarh	18.27
6.	Delhi	2.05
7.	Goa	0.88
8.	Gujarat	35.77
9.	Haryana	4.56
10.	Himachal Pradesh	33.84
11.	Jammu & Kashmir	111.05
12.	Jharkhand	18.18
13.	Karnataka	24.7
14.	Kerala	6.11
15.	Madhya Pradesh	61.66
16.	Maharashtra	64.32
17.	Manipur	10.63
18.	Meghalaya	5.86
19.	Mizoram	9.09
20.	Nagaland	7.29
21.	Odisha	25.78
22.	Punjab	2.81
23.	Rajasthan	142.31
24.	Sikkim	4.94
25.	Tamil Nadu	17.67
26.	Telangana	20.41
27.	Tripura	2.08
28.	Uttar Pradesh	22.83
29.	Uttarakhand	16.8
30.	West Bengal	6.26
31.	Uts	0.79
	Total	**748.99**

TABLE 4.3 Commissioning Status of Grid Connected Solar Projects as of 31-12-2019 (MNRE, 2020)

SL. No.	State/UT	Cumulative Capacity till 31-03-2019 (MW)	Capacity Added in 2019-20 till 31-12-2019 (MW)	Cumulative Capacity till 31-12-2019 (MW)
1.	Andaman & Nicobar	11.73	0.46	12.19
2.	Andhra Pradesh	3085.68	473.34	3559.02
3.	Arunachal Pradesh	5.39	0.22	5.61
4.	Assam	22.4	18.83	41.23
5.	Bihar	142.45	6.9	149.35
6.	Chandigarh	34.71	2.28	36.99
7.	Chhattisgarh	231.35	0	231.35
8.	Dadra & Nagar Haveli	5.46	0	5.46
9.	Daman & Diu	14.47	2.09	16.56
10.	Delhi	126.89	29.23	156.12
11.	Goa	3.89	0.89	4.78
12.	Gujarat	2440.13	323.42	2763.55
13.	Haryana	224.52	24.75	249.27
14.	Himachal Pradesh	22.68	9.89	32.57
15.	Jammu & Kashmir	14.83	4.47	19.3
16.	Jharkhand	34.95	3.45	38.4
17.	Karnataka	6095.55	1179.37	7274.92
18.	Kerala	138.59	3.16	141.75
19.	Lakshadweep	0.75	0	0.75
20.	Madhya Pradesh	1840.16	397.33	2237.48
21.	Maharashtra	1633.54	29.88	1663.42
22.	Manipur	3.44	1.14	4.58
23.	Meghalaya	0.12	0	0.12
24.	Mizoram	0.5	1.02	1.52
25.	Nagaland	1	0	1
26.	Odisha	394.73	3.11	397.84
27.	Puducherry	3.14	2.37	5.51
28.	Punjab	905.62	41.48	947.1
29.	Rajasthan	3226.79	1617.42	4844.21
30.	Sikkim	0.01	0.06	0.07

TABLE 4.3 *(Continued)*

SL. No.	State/UT	Cumulative Capacity till 31-03-2019 (MW)	Capacity Added in 2019-20 till 31-12-2019 (MW)	Cumulative Capacity till 31-12-2019 (MW)
31.	Tamil Nadu	2575.22	1213.14	3788.36
32.	Telangana	3592.09	28.66	3620.75
33.	Tripura	5.09	4.32	9.41
34.	Uttar Pradesh	960.1	85	1045.1
35.	Uttarakhand	306.75	8.74	315.49
36.	West Bengal	75.95	33.46	109.41

4.3.2 WIND ENERGY

India is in 4th place for the largest installed wind power capacity. China is in 1st place, and U.S. and Germany are in 2nd and 3rd place, respectively, for the largest installed wind capacity. The wind power installed capacity in India as of March 2017 is approximately 32 GW. The wind energy potential of 302 GW at a hub height of 100 m has been recently analyzed by the National Wind Energy Institute (NIWE) in India. Wind energy has spread through the southern, western, and northern parts of India. In Andhra Pradesh, Gujarat, Karnataka, Madhya Pradesh, Maharashtra, Rajasthan, Tamil Nadu, and Telangana, the wind power capacity is being focused on. The wind power potential is located in the States of Andhra Pradesh, Karnataka, Madhya Pradesh, Maharashtra, Rajasthan, Tamil Nadu, and Telangana. With about 7.5 GW, Tamil Nadu has the highest installed capacity in the country, and its state regulations are very favorable for the production of wind power. Maharashtra has the second-highest installed capacity of approximately 5 GW, followed by Gujarat, with the country's third-highest installed capacity of approximately 4 GW (Bakshi, 2002).

Since the wind is an intermittent, site-specific energy source, a thorough evaluation of wind power is expected before deciding on potential sites. The government has installed more than 800 wind monitoring stations across the country through the National Wind Energy Institute (NIWE) and 50 m, 80 m, 100 m, and 120 m above ground level. According to the latest estimates, the nation's gross intensity was 30,225 GW and 69,550 GW at 100 meters and 120 meters above sea level, respectively. Figure 4.7 and Table 4.4 shows that wind potential is largely available in seven windy Indian states

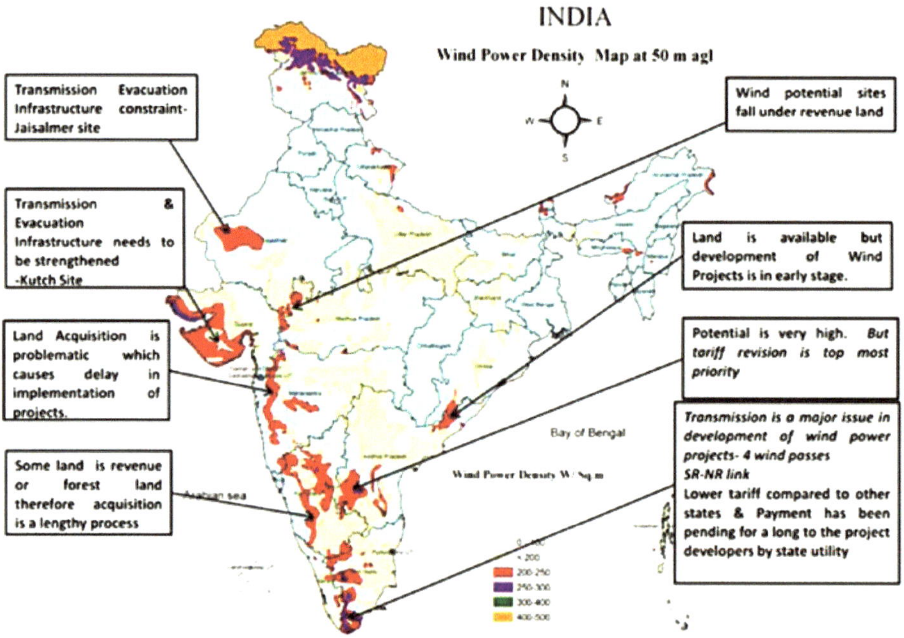

FIGURE 4.7 Wind power potential of India

Source: Reprinted with permission from Chaurasiya et al., 2019. https://creativecommons.org/licenses/by/4.0/

TABLE 4.4 Wind Power Potential in India at 100 Meters and 120 Meters, Above Ground Level (MNRE, 2020)

Sl. No.	State	Wind Power Potential at 100 Meters Above Ground Level (GW)	Wind Power Potential at 120 Meters Above Ground Level (GW)
1.	Andhra Pradesh	44.23	74.9
2.	Gujarat	84.43	142.56
3.	Karnataka	55.86	124.15
4.	Madhya Pradesh	10.48	15.4
5.	Maharashtra	45.39	98.21
6.	Rajasthan	18.77	127.75
7.	Tamil Nadu	33.8	68.75
	Total (7 Windy States)	**292.97**	**651.72**

As of 31-12-2019, the installed capacity of grid-interactive wind power in India is 37.50 GW, and Table 4.5 indicates the state-wise installed capacity (in MW).

TABLE 4.5 State-Wise Wind Power Installed as of 31-12-2019 (MNRE, 2020)

Sl. No.	State	Wind Power (MW)
1.	Andhra Pradesh	4092.45
2.	Gujarat	7359.22
3.	Karnataka	4753.4
4.	Kerala	62.5
5.	Madhya Pradesh	2519.89
6.	Maharashtra	5000.33
7.	Rajasthan	4299.72
8.	Tamil Nadu	9285.265
9.	Telangana	128.1
10.	Others	4.3
	Total (MW)	37505.175

4.3.3 OCEAN ENERGY

Wave, thermal, tidal energy, and ocean currents are the primary sources of ocean energy. As India is a tropical region, a constant temperature difference between surface water and the deep ocean is available. You can use this gradient to simultaneously produce power and freshwater. A larger-capacity floating offshore plant must now be attempted to meet coastal mainland water requirements (Chul & Jo, 2008). The capacity is approximately 40,000 MW along the 6,000 Km of coastline. However, this energy is less powerful than in the more northern and southern latitudes available. The research and development (R&D) of wave energy exploration began in India in 1982 at the Ocean Engineering Center, Indian Institute of Technology, Madras. Principal calculations reveal an annual wave energy potential of between 5 MW and 15 MW per meter along the Indian coast, which results in a theoretical potential of about 40,000–60,000 MW on the shores of around 6,000 Km. However, there is expected to be a considerably lower realistic and economic opportunity. According to a report undertaken by the Indian Technology Institute, Chennai, and Risk and Infrastructure Solutions Limited (CRISIL) in December 2014, the tidal capacity was projected to be nearly 12,455 MW. The possible areas with low/medium tidal wave intensity are Sunderbans in West Bengal, the Gulf of Khambat, Palk Bay–Mannar and Hoogly River, South Haldia and Gulf of Kutch, and Southern regions in Gujarat. Tidal energy is currently in the process of R&D and is not commercially deployed in India. As capital costs range from Rs. 30 crores to Rs. 60 crores per MW, previous tidal power attempts have not been fruitful (G.O.I. Ministry of New and Renewable Energy, 2021). Now in India, ocean energy is accepted as RE this step would improve ocean energy in India (PIB, India, 2020).

4.3.4 GEOTHERMAL ENERGY

In 1970, India started exploring and analyzing geothermal fields. The GSI (Geological Survey of India) has discovered 350 geothermal energy locations across the country. The Puga valley in Ladakh is the most promising of these. India's geothermal energy potential is projected to be about 10,000 MW. India's geothermal regions are the Himalayas, Cambay, the West Coast, Sohana, Son-Narmada-Tapi (SONATA), Mahanadi, and the Godavari. India's geothermal provinces (Figure 4.8) can produce around 10,600 megawatts of electricity. However, geothermal energy is only now being discovered and produced for a variety of purposes, and it is beginning to show signs of potential hope. When looking at India's energy future, green energy would play a significant role. Many of the discovery work in India is done by organizations including the GSI and the National Geophysical Research Institute of Hyderabad (NGRI).

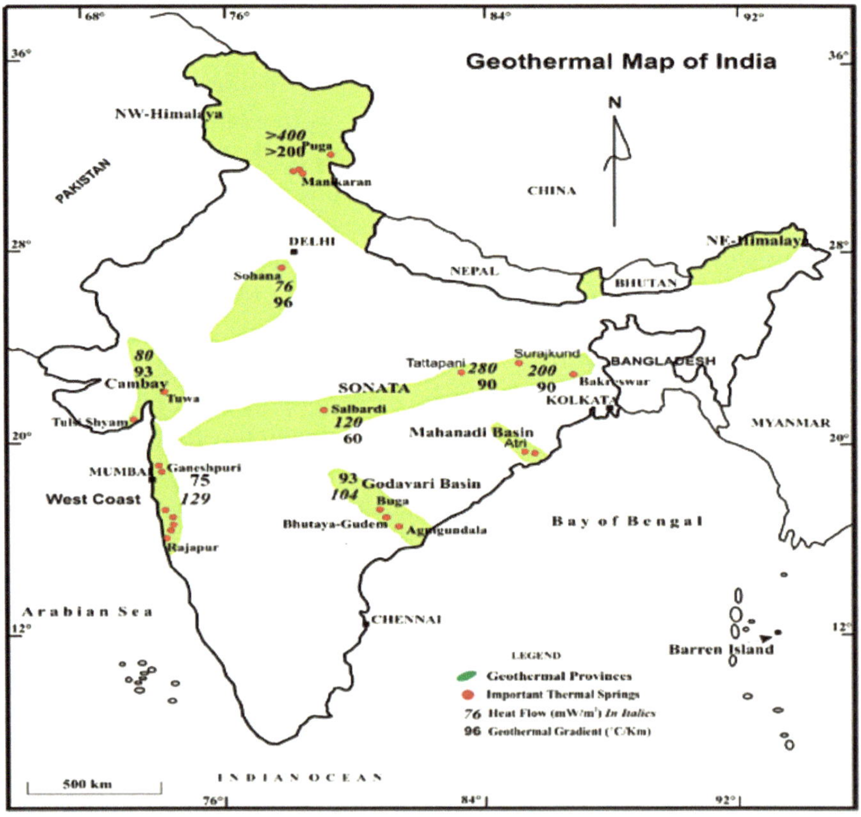

FIGURE 4.8 Geothermal provinces in India

Source: Reprinted from Chandrasekharam, 2000.

The deepest exploratory geothermal wells drilled in India are listed in Table 4.6.

TABLE 4.6 Geothermal Wells Depths Drilled in Different Regions of India (Vaidya et al., 2015)

Region	Depth
Chhattisgarh, Tatapani, Sarguja	Near 620 m
Himachal Pradesh and Manikaran	Near 700 m
Jammu & Kashmir, Chumthang, and Ladakh	Near 220 m
Tapoban	Near 728 m
Manikaran, Jammu & Kashmir, Puga, and Ladakh	Near 385 m

4.3.5 BIOMASS ENERGY

Biomass is characterized as bio-residue made available by vegetation based on water, forest, or organic waste, by crop production products, and by waste from the agricultural or food industry. There are various biomass resources available in various forms in India. In terms of how they are found in nature, they can be classified as grasses, woody plants, vegetables, plants, green manure, and aquatic plants. Algae and jatropha are now used for the manufacture of biodiesel. Residually, the core and different biomass sources of energy can be categorized as farm crops, electric plantations, and urban and industrial waste. The various classifications of biomass available in India are given in Figure 4.9.

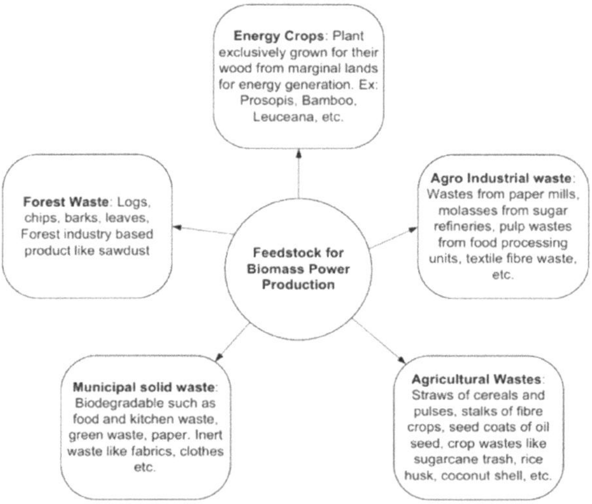

FIGURE 4.9 Biomass classifications are available in India (Kumar et al., 2015).

Source: Reprinted with permission from Kumar et al., 2015. Copyright 2015. Elsevier

Thermo-chemical and bio-chemical are two major processes that are currently in use for the transformation of biomass into energy. The choice of biomass conversion technologies depends on how the energy used, such as heat generation combustion, mechanical energy, electrical energy, and liquid fuels for use as transport fuels, etc., are obtained through pyrolysis, fermentation, and mechanical extractions. By gasification, biomass was removed to form syngas. Various electricity-generating schemes in different countries of India are being established to meet the energy needs of biomass gasification. The states generate energy using bagasse cogeneration, which uses sugar mill waste. A variety of power generation projects in India have already proven promising, focusing on gasification-based cogeneration of rural electrification plants. Not only have these plants solved the problem of rural electrification for remote villages where the cost of conventional electrification networks can be prohibitive, but the cost of power generation has also been comparatively low (Kumar et al., 2015). Bio-energy potential in Northeastern Region states is enlisted in Table 4.7.

TABLE 4.7 North–Eastern Region States Bioenergy Potential (MNRE, 2020)

SL. No.	States/Uts	Biomass Power (MW)	Waste to Energy (MW)
1.	Arunachal Pradesh	8	–
2.	Assam	212	8
3.	Manipur	13	2
4.	Meghalaya	11	2
5.	Mizoram	1	2
6.	Nagaland	10	–
7.	Sikkim	2	–
8.	Tripura	3	2

In the region, over 500 biomass power and bagasse cogeneration projects with a total capacity of 9,806 MW have been installed for grid feeding power. Maharashtra, Karnataka, Uttar Pradesh, Tamil Nadu, and Andhra Pradesh are states that have taken a leadership role in the implementation of bagasse cogeneration projects. Table 4.8 comprises the state-wise installed capacity of biomass IPP/bagasse cogeneration/non-bagasse cogeneration in India. Chhattisgarh, Madhya Pradesh, Gujarat, Rajasthan, and Tamil Nadu are the leading states for biomass energy projects. Current Status of biomass is as follows:

- A total capacity of 10,145 MW has been installed in the biomass power and cogeneration sector.
- Installed capacity of biomass IPP – 1,826 MW.
- Installed capacity of bagasse cogeneration – 7,547 MW.
- Installed capacity of non-bagasse cogeneration – 772 MW.

TABLE 4.8 State-Wise Installed Capacity of Biomass IPP/Bagasse Cogeneration/ Non-Bagasse Cogeneration in India as of 31-10-2020 (information as received from SNAs) (Ministry of New and Renewable Energy, 2021b)

Name of State/UT	Biomass IPP	Bagasse Cogeneration	Non-Bagasse	Cumulative Installed Capacity (as of 31-12-2019)
Andhra Pradesh	171.2	206.9	105.57	483.67
Bihar	12	100.5	12.2	124.7
Chhattisgarh	222.4	20	2.5	244.9
Gujarat	44.5	20.8	12	77.3
Haryana	19.4	102	89.26	210.66
Karnataka	137.3	1729.8	20.2	1887.3
Madhya Pradesh	92.5	0	14.847	107.347
Maharashtra	217	2,351	16.4	2584.4
Punjab	138.5	161	173.95	473.45
Rajasthan	114.3	4.95	2	121.25
Tamil Nadu	218.7	750.4	43.55	1012.65
Telangana	60.1	98	2	160.1
Uttarakhand	0.12	72.6	57.5	130.22
Uttar Pradesh	28	1929.5	159.76	2117.26
West Bengal	300	–	19.92	319.92
Odisha	50.4	–	8.82	59.22
Himachal Pradesh	–	–	9.2	9.2
Kerala	–	–	2.27	2.27
Meghalaya	–	–	13.8	13.8
Jharkhand	–	–	4.3	4.3
Assam	–	–	2	2
Manipur	–	–	–	0
Nagaland	–	–	–	0

4.4 POLICY ADVANTAGE

4.4.1 CHINA

The energy market in China is a combination of market and administrative initiatives, including targets for administrative planning. Under the 12th Five-Year Plan, the National Energy Administration (NEA) set ambitious wind and solar targets backed by subsidized feed-in tariffs, and the industry frequently met those targets, which were then often revised upwards by policymakers (Song, 2020). Capacity goals have been viewed by wind and solar industry analysts as a leading indicator of policy aspirations for RE. NEA, however, set 2020 goals for wind and solar with the 13th Five-Year Plan that indicated steadier progress and refused to change the targets as the market outpaced them.

China has implemented green certificates and a sustainable obligation to encourage the adoption of renewables. Although these appear similar to markets in the U.S. and U.K., their purpose and function vary significantly, a point frequently overlooked outside of China in reporting. To reduce the government's subsidy payment requirement, the green certificate market was developed mostly by persuading companies or individuals to purchase certificates, thus transferring the subsidy obligation to current projects rather than creating a market for projects that will supply additional RE beyond those already in place. For this reason, the green market for certificates has never taken off. The renewable obligation (RO), which passed three drafts in 2018 before its implementation, also varies from similar mandates in the United States and Europe. Instead of setting goals for provinces or industries to target over the long term, China's RO defines demand targets for just three years (including the current year). Indeed, the regional responsibility for 2020 was updated in June of this year to better reflect production. This implies that the responsibility is the same as the regulatory planning quota and that it does little to stimulate long-term market-based clean energy projects (Hove, 2020).

While many of the short-term initiatives to date have concentrated on administrative planning, in the medium term, the government sees spot markets and market measures as crucial for streamlining investment in the power sector, bringing down electricity costs and prices, and improving RE uptake. New wind and solar investments are primarily channeled through competitive tenders for either "grid parity" project contracts—long-term contracts for local coal grid tariffs—or through competitive auctions for remaining feed-in tariff funds (Compilation and Translation Bureau, 2016).

4.4.2 INDIA

The Indian government sets ambitious RE objectives for the short to medium period. The nation aims to have 175 GW of installed RE capacity by 2022. The Government of India declared in 2018 that RE will be increased by 227 GW before 2022 and 275 GW by 2027. On September 23, 2019, at the United Nations Climate Summit in New York, India's Prime Minister announced a new goal of 450 GW of renewable electricity production without defining a deadline (International Energy Agency, 2020).

In the Indian area of power generation, transport, and heat renewables, the dedicated Ministry of New and Renewable Energies (MNRE) will be responsible. The MNRE includes R&D, research, registration of the National Institute of Solar Energy and the National Institutes of wind energy, standardization, skills advancement, resource valuation, and awareness programs. The MNRE also includes bio-energies for electricity, including biogas. To provide loans to ventures in the fields of solar power and energy conservation, the Indian Renewable Energy Development Agency (IREDA) operates as a non-banking financial entity in the MNRE (2020).

Solar Energy Corporation of India (SECI) is responsible for implementing various MNRE support schemes, such as the solar park scheme and the grid-connected solar roofing scheme. SSS-NIRE is an autonomous institution of the MNRE, an emerging research and development center with a mandate to concentrate on bioenergy and to develop advanced renewables and biofuel technologies.

The Ministry of Power (MoP) oversees India's electricity market, including renewables for the generation of power. The Minister of Power also oversees the MNRE and is responsible for RE. India relies on renewable purchase obligations (RPOs), renewable electricity certificates (RECs), the rapid depreciation of RE assets for commercial and industrial users, and, most recently, competitive tenders for utility-scale renewables (Shrimali & Tirumalachetty, 2013). With long-term power buy deals comprising fixed-price contracts to meet the 2022 objective, the company has initiated open solar photovoltaic auctions (2010) and wind auctions (2017). The MNRE declared that 25–30 GW of solar PV would be tendered each year until the end of 2021 to reach the 100 GW level by 2022 (India's solar generation potential was 32.5 GW in the year 2019).

4.5 SWOT ANALYSIS ON INDO-CHINA RENEWABLE ENERGY (RE) HUB

4.5.1 THE STRENGTHS

4.5.1.1 POLICY ADVANTAGE

China and India are already well on track to meet their Paris Agreement climate goals. China plans to raise its RE capacity by 38% in 2020 compared to 2015 levels, increasing the installed capacity to 680 gigawatts (GW) and about $361 billion in funding in RE. For context, Germany, which ranked first in its RE policy in the 2016 Allianz Climate and Energy Monitor, presently has around 100 GW of RE completed. There is about 0.04 GW of capacity in a new park of 10 wind turbines.

India is also increasing its potential for RE at a quick rate. In 2016, solar and wind installations achieved the annual target by 43% and 116%, respectively. India plans to reach 175 GW of installed renewables for the year 2022. India is expected to easily achieve its climate goals with market forces put into motion by a strong policy intent (Liesch et al., 2017). China is canceling proposals for new fossil-based power plants and quickly shutting down old coal-fired power plants, whereas India is exploring plans to stop building new coal-fired energy plants beyond 2022 (Liesch et al., 2017). In China, RE policy is in the developing stages, and national motivation is critical for the further implementation of RE targets. The Chinese Center for Renewable Energy Development (CRED) has been tasked with the drafting of a new regulation, the Law on Renewable Energy Development. The National Development and Reform Commission is set up to support and provide directions for the growth of RE. The legislation aims to meet short-term energy needs while reinforcing the priorities of long-term sustainable growth (NREL, 2004). In several administrative acts of RE promotion, India has pioneered in the world, such as:

- Electricity regulatory commission in the liberalized market–1991;
- Compulsory environmental audits for power plants–1992;
- Energy conservation bill–2000;
- Renewable energy promotion bill–2005 (NREL, 2004).

4.5.1.2 GEOGRAPHIC ADVANTAGE

China has a vast geographical capacity for variable renewable electricity (VRE). The maximum capacity that can be installed for solar power is 12,936 GW. The maximum capacity that can be installed for onshore wind power is 575–4,909 GW, and for offshore wind is 559–932 GW. The geographic ability and mean capacity factor (CF) measured at each asset stage of the VRE also provide the necessary information for policymakers, academic peers, and private investors to determine the performance of VRE capital. For example, they may be used to guide decision-making for VRE investments in project financing. They also support the creation of efficient subsidy schemes required to meet China's goals for RE. The best opportunities (both high mean CF and high potential) are located in Inner-Mongolia (for wind) and North China (for solar) (NREL, 2004).

India, with its strategic geographical position, has ample non-conventional energy supplies. Geographically, India is the perfect land for solar energy. As of 30 September 2020, 36.17% (including significant hydro) of India's installed power generating capacity comes from renewable sources. According to the 2027 blueprint, India plans to have 275 GW of clean energy. India has also made considerable strides in terms of policy and has consolidated the structure for achieving the goal of transforming the usage of green energy sources as a common and prevalent mode of energy use. Significant untapped RE potential from India will contribute significantly to alleviating power shortages. They can also enhance energy security, contribute to regional growth, boost access in remote (rural) areas, diversify fuel sources, and provide environmental benefits. Recognizing these advantages, India's policymakers have paid a great deal of attention to clean energies, setting aggressive industry targets. Meeting these targets would take considerable financial expenditures and strategic effort to resolve the several challenges facing the various industries of RE (Sargsyan et al., 2011).

4.5.2 THE WEAKNESSES

China's green energy growth strategy is starting to reveal certain drawbacks that may better be reflected by the following: (i) the lack of continuity and continuity of policies; the coordinating role has not been implemented; (ii) the promotion program is inadequate, and the system is not complete; (iii) government subsidies do not cover adequate areas; the subsidies are limited and not smooth; (iv) the investment and financing mechanism is not sound and complete; the main investment body is more than single; domestic undertakings are not yet adequately invested; as a real investment man for industrialized growth, undertakings are not incorporated into the development chains; and (v) China is currently investing in scientific R&D of RE sources. This has plummeted economic growth into a fairly passive climate (Peidong et al., 2009).

India is a developing nation, and India's RE development faces some limitations that should be rectified for India to achieve its target set by the Indian government. The major limitation of India for the growth of the RE sector are – (i) lack of sufficiently trained and technical manpower available; and (ii) lack of infrastructure to incorporate. Infrastructure shortage typically includes highways, grid access, Communication, and expenses. The lack of proven reliability of current technologies in India is a challenge for the implementation of the new technology; (iii) the shortage of knowledge and connectivity is seen as a significant obstacle to the introduction of emerging renewable technologies; (iv) there is a lack of awareness and management around emerging energy technology; and (v) lack of understanding among end-users (Madurai Elavarasan et al., 2020).

4.5.3 THE OPPORTUNITIES

China has adopted several supporting steps to positively encourage the accelerated growth of the RE sector. China's Green Energy Development Plan will place green energy in a strategic role, improve scientific R&D of green energy, increase investment in green energy, innovate the system, and encourage the development of green energy. Chinese government officials emphasized that the green energy sector is a major economic growth platform, that it is the strategic focal point of a new round of international competitiveness, that the current global financial crisis has created prospects for the creation of the green energy sector and the development of the green energy sector. China's government attaches considerable priority to the production of RE. The strategic role of green energy is becoming more and more popular, and it will undoubtedly be further established. Promote the production and use of RE in China, will provide adequate space for the growth of the RE industry. There are currently favorable conditions and prospects for the growth of the RE sector in China. The planet is moving towards a low carbon period from industrialization or a high carbon age, and the robust growth of the RE sector has become a worldwide movement and consensus. In the international situation, the Chinese must pursue the pattern of the low-carbon period and actively support the growth of the green era. The energy sector and related foreign organizations not only meet the duty and obligation to offer financial and technical assistance to emerging countries for the resolution of developed nations. While there are many hurdles, they are all financial commitments and obligations combined with China's long-term opening-up trend and multilateral ties with other countries that are conducive to the growth of China's RE sector (Yang et al., 2018).

India has one of the most extensive clean energy expansion projects in the world. Support from the government and international investors has fueled this expansion. As a result of a highly favorable policy, climate, a steady flow of money, declining costs, and advanced technologies, the RE market in India has expanded very rapidly in the last five years. By announcing an ambitious goal of 175 GW from renewables by 2022, the government made clear its plan to drive toward a low-emission power grid in 2015. In 2015, the government made clear its intention to move towards a low-emission electricity system by declaring an ambitious target of 175 GW from renewables by 2022. The message was straightforward; broader and affordable electricity availability but at a minimum environmental expense. The renewables industry reacted to the government's demand for climatic conditions growth by rapidly ramping up production, India increased the share of RE sources in its power generation industry from 6% to 10%, with an annual growth rate of around 17.5% between 2014 and 2019. India is now among the top five RE producers in the world, with an installed RE capacity of 83 GW, including 31 GW under construction and 35 GW tenders, and is on track to exceed its original target. In fact, by 2022, 225 GW of renewables will be installed, with a goal of 40% RE by 2030 (Sinha, 2020).

4.5.4 *THE THREATS*

Major threats to the development of RE in India and China are the geographical mismatch between the areas where RE is produced and the areas where energy demand is high, the lack of development of power transmission infrastructure, and financial conditions.

4.5.4.1 *GEOGRAPHICAL THREATS*

The region's geographical location and environmental conditions can be an obstacle to the production of RE. For example, in India and China, where solar and wind power are intermittent, solar energy is sporadic on the earth's surface, depending on the position of the earth. This limits people in these areas from the use of solar energy as it will not be reliable. Moreover, as the population increases, natural resources and RE resources begin to become restricted. Thus, RE options have been costly and therefore costly for the production of RE, which is therefore inaccessible to customers (Yueqin & Yisi, 2021). In reality, this hampers the production in some countries of RE technologies.

4.5.4.2 *LACK OF INFRASTRUCTURE*

The lack of physical infrastructure for transmission and distribution lines, as well as the machinery and utilities required for power firms, is a significant infrastructural obstacle for the production of RE in India and China. Some of the machinery in these countries is normally not easily available and is often imported from industrialized nations. Because imported machinery is costly compared to locally produced, RE generation in India and China would be expensive and even inaccessible. Another major infrastructural obstacle to the expansion of RE technology is the lack of grid connectivity, particularly in the wind power sector. In such situations, energy (whether electrical or mechanical) is transmitted from the point of origin to the point of usage, during which there is a high degree of transmission loss. As a result, many consumers are losing their interest in RE and are not ready to invest in them because of the threat of damage (Energy & Fran, 2021).

4.5.4.3 FINANCIAL THREAT

Capital investment, transaction costs, economic status, and availability of incentives and subsidies are key factors determining the pace of implementation of RE technologies. The initial capital cost of RE is comparatively high compared to traditional energy sources, which in turn increases the cost of generating RE. Since many producers want to keep their initial investment costs down while increasing earnings, high investment costs remain a major hurdle to the development of sustainable RE initiatives (Online & Aes, 2020).

4.6 CONCLUSION

India and China have now become RE hubs for the manufacturing of renewable machinery and RE generation. Both countries now shifting toward renewable power generation to cut down the carbon emissions, and as well as to meet future energy requirement when fossil fuel are reduced day by day. There is a lot of possibilities to generate power through renewable source in both countries. Government policies also promote the use of RE in various fields and invest in the R&D of new renewable technology in both countries. Each government sets goals for RE to meet the future requirement and form a separate ministry to provide direction and support for the growth of RE. Major conclusion remarks for India and China are enlisted below:

- The Indian Government is constantly working to achieve the 2030 target to produce 450 GW of renewable energy.
- China's major target is to reduce CO_2 intensity by over 65% before the year 2030.

- China's ambitious target is to reach carbon neutrality up to the year 2060.
- The government of India forms a separate ministry presently named as Ministry of New and Renewable Energy (MNRE) which works on all matters related to renewable energy.
- India and China are members of the Clean Energy Ministerial (CEM), a high-level global platform to support policies and strategies that accelerate clean energy technologies, share valuable lessons and practice guidelines, and facilitate the transition to a sustainable clean energy economy.
- India and China promote local manufacturing, and export hubs for renewable energy. The government of both countries also provides loans and subsidies for promoting manufacturing plants related to renewable energy.

KEYWORDS

- **biomass**
- **energy**
- **energy security**
- **renewable energy**
- **solar energy**
- **sustainability**
- **sustainable development**
- **SWOT analysis**

REFERENCES

Bakshi, R., (2002). Wind energy in India. *IEEE Power Eng. Rev., 22*, 16–18. https://doi.org/10.1109/MPER.2002.1029967.

Chandrasekharam, D. 2000. "*Geothermal Energy Resources of India: Country Update.*" World Geothermal Congress 2000, no. April: 25–29. https://www.geothermal-energy.org/pdf/IGAstandard/WGC/2000/R0558.PDF

China Mineral Resources, (2015). *China Mineral Resources* (pp. 52–53). Minist. L. Resour. People's Repub. China.

Chul, P., & Jo, H., (2008). *Recent Development of Ocean Energy in Korea*, 911–916.

Compilation and Translation Bureau, (2016). *The 13th Five-Year Plan for Economic and Social Development of the People'S Republic of China* (pp. 97–99). Cent. Compil. Transl. Press.

Donatini, F., (2018). Geothermal power. In: *Power Engineering Advances and Challenges* (pp. 179–215). CRC Press. https://doi.org/10.1201/9781315202105-7.

Dwijen Vaidya, Manan Shah, Anirbid Sircar, Shreya Sahajpal, Shubhra Dhale, (2015) " Geothermal Energy: Exploration Efforts In India ", International Journal of Latest Research in Science and Technology . Vol. 4, Issue 4, pp 61-69.

Energy, C., & Fran, T. A., (2021). *Renewable Energies IN China: From A Proactive Political Approach to Bottlenecks in the Field*, 4–7.

Hähnlein, S., Bayer, P., Ferguson, G., & Blum, P., (2013). Sustainability and policy for the thermal use of shallow geothermal energy. *Energy Policy, 59*, 914–925. https://doi.org/10.1016/j.enpol.2013.04.040.

Hove, A. (2020). *Current Direction for Renewable Energy in China*. Oxford Energy Comment.

International Energy Agency, (2020). *India 2020 Energy Policy Review* (pp. 1–304). Int. Energy Agency.

Jingjing, L., Xing, Z., DeLaquil, P., & Larson, E. D., (2001). Biomass energy in China and its potential. *Energy Sustain. Dev., 5*, 66–80. https://doi.org/10.1016/S0973-0826(08)60286-0.

Kumar, A., Kumar, N., Baredar, P., & Shukla, A., (2015). A review on biomass energy resources, potential, conversion and policy in India. *Renew. Sustain. Energy Rev., 45*, 530–539. https://doi.org/10.1016/j.rser.2015.02.007.

Kumar, J, C. R., & Majid, M. A., (2020). Renewable energy for sustainable development in India: Current status, future prospects, challenges, employment, and investment opportunities. *Energy. Sustain. Soc., 10*, 2. https://doi.org/10.1186/s13705-019-0232-1.

Liesch, T., Ruiz-Vergote, S., Tewari, R., Höhne, N., Burck, J., & Freundt, M., (2017). *Deep Dive: What Do Current Policy Developments in China, India, and the US Mean for Investing in Renewables?* (pp. 1–20).

Madurai, E. R., Afridhis, S., Vijayaraghavan, R. R., Subramaniam, U., & Nurunnabi, M., (2020). SWOT analysis: A framework for comprehensive evaluation of drivers and barriers for renewable energy development in significant countries. *Energy Reports, 6*, 1838–1864. https://doi.org/10.1016/j.egyr.2020.07.007.

Ministry of New and Renewable Energy, (2021a). *Solar Energy Overview*, 1–2.

Ministry of New and Renewable Energy, (2021b). *Bio Energy Overview*. 2021, pp. 1–3.

Ministry of New and Renewable Energy, G. of I., (2021). *New Technologies Ocean Energy.* Routledge. https://doi.org/10.4324/9781315210698-9.

MNRE, (2001). *Jawaharlal Nehru National Solar Mission and Other Programs in Renewable Energy Supported by MNRE*. Annual report, pp. 14–16.

MNRE, (2020). *Ministry of New and Renewable Energy Annual Report 2019-20* (pp. 160–180). Minist. New Renew. energy.

NREL, (2004). *Renewable Energy Policy in China: Overview*. NREL Int. Programs Fact Sheet.

Online, E., & Aes, J. T. H. E., (2020). *Barriers to Renewable Energy Technologies Development*, pp. 1–12.

Pandve, H., (2009). India's national action plan on climate change. *Indian J. Occup. Environ. Med. 13*, 17. https://doi.org/10.4103/0019-5278.50718.

Peidong, Z., Yanli, Y., jin, S., Yonghong, Z., Lisheng, W., & Xinrong, L., (2009). Opportunities and challenges for renewable energy policy in China. *Renew. Sustain. Energy Rev., 13*, 439–449. https://doi.org/10.1016/j.rser.2007.11.005.

PIB, India, (2020). *Press Information Bureau Government of India Ministry of New and Renewable Energy.*

Prem Kumar Chaurasiya, Vilas Warudkar, Siraj Ahmed, (2019). Wind energy development and policy in India: A review, Energy Strategy Reviews, Volume 24, Pages 342-357. https://doi.org/10.1016/j.esr.2019.04.010.

Sargsyan, G., Bhatia, M., Banerjee, S. G., Raghunathan, K., & Soni, R., (2011). *Unleashing the Potential of Renewable Energy in India.* Unleashing the potential of renewable energy in India. https://doi.org/10.1596/978-0-8213-8780-1.

Shrimali, G., & Tirumalachetty, S., (2013). Renewable energy certificate markets in India—A review. *Renew. Sustain. Energy Rev., 26*, 702–716. https://doi.org/10.1016/j.rser.2013.06.034.

Sinha, S., (2020). *Why India is the New Hotspot for Renewable Energy Investors* (pp. 1–5). World Econ. Forum.

Sino-German Energy Transition Project, (2020). *China Energy Transition Status Report 2020.*

Song, X., (2020). Challenges of sustainable development in China. *Sustain. Dev., 10*, 517–526. https://doi.org/10.12677/SD.2020.104065.

Wang, S., Yuan, P., Li, D., & Jiao, Y., (2011). An overview of ocean renewable energy in China. *Renew. Sustain. Energy Rev., 15*, 91–111. https://doi.org/10.1016/j.rser.2010.09.040.

Wang, Y., Liu, Y., Dou, J., Li, M., & Zeng, M., (2020). Geothermal energy in China: Status, challenges, and policy recommendations. *Util. Policy, 64*, 101020. https://doi.org/10.1016/j.jup.2020.101020.

Yadav, K., & Sircar, A., (2020). Geothermal energy provinces in India: A renewable heritage. *Int. J. Geoheritage Park.* https://doi.org/10.1016/j.ijgeop.2020.12.002.

Yang, D., Wang, X., & Kang, J., (2018). SWOT analysis of the development of the green energy industry in China: Taking the solar energy industry as an example. *Proc. – 2018 2nd Int. Conf. Green Energy Appl.: ICGEA*, 2018 103–107. https://doi.org/10.1109/ICGEA.2018.8356320.

Yueqin, J., & Yisi, C., (2021). *Renewable Energy in China: Market Barriers and Policy Options,* 1–7.

Zhang, Y. L., Lin, Z., & Liu, Q. L., (2014). Marine renewable energy in China: Current status and perspectives. *Water Sci. Eng., 7*, 288–305. https://doi.org/10.3882/j.issn.1674-2370.2014.03.005.

Zhang, Y., Ren, J., Pu, Y., & Wang, P., (2020). Solar energy potential assessment: A framework to integrate geographic, technological, and economic indices for a potential analysis. *Renew. Energy, 149*, 577–586. https://doi.org/10.1016/j.renene.2019.12.071.

CHAPTER 5

Japan's Nuclear Energy Crisis: Lessons for Energy Security with Sustainable Clean Energy

RAMKISHORE SINGH,[1] MANISH GUPTA,[2] ASHUTOSH VERMA,[1] MANISH VYAS,[3] and DHARAM BUDDHI[1]

[1]*Department of Research and Innovation, Lovely Professional University, Phagwara, Punjab, India*

[2]*Mittal School of Business, Lovely Professional University, Phagwara, Punjab, India*

[3]*School of Pharmaceutical Sciences, Lovely Professional University, Phagwara, Punjab, India*

5.1 INTRODUCTION

Since the dawn of time, energy has been one of the most basic requirements for human growth and society's long-term viability. With the advancement of humanity, the pattern of energy consumption, sources, and methods for converting energy from its most basic form (naturally accessible biomass on the earth's most upper surface) to its highest-grade form (electricity) have evolved. Prior to 1800, civilization depended nearly entirely on biomass as a major energy source for conducting energy-intensive activities such as cooking, lighting, and heating. During the period, traditional biomass was the only source that could meet the world's entire primary energy demand of around 5,550 TWh. Changes in global society's lifestyle have been noticed as time has passed. The past 200 years have seen a significant change in the human lifestyle, which has resulted in a massive rise in human energy

The South Asian Energy Corridor: Toward Sustainability and Security. Amritanshu Shukla, PhD, Atul Sharma, PhD & Saurabh Mishra, PhD (Eds.)

requirements, which has been met by recognized fossil fuels. Traveling was mostly done on foot or with animals if one could afford them until the 19th century. For lighting the interiors, candles or oil derived mainly from animal fats were employed. The daily activities were in perfect harmony with nature. People used to retire to bed shortly after the sun set since there were few cheap alternatives for artificial illumination inside. As a result, most of the activities that required illumination, such as farming, trade, traveling, and other professions, were carried out throughout the day in natural sunshine.

People managed to find new methods to make their lives fashionable and much more pleasant than they were previously in the 18th century by the beginning of the 19th century. In addition, many new energy-powered technologies were created; unfortunately, these technologies were inefficient. Humans discovered and began using various sources of fossil fuels as the need for high-grade energy for powering newly created sophisticated devices grew. Several other fossil fuels-driven automatic machines and equipment technologies, such as motorbikes, automobiles for transportation, steam engines, and turbines for power generation, cooking devices, and a few more powerful machines for industrial applications, were developed and used at the same time. With the Industrial Revolution in the 19th century, significant changes in lifestyle occurred. Industrial installations in and near metropolitan regions have created a significant number of job possibilities, attracting individuals not just from urban areas but also from rural places. People from rural regions migrated to cities and towns in search of a better and more comfortable living, resulting in a huge shift in rural and urban populations. Global industrialization occurred as a consequence of rapid technological advances and the discovery of high-energy-density fossil fuel deposits, resulting in a significant shift in society's energy consumption pattern. Traditional biomass's contribution to energy requirements remained about 10% of overall primary energy consumption at the time. Furthermore, from the beginning of the 19th century, worldwide primary energy consumption has risen by more than 30 times. This paradigm shift in energy consumption quantity and pattern is the consequence of changing lifestyles, rising living standards, industrialization, and a constantly increasing population. Furthermore, by the mid-20th century, the proportion of fossil fuels (coal, oil, and gas) in the global energy mix had risen considerably, accounting for about 80% of global society's total primary energy needs. We now use nearly 4,000 times the amount of fossil fuel utilized in the 1700s, after the development of the coal-fired steam engine. The shift in global energy supply sources from 1800 to 2019 may be shown clearly in Figure 5.1.

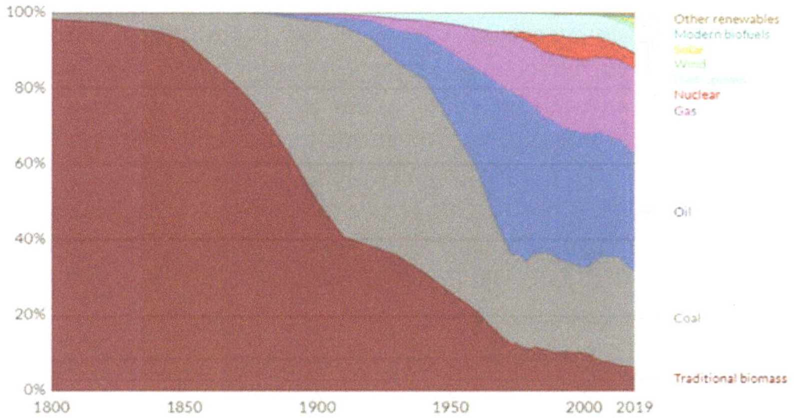

FIGURE 5.1 Global energy consumption by source (%) since 1800.

Source: Reprinted from Our World in Data https://ourworldindata.org/energy-mix.

The reckless and aggressive extraction of fossil fuels has placed the world's civilization in jeopardy. Excessive combustion of fossil fuels puts strain on limited fuel sources while also constantly pumping CO_2 into the atmosphere and rising CO_2 concentrations in the environment together with other greenhouse gases (GHGs). Total CO_2 emissions in 2019 are projected to be about 1,00,000 times higher than they were in 1,800 (*see* Figure 5.2). CO_2 emissions in 1,800 were less than 0.03 billion tons (https://ourworldindata.org/grapher/annual-co-emissions-by-region?stackMode=absolute&time=1800.latest). The increasing concentration of CO_2 and other GHGs in the environment has caused global warming by raising the temperature of the earth's atmosphere.

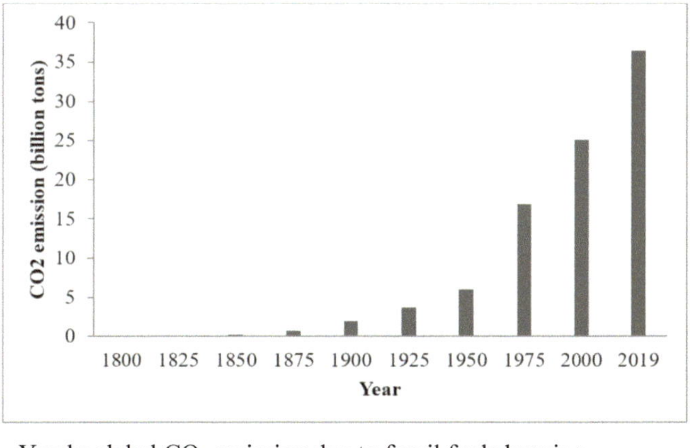

FIGURE 5.2 Yearly global CO_2 emission due to fossil fuels burning.

Source: Reprinted from Our World in Data. https://ourworldindata.org/energy-mix

In recent years, a group of international scientists affiliated with the IPCC has conducted significant studies on forecasting future rates of global warming. The IPCC made one of the forecasts, which was published in 2014 (Intergovernmental Panel on Climate Change (IPCC), 2014). The study contains global warming projections for the best-case and worst-case scenarios (Figure 5.3). Estimates were made for the end of the 21st century. Giorgetta et al. (2013) simulated another for the worst-case scenario, which represents a doubling of CO_2 and an increase in mean global temperature of approximately 3–4°C by 2,100 (Elias, 2018).

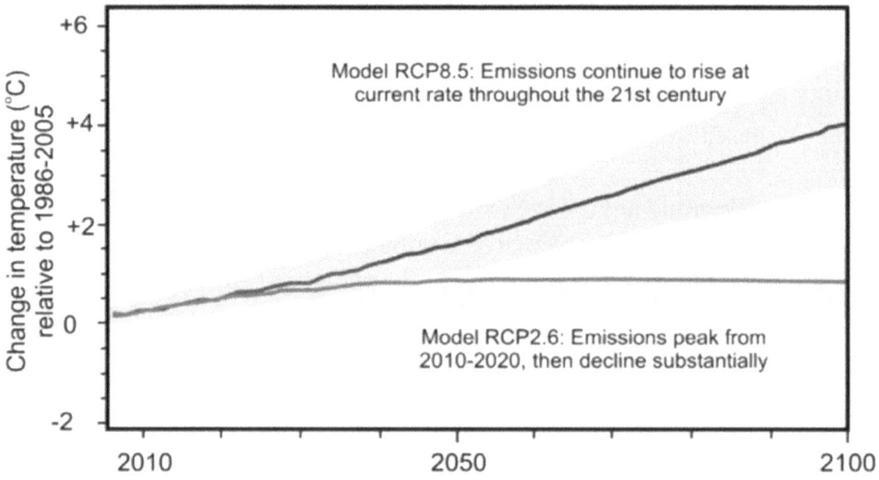

FIGURE 5.3 IPCC scenarios predicting future global average temperatures: best- and worst-case scenarios (IPCC, 2014).

Source: Adapted from IPCC, 2014

If the pace of fossil fuel combustion is not decreased and stabilized at its present level, the increase in global mean temperature may continue. The melting of permafrost in the Arctic and subarctic areas has been sparked by the increasing global mean temperature caused by the enormous quantity of CO_2 put into the earth's atmosphere every year. Furthermore, as temperatures rise, the soil profile is progressively altered. In the highest latitudes, where global warming is magnified, the impact is more severe. Furthermore, climate change may be extremely disruptive to social and economic life, resulting in severe problems such as catastrophic drought, heat waves, and bigger storms that may be life-threatening.

5.2 NUCLEAR POWER A CO_2 NEUTRAL OPTION

The effect of fossil fuel-driven climate change has been recognized by the global scientific community and governments, who have emphasized the use of alternative CO_2-neutral fuels. CO_2 neutral, renewable, and low-carbon energy sources may reduce CO_2 emissions as much as feasible, as quickly as possible. As a result, worldwide technological development and awareness-building initiatives have been launched to make use of discovered alternative CO_2 neutral fuels in a cost-effective, technically viable, and long-term way.

Nuclear energy has been recognized as one of the most promising alternatives for providing energy to humanity as an alternative fuel source. Nuclear energy has been seen as a critical component of climate change mitigation among renewable and CO_2-neutral energy sources. However, whether or not to utilize nuclear power for energy generation is constantly a hot topic of discussion. The first suggestion for utilizing nuclear power as a climate-protection alternative was made in early 2000 (Sailor, 2000), but reservations about including nuclear as a climate-change mitigation option were voiced repeatedly. Scientists are expressing their views on nuclear power in the media and directly to governments, both in favor (Hansen, 2019) and against it (Dorfman, 2019). However, the bulk of scientific research differs on whether nuclear power can be used to reduce CO_2 emissions (Mez, 2012). However, some argue that nuclear power facilities will be 10 times less harmful than coal power plants and may be utilized to replace them (Knapp, Pevec, & Matijevi, 2010; Knapp & Pevec, 2018), while others argue that the existing nuclear fleet should be kept operational (Roth & Jaramillo, 2017). Buongiorno et al. (2019) recently highlighted nuclear power's promise as a low-carbon solution, although cost reduction is a significant issue. At the same time, Parsons et al. (2019) conclude that nuclear energy is required to fulfill the Paris Agreement's climate objectives. Furthermore, the International Atomic Energy Agency (IAEA) identified a broad variety of nuclear power scenarios and features that may be regarded as climate mitigation technology (International Atomic Energy Agency, 2018). Sovacool et al. (2020) examined the relationship between CO_2 emissions and power production during a 25-year period in 123 countries and concluded that large-scale national nuclear projects had no effect on CO_2 emissions. The discussion is still going on and may go on for a little longer since there is no clear agreement on the good or bad elements.

5.2.1 NUCLEAR ENERGY POTENTIAL AND RESOURCES

Nuclear energy produces usable energy, such as electricity, from radioactive materials such as uranium and thorium. In nuclear reactors, such materials undergo a fission chain reaction, which releases thermal energy by breaking the atoms of radioactive elements. Furthermore, the released thermal energy is converted to steam, which is utilized to power turbines and generate electricity. The method of generating electricity is quite similar to that of conventional thermal power plants, with the exception that the heat generated by fossil fuels is replaced by a fission-fusion reaction in a nuclear reactor. Nuclear reactors now employ uranium 235 isotopes as fuel and use the thermal neutron spectrum. The fission of uranium 235 produces heat and, ultimately, electricity. However, natural uranium includes mostly isotope uranium 238 (about 99.3%) and a small amount of isotope uranium 235 (only 0.7%), which is utilized as nuclear reactor fuel. Furthermore, the bulk proportion of uranium ore in most rock mining, with the exception of a few Canadian ones, is between 0.03% and 034%. As of 2017, about 8 million tons of uranium have been found, with 3.2 million tons classified as assailable. The cost of recovery was predicted to be less than 260 dollars per kilogram of uranium (OECD NEA, IAEA, 2019). Australia, Canada, and Kazakhstan are significant uranium producers, accounting for about 67% of total world output. Only Kazakhstan generates approximately 40% of total worldwide uranium output, making it the world's biggest producer (WNA, 2019).

In addition, fast breeder reactors were created in response to the restricted supply of the isotope uranium 235 in conventional nuclear reactors. The fast breeder reactors work by transmuting uranium-238 into the fissile isotope plutonium-239, which may be used as reactor fuel in the same way as uranium-235. However, the bulk of fast breeder reactors performed poorly during operation. The BN-600 fast breeder reactor in Beloyarsk, Russia, is the sole exception, with a lifetime load factor of around 75%. Pistner & Englert (2017) have published a thorough review of fast breeder reactors that have been installed since 1947. One of the most promising ideas so far has been the sodium fast reactor. The reactor's main coolant is sodium, which is activated as it passes through the reactor core. Sodium-24 (half-life of 15 hours) is produced throughout the process by catching neutrons, while sodium-22 is produced via the (n, 2n) reaction (half-life of 2.6 yrs.). However, owing to the strong reactivity of sodium with air and water, these reactors are likely to have malfunctions and extra risks for accidents, resulting in a sodium leak, either to the tertiary loop (which usually includes a steam turbine and water as a fluid) To minimize the risk of fire, the reactor containment is filled with an inert gas like Argon. Sodium leaks and fires, on the other hand, were unavoidable in any of the French SFRs (Schneider, 2009), BN 350, or BN 600 systems (Poplavskii et al., 2004). The largest sodium leak-caused fire at Monju SFR in Japan forced the plant's closure in 1995 (Kondo, 1998). As a result, additional safety measures in the steam generators are needed, such as limiting oxygen input and promptly detecting leakage.

5.2.2 CO_2 EMISSIONS FROM NUCLEAR POWER PLANTS

As mentioned previously in this chapter, there are significant variations of opinion about nuclear power plants' low CO_2 emissions. Many studies have recently been performed to assess the direct and indirect CO_2 emissions from nuclear power facilities. One research found CO_2 emissions ranging from 1.4 g CO_2/kWhe to 288 g CO_2/kWhe, with a mean value of 66 g CO_2/kWhe. Different grades of uranium ore, mining methods and procedures, fuel manufacture, and power sources utilized in mining all contribute to the variance in emission values per kWh (Sovacool, 2008). In the European setting, Beerten et al. (2009) predicted 32 g CO_2/kWhe. The energy expenditure required for mining and fuel manufacturing is linked to nuclear lifetime CO_2 emissions. In addition to the CO_2 costs from the nuclear lifespan of 9–70, Jacobson (2009) conducted life cycle studies for various energy generating technologies and found 59–106 g CO_2/kWhe as opportunity cost emissions for nuclear power. He also said that overall nuclear energy prices range from 68 to 180 g CO_2/kWhe. According to Wallner et al. (2011), nuclear power plants should only be utilized if they produce more energy than is used in the mining and enrichment of fuel. For low grade ores, he reported net negative energy. Nikolaus et al. (2021) recently reported a 2–3% decrease in total world GHG emissions. If stated plans for additional nuclear plants and lifespan extensions are executed, further reductions are anticipated through 2040. However, the supply of uranium-235 is technically limited, and the timely availability of new nuclear technologies that can operate on uranium-238 does not seem to be feasible.

5.3 JAPAN NUCLEAR ENERGY CRISES

Almost no power generation technique is completely risk-free. With every technology, there are always certain dangers involved. These dangers may be financial, societal, or unintentional. Nuclear power also has minimal dangers. If the process is not regulated, the power intensity of fission-fusion chain reactions in radioactive fuels is much more hazardous than any other kind of risk. As mentioned previously in this chapter, the use of nuclear power plants as a climate change mitigation option has long been a source of controversy, and there have been a few instances of mishaps in nuclear power plants, either due to human mistake or natural catastrophe.

An earthquake with a magnitude of 9.0 on the Richter scale was just detected off the coast of Japan on March 11, 2011. Along Japan's north-eastern shore, the earthquake triggered a massive and destructive tsunami that reached a height of 10 meters. At Miyako, Iwate, the tsunami reached a height of nearly 30 meters. It was an unusual double complex earthquake with a 3-minute duration (Jibran et al., n.d.). The earthquake-caused tsunami had such a large impact that it shifted a 650 km seabed from north to south, 10–20 m horizontally; the local coastline was 0.5 m away; a 560 km^2 area was flooded; approximately 19,000 people were killed; and Furthermore, the catastrophic earthquake and tsunami forced the automatic closure of 11 of the country's 55 nuclear facilities. Due to their proximity to the epicenter, two nuclear power facilities, Fukushima and Onagawa, were severely damaged. The destruction resulted in a huge discharge of radioactive material, much more than Chernobyl in 1986 and far worse than the Three Mile Island disaster in 1979 (Timeline: Japan Power Plant Crisis, 2011). Furthermore, the natural catastrophe damaged the plant's emergency supply systems, which were required to keep the reactors cold. Fuels melted, hydrogen exploded, and radioactive material was released into the environment as a result of the emergency supply system failure. General Electric, Toshiba, and Hitachi designed and delivered the early 1960s Mark I containment boiling water reactors (BWR) at the Fukushima nuclear power facilities. The Mark I containment design is a drywell in the form of an inverted common incandescent light bulb that houses the reactor vessel and main pipework, which is connected to a torus-shaped suppression chamber by many massive vent pipes. Underneath the drywell lies the suppression chamber. Between 1971 and 1975, a few reactors at Fukushima began commercial operation, while more were added subsequently. The plant's overall power generation capacity was about 2,344 MWe (Mark, Richard, & Mary, 2011). Because the interruption was so significant and unexpected, no safety precautions could be implemented. However, attempts were undertaken to cool the reactor cores using seawater (Mark, Richard, & Mary, 2011), and boron was added to the seawater to absorb neutrons, allowing the nuclear chain reaction to resume without overheating the reactors (Seoul to Send Boron in Bid to Cool Reactors, 2011). Increased pressure in the main containments of the units and chambers necessitated evacuation to alleviate pressure in the environment and ultimately release radiation from the reactor cores (Fukushima, 2012). The massive explosions in the reactor building were thought to have occurred as a consequence of a chemical reaction between the fuel's zirconium coating and high-temperature steam, which generated significant quantities of hydrogen in the reactors. The total quantity of hydrogen generated in the chemical process was calculated to be between 2,400 and 3,000 kg (Higgins, 2012). Due to the explosion in the plant, all attempts to cool the chambers were futile, and the cooling water in the refueling pool became polluted (Lessons Learned from the Fukushima Nuclear Accident for Improving Safety and Security of U.S. Nuclear Plants, 2016). All of these resulted in substantial radioactive material being released into the facility. As a result, radioactive pollution in waters around the facility, agricultural goods, and neighboring prefectures exceeded the permissible limit (Gilmour et al., 2016). On the 12th day after the disaster, the limit of radioactive contamination in Tokyo's drinking water was more than double the acceptable threshold for infants. The radioactive depositions in regions outside a 20-kilometer radius, which functioned as an evacuation zone surrounding the Fukushima facility, were much higher than those seen in Chernobyl (Fairlie, 2013).

Initially, a precautionary evacuation was ordered on the same day as the accident in a radius of less than 3 km from the facility, which was later expanded to 10 km and then 20 km when the amount of radioactive contamination was detected. The condition worsened on the fifth day following the accident, with the dosage rate reaching 1,085 mrem/h. Furthermore, the dosage rates in the factory were up to 400 times greater, making recovery attempts much more difficult (The Fukushima Daiichi Accident, 2015). It's worth noting that the worldwide prolonged dosage limit for the general population is 100 mrem for a year. Winds also contributed to soil pollution in the region (Chronological Fact Sheet on 2011 Crisis at Fukushima Nuclear Power Plant, 2011), as well as contamination of food items such as vegetables, meat, and dairy products, as well as water. As a result, the local community was forced to relocate 40 kilometers away from the Fukushima site, affecting approximately 1,00,000 residents and forcing them to live an indefinite distance from their homes while also preventing the health effects of radioactive doses that exceeded Japan's regulatory limits. In the nation, unusually short-term fatalities and radiation-related illnesses were plainly noticed. After almost nine months of mishaps, the Japanese government announced a cold shutdown of the damaged Fukushima reactors on December 16, 2011, allowing some people to return to the least polluted evacuation zones with the lowest risk of additional radioactive leaks (Justin, 2011). The decommissioning and full shutdown of the facility was anticipated to take approximately a year.

Other nuclear power plant accidents have occurred in the past, but the loss of human lives, financial losses associated with managing and implementing safety measures, and community displacements were not as severe as those experienced in Japan as a result of the Fukushima nuclear catastrophe. Scientists, policymakers, investors, and other stakeholders in nuclear power across the world have altered their minds as a result of these events. Previously, nuclear power was thought to be one of the CO_2 neutral and long-term energy supply alternatives for humanity, with reasonable safety precautions. However, after the Fukushima nuclear catastrophe, these precautions seem to be insufficient to protect society from the deaths caused by leakage of radioactive fuel used in nuclear power plants, which resulted in an uncontrolled chemical chain reaction of radioactive elements.

Other RE sources, which seem to be more promising and economically viable, are now receiving greater attention from the worldwide community of scientists and policymakers. Nuclear power plants pose significant safety and environmental risks. Current nuclear power facilities use better technology and tighter safety regulations, lowering the danger of future accidents considerably. However, catastrophic "catastrophes" remain a possibility. Nuclear mishaps in the past have resulted in fatalities, major diseases, and severe environmental damage, and their consequences have lasted for decades. Even if all safety precautions are taken, there is no assurance that a nuclear power station will not have an accident. Threats of terrorism are another issue that has to be handled (MacKenzie, 2011).

5.4 ENERGY SECURITY WITH SUSTAINABLE CLEAN ENERGY

RE sources have gotten a lot of attention in recent years because of their near-zero GHG emissions, plentiful availability, and promise for providing electricity to the world's population. As a result of increased reliance on power-driven automatic electronics-based technologies for routine activities, the sector is receiving a lot of attention, which has resulted in a high demand for energy that will almost certainly increase significantly in the near future, as well as environmental concerns, particularly the risk of drastic climate change; need of affordable energy for rural populations; as well as advancements in technologies and reduction in costs.

Hydropower, biomass energy, solar energy, wind energy, geothermal energy, tidal and ocean energy are all examples of possible RE sources. Furthermore, researchers and innovators have developed technologies that have proven to be extremely useful and beneficial in the modernization of RE resources, particularly biomass, geothermal, small and micro-hydropower, low temperature solar energy applications, mini- and micro-wind turbines, and the conversion of vegetable and animal waste to com Further developments in many other sophisticated power-producing technologies, such as solar thermal and photovoltaic technologies, as well as supportive policies from governments, have given people and large corporations across the world a substantial boost in terms of implementation.

5.4.1 POTENTIAL OF RENEWABLE ENERGY (RE) SOURCES

In comparison to the potential, the contribution of clean and RE sources to primary energy consumption is extremely low. Several estimates of global and regional RE potential have been made and are published in the literature. All agreed that RE's entire worldwide technological potential is considerably more than present global energy consumption (Verbruggen, Moomaw, & Nyboer, 2011). Theoretical and technological potentials of RE were evaluated in one of the studies, revealing much more availability than the present world primary energy consumption (*see* Table 5.1). Even if the lowest technological potential of solar energy is included in the estimate, only solar can provide nearly 100 times the world's primary energy consumption.

TABLE 5.1 Theoretical and Technical Renewable Energy Potentials versus Utilization in 2015 (Teske, Nagrath, Morris, & Dooley, 2019)

Renewable Energy Resource	Theoretical Potential (Annual Energy Flux) [EJ/year] IPCC 2011	Technical Potential [EJ/year] Global Energy Assessment 2012, Chap. 11, p. 774	Utilization in 2015 [EJ/year] IEA-WEO 2017
Solar energy	3,900,000	62,000–2,80,000	1.3
Wind energy	6,000	1,250–2,250	1.9
Bioenergy	1,548	160–270	51.5

TABLE 5.1 *(Continued)*

Renewable Energy Resource	Theoretical Potential (Annual Energy Flux) [EJ/year] IPCC 2011	Technical Potential [EJ/year] Global Energy Assessment 2012, Chap. 11, p. 774	Utilization in 2015 [EJ/year] IEA-WEO 2017
Geothermal energy	1,400	810–1,545	2.4
Hydropower	147	50–60	13.2
Ocean energy	7,400	3,240–10,500	0.0018
Total		76,000–2,94,500	(Total Primary Energy Demand 2015) 555 EJ/year

Furthermore, the IPCC determined that clean and RE's worldwide technological potential would not restrict the sector's continuing growth (Verbruggen, Moomaw, & Nyboer, 2011). However, due to land availability and other resource restrictions, the exploitable potential of clean, sustainable energy may be decreased by a portion.

5.4.2 GROWTH IN CLEAN AND RENEWABLE ENERGY (RE) SECTOR

Joachim Nitsch estimated that solar contributed 0.1% of total world energy consumption in 1994, with installed capacities of 2,000 MW, 350 MW, and 80 MW for wind, solar thermal power, and solar photovoltaic power, respectively. A total of 200 km^2 of collectors, as well as 30,000 contemporary straw- and wood-fueled heating units and biogas installations, were recorded by the author (Joachim, 1994).

Until 2019, the total installed capacity of wind and solar power on the globe was estimated to be significantly more than 600 GW and more than 580 GW, respectively. In China, the United States, and India, the total installed capacity of solar power was more than 200 GW, 62 GW, and 35 GW, respectively, and the installation of the technologies has grown significantly in the past two years. In the same year, the aforementioned nations reported wind power installed capacity of about 210 GW, over 100 GW, and 37.5 GW, respectively (https://ourworldindata.org/renewable-energy). In the past 10 years, the percentage of RE in yearly power production in the United States has risen from 10% to 20%, more than doubling. Several reasons, such as lower technological costs and favorable regulations, have contributed to the sector's development in the United States, with RE's share projected to rise to over 30% by 2030 and up to 65% by 2050 (Wesley, Nathaniel, &

Trieu, 2021). Between 1994 and 2015, research on the effect of financial development on RE use was performed in 34 upper-middle-income developing nations. The research concludes that although economic development has a negative influence on RE use, consumer prices have little impact. The research also shows that increased demand for environmentally friendly RE sources is triggered by financial development (Muhammad et al., 2021). Furthermore, according to another research, RE is a complement to non-RE in high-income nations, while renewable and non-RE are replacements in middle-income countries (Nadia & Seema, 2021). RE has been recognized as a favorable precondition for the growth of nations. Nevertheless, the government and corporate houses must take deliberate measures and coordinate their efforts (Elena et al., 2021).

Several economically feasible and user-friendly methods for effectively capturing and using RE have been created in recent years. Such advancements in technical development have accelerated the adoption of RE technology and accelerated the sector's growth. Because the developed technologies are modular, they are suitable for both large and small-scale applications; they are flexible in meeting forecast demand; short lead times reduce risk and financing charges; they have the potential for market expansion and rapid dissemination, as well as favorable land-use characteristics. Currently, conventional biomass, hydropower, nuclear, wind, and solar provide about 6.4%, 6%, 4%, 2%, and 1% of worldwide primary energy consumption, respectively. Traditional biomass is mostly utilized for cooking and heating, especially in poor nations' rural regions. Large-scale hydropower has reached its economic capacity and has little room for growth, and the choice is often constrained by financial, environmental, and social factors. The combined contribution of sustainable clean RE sources such as geothermal, wind, solar, and biofuels is projected to be about 4.66% of total world energy consumption (https://ourworldindata.org/energy-mix) (Figure 5.4).

5.4.3 COST AND SUSTAINABILITY OF CLEAN AND RENEWABLE ENERGY (RE)

The best thing about installing RE power generation technologies is that the cost per unit of output is steadily dropping as a consequence of mass manufacturing and the availability of lower-cost raw materials. In addition, the cost of fuel is either nil (wind, sun, geothermal, and tidal) or cheap (in the case of biomass or waste). On the other hand, the price of fossil fuels is steadily rising. In the last 10 years, the cost of RE technology has plummeted, and the majority of them are now economically feasible. The unit life cycle costs of energy from geothermal, wind, solar thermal, and solar photovoltaic were about 15, 38, 58, and 90 cents/kWh, respectively, until 2000 (The World Bank, 2005), when they were reduced by a factor of 50%, around 1/5th, 1/6 the price has continued to fall till today. The current cost per unit for concentrated solar power with storage is projected to be about 9.0/kWh; however, by 2030, this is expected to decrease to 5.0/kWh (Shahabuddin et al., 2021). Solar energy is one of the alternatives accessible in abundance and free of charge for the energy sector's sustainability as well as effective climate change mitigation. The source is thought to be the most promising source of RE. However, the percentage of the projected potential of the source that has been utilized to far has not exceeded 2–3%, indicating the necessity to take drastic measures to harness the potential and reduce GHG emissions.

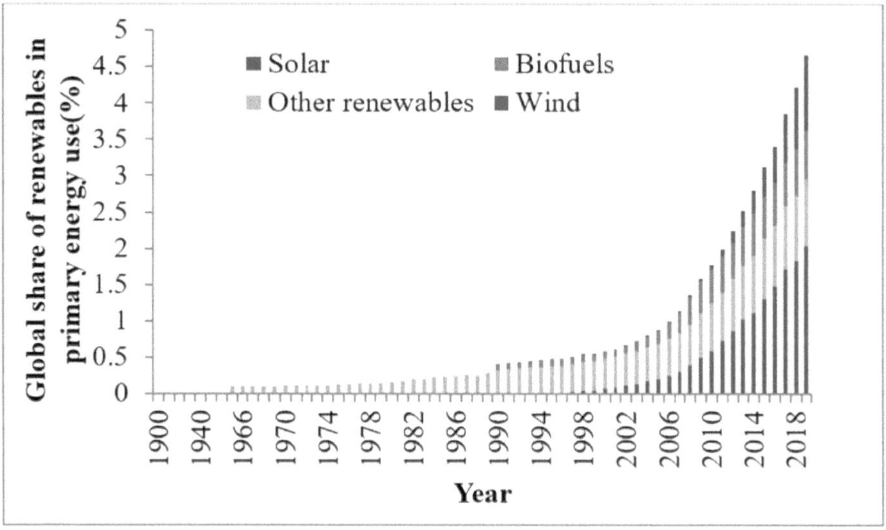

FIGURE 5.4 Growth in global share of sustainable clean energy.

Source: Reprinted from Our World in Data https://ourworldindata.org/energy-mix.

5.4.4 *RISKS AND ADVANTAGES OF NUCLEAR POWER*

The intermittent nature of the source and intensity fluctuation throughout the day are the main challenges that technocrats, policymakers, and scientists confront in expediting the process of implementation and usage. The two commercially accessible solar power technologies, solar photovoltaic and solar concentrating electricity, come with storage options such as electrochemical and thermal, allowing users to depend on the technologies. Furthermore, the grid-tied alternative, especially in the case of solar PV, has built confidence among consumers and developers, and as a result, major corporations across the world are showing a strong desire to invest in the installation and development of MW power plants. In addition, the cost of implementation has dropped dramatically in the past 4–5 years, making it more accessible. Furthermore, via thermolysis, thermochemical cycle, hydrocarbon cracking, reforming, and solar gasification, concentrated solar power may be used to convert biomass/waste into usable liquid or gaseous fuels (Shahabuddin et al., 2021). Solar-assisted power generation (SAPG) is also a viable option for producing energy that is clean, dependable, and efficient. This is a new technique for generating low-cost, high-efficiency solar/coal hybrid electricity. Non-concentrating solar collectors are used for preheating in coal-fired thermal power plants in this technique. One recent research suggested a SAPG to produce 12.6 MWe of extra output power with a 0.57% improvement in coal thermal system conversion efficiency. Exergy loss was decreased by 36.6% because of the integrated preheating system. Furthermore, the proposed system was estimated to cost only 47.4% of the conventional SAPG (Yu, Yingying, & Junjie, 2021), and the flexibility of design and customization of capacity as per the users' needs without a significant increase in cost, particularly in the case of solar photovoltaic, is an intriguing feature of the technology that allows RE applications also have increased conversion efficiency, dependability, and durability. Furthermore, short lead times, a high location-specific potential, attractive land-use characteristics, and ease of decommissioning are all driving factors for nuclear power. Unlike nuclear power plants, geopolitical risk has a significant and positive effect on the diffusion of RE (Osama, 2021). There are no issues of severe contamination risks, unexpectedly longer decommissioning times, resources, and life-threatening uncontrollable problems, if any unexpected natural disasters occur. Also, there is always the possibility of a terrorist attack on a nuclear power station, which may have catastrophic consequences, while the effect of such an event on clean energy power plants may be less severe, and recovery and repair are relatively much simpler and can be done rapidly.

KEYWORDS

- **CO_2 emissions**
- **energy security**
- **Japan's nuclear energy crises**
- **nuclear energy**
- **nuclear power plants**
- **renewable energy sector**
- **sustainable clean energy**

REFERENCES

Beerten, J., Laes, E., Meskens, G., & D'haeseleer, W., (2009). Greenhouse gas emissions in the nuclear life cycle: A balanced appraisal. *Energy Pol., 37*(12), 5056–5068.

Buongiorno, J., Corradini, M., Parsons, J., & Petti, D., (2019). Nuclear energy in a carbon-constrained world: Big challenges and big opportunities. *IEEE Power Energy Mag., 17*(2), 69–77.

Chronological Fact Sheet on 2011 Crisis at Fukushima Nuclear Power Plant, (2011). Retrieved from. https://www.nirs.org/wp-content/uploads/reactorwatch/accidents/Fukushimafactsheet.pdf (accessed on 12 August 2023).

Dorfman, P., (2019). *Letter: Global Investment in Nuclear is in Stark Decline*. Financial Times. Retrieved from: https://www.ft.com/content/75ea3180-0a02-11ea-bb52-34c8d9dc6d84 (accessed on 12 August 2023).

Elena, S., Sebastian, K., Eugeny, K., & Viktor, S., (2021). Governmental support, and renewable energy production: A cross-country review. *Energy, 230*, 120903, ISSN 0360-5442, https://doi.org/10.1016/j.energy.2021.120903.

Elias, S. A., (2018). Climate change and energy. In: Dominick, A. D., & Michael, I. G., (eds.), *The Encyclopedia of the Anthropocene* (Vol. 1, pp. 457–466). Oxford: Elsevier.

Fairlie, I., (2013). *Assessing Long-Term Health Effects from Fukushima's Radioactive Fallout.* Retrieved from: http://www.ianfairlie.org/news/assessing-long-term-health-effects-from-fukushimas-radioactive-fallout/ (accessed on 12 August 2023).

Fukushima, (2012). Nuclear Accident Analysis Report.

Gilmour, S., Miyagawa, S., Kasuga, F., & Shibuya, K., (2016). Current measures on radioactive contamination in Japan: A policy situation analysis. *PLoS One, 11*(3), e0152040. https://doi.org/10.1371/journal.pone.0152040.

Giorgetta, M. A., Jungclaus, J., Reick, C. J., Legutke, S., Bader, J., et al., (2013). Climate and carbon cycle changes from 1850 to 2100 in MPI-ESM simulations for the coupled model intercomparison project phase 5. *Journal of Advances in Modeling Earth Systems, 15*, 572–597.

Hansen, J., (2019). *Letter: EU Must Include Nuclear Power in its List of Sustainable Sources*. Financial Times.

Higgins, J., (2012). *The Next Meltdown? Responding to a Nuclear Accident in the Developing World* – Homeland Security Affairs. Retrieved from: https://www.hsaj.org/articles/205 (accessed on 12 August 2023).

https://ourworldindata.org/energy-mix (accessed on 12 August 2023).

https://ourworldindata.org/grapher/annual-co-emissions-by-region?stackMode=absolute&time=1800.latest (accessed on 12 August 2023).

https://ourworldindata.org/renewable-energy (accessed on 12 August 2023).

Intergovernmental Panel on Climate Change (IPCC), (2014). Climate change 2014: Synthesis report. In: Core Writing Team, Pachauri, R. K., & Meyer, L. A., (eds.), *Contribution of Working Groups I, II and III to the Fifth Assessment Report of the Intergovernmental Panel on Climate Change* (p. 151). Geneva, Switzerland: IPCC.

International Atomic Energy Agency, (2018). *Energy, Electricity and Nuclear Power Estimates for the Period up to 2050, 2018 Edition.* No. 1 in Reference Data Series. International Atomic Energy Agency, Vienna.

IPCC, (2014). *Climate Change 2014: Synthesis Report.* Contribution of Working Groups I, II and III to the Fifth Assessment Report of the Intergovernmental Panel on Climate Change. Geneva, Switzerland, IPCC, 2014.

Jacobson, M. Z., (2009). Review of solutions to global warming, air pollution, and energy security. *Energy Environ. Sci., 2*(2), 148–173.

Japan 2011 Earthquake: U.S. Department of Defense (DOD) Response, (2011). Retrieved from: https://www.files.ethz.ch/isn/128009/159781.pdf (accessed on 12 August 2023).

Japan Finds Costs Are Ballooning for Dismantling Fukushima, (2016). Retrieved from: https://www.spokesman.com/stories/2016/oct/25/japan-finds-costs-are-ballooning-for-dismantling-f/ (accessed on 12 August 2023).

Jibran, I., Fares, M. H., Abdel-Mohsen, O. M., & Evan, K. P. (2021). *Chapter 20 Assessment of Radiation Pollution from Nuclear Power Plants.* Pollution Assessment for Sustainable Practices in Applied Sciences and Engineering. In: Abdel-Mohsen, O., Mohamed Evan, K., Paleologos Fares, M. (eds.). Howari. Published by: Butterworth-Heinemann: An imprint of Elsevier. pp. 1027–1053. https://doi.org/10.1016/B978-0-12-809582-9.00020-7.

Joachim, N., (1994). The global potential of renewable energy sources. *Structural Engineering International, 4*(2), 72–75. doi: 10.2749/101686694780650995.

Justin, M., (2011). *Fukushima is in Cold Shutdown, Says Japanese Prime Minister.* Retrieved from: https://www.theguardian.com/world/2011/dec/16/fukushima-cold-shutdown-japanese-pm1 (accessed on 12 August 2023).

Knapp, V., & Pevec, D., (2018). Promises and limitations of nuclear fission energy in combating climate change. *Energy Pol., 120*, 94–99.

Knapp, V., Pevec, D., & Matijevi´c, M., (2010). The potential of fission nuclear power in resolving global climate change under the constraints of nuclear fuel resources and once-through fuel cycles. *Energy Pol., 38*(11), 6793–6803.

Kondo, S., (1998). History and perspective of fast breeder reactor development in Japan. *Energy, 23*(7, 8), 619–627.

Lessons Learned from the Fukushima Nuclear Accident for Improving Safety and Security of U.S. Nuclear Plants, (2016). https://doi.org/10.17226/21874.

MacKenzie, D., (2011). *Briefing: How Nuclear Accidents Damage Human Health.* New Scientist. Retrieved from: https://www.newscientist.com/article/dn20244-briefing-how-nuclear-accidents-damage-human-health/ (accessed on 12 August 2023).

Mark, H., Richard, J. C., & Mary, B. N., (2011). *Fukushima Nuclear Disaster*. Congressional Research Service. https://fas.org/sgp/crs/nuke/R41694.pdf (accessed on 12 August 2023).

Mez, L., (2012). Nuclear energy–Any solution for sustainability and climate protection? *Energy Pol., 48*, 56–63.

Moomaw, W., Yamba, F., Kamimoto, M., Maurice, L., Nyboer, J., Urama, K., Weir, T. (2011). Introduction. In: *IPCC Special Report on Renewable Energy Sources and Climate Change Mitigation* [Edenhofer, O., Pichs-Madruga, R., Sokona, Y., Seyboth, K., Matschoss, P., Kadner, S., Zwickel, T., Eickemeier, P., Hansen, G., Schlömer, S., & von Stechow, C. (eds.)], Cambridge University Press, Cambridge, United Kingdom and New York, NY, USA.

Muhammad, S., Betül, A. T., Sevgi, S. S., & Xuan, V. V., (2021). The effect of financial development on renewable energy demand: The case of developing countries. *Renewable Energy,* ISSN 0960-1481, https://doi.org/10.1016/j.renene.2021.06.121.

Nadia, D., & Seema, N., (2021). Does transitioning towards renewable energy accelerate economic growth? An analysis of sectoral growth for a dynamic panel of countries. *Energy*, 121290, ISSN 0360-5442, https://doi.org/10.1016/j.energy.2021.121290.

Nikolaus, M., Nikolaus, A., Klaus, G., Wolfgang, K., Wolfgang, R., & Wolfgang, L., (2021). Nuclear energy - The solution to climate change? *Energy Policy, 155*, 112363.

OECD NEA, IAEA, (2019). *Uranium 2018: Resources, Production and Demand.* Tech. Rep. Nuclear Energy Agency/Organization for Economic Co-operation and Development, Paris.

Osama, D. S., (2021). Is the geopolitical risk an incentive or obstacle to renewable energy deployment? Evidence from a panel analysis. *Renewable Energy, 178*, 377–384. ISSN 0960-1481. https://doi.org/10.1016/j.renene.2021.06.089.

Parsons, J., Buongiorno, J., Corradini, M., & Petti, D., (2019). A fresh look at nuclear energy. *Science, 363*(6423), 105, 105.

Pistner, C., & Englert, M., (2017). Neue Reaktorkonzepte - eine Analyse des aktuellen Forschungsstands. In: *Im Auftrag der Schweizerischen Energie-Stiftung*. Tech. rep. ¨Oko Institut E.V., Darmstadt.

Poplavskii, V. M., Bagdasarov, Y. E., Kamaev, A. A., Tsikunov, A. G., Zybin, V. A., & Ivanenko, V. N., (2004). Danger of burning of sodium coolant. *Atom Energy, 96*(5), 327–331.

Roth, M. B., & Jaramillo, P., (2017). Going nuclear for climate mitigation: An analysis of the cost effectiveness of preserving existing U.S. nuclear power plants as a carbon avoidance strategy. *Energy, 131*, 67–77.

Sailor, W. C., (2000). Nuclear power: A nuclear solution to climate change? *Science, 288*(5469), 1177, 1178.

Schneider, M., (2009). Fast breeder reactors in France. *Sci. Global Secur., 17*(1), 36–53.

Seoul to Send Boron in Bid to Cool Reactors, (2011). The Japan Times. Retrieved from: https://www.japantimes.co.jp/news/2011/03/17/national/seoul-to-send-boron-in-bid-to-cool-reactors/#.WlHNPyOB2LI (accessed on 12 August 2023).

Shahabuddin, M., Alim, M. A., Tanvir, A., Mofijur, M., Ahmed, S. F., & Greg, P., (2021). A critical review on the development and challenges of concentrated solar power technologies. *Sustainable Energy Technologies, and Assessments, 47*, 101434, ISSN 2213-1388, https://doi.org/10.1016/j.seta.2021.101434.

Sovacool, B. K., (2008). Valuing the greenhouse gas emissions from nuclear power: A critical survey. *Energy Pol., 36*(8), 2950–2963.

Sovacool, B. K., Schmid, P., Stirling, A., Walter, G., & MacKerron, G., (2020). Differences in carbon emissions reduction between countries pursuing renewable electricity versus nuclear power. *Nature Energy, 5*(11), 928–935.

Teske, S., Nagrath, K., Morris, T., & Dooley, K., (2019). Renewable energy resource assessment. In: Teske, S., (ed.), *Achieving the Paris Climate Agreement Goals*. Springer, Cham. https://doi.org/10.1007/978-3-030-05843-2_7.

The Fukushima Daiichi Accident, (2015). Retrieved from: https://www-pub.iaea.org/MTCD/Publications/PDF/Pub1710-ReportByTheDG-Web.pdf (accessed on 12 August 2023).

The World Bank, (2005). *Renewable Energy Potential in Selected Countries Volume I: North Africa, Central Europe, and the Former Soviet Union and Volume II: Latin America Final Report.*

Timeline: Japan Power Plant Crisis, (2011). BBC News. Retrieved from: https://www.bbc.com/news/science-environment-12722719 (accessed on 12 August 2023).

Wallner, A., Wenisch, A., Baumann, M., & Renner, S., (2011). *Energiebilanz der Nuklearindustrie.* Tech. rep. Austrian Energy Agency/¨Osterreichisches ¨Okologieinstitut, Vienna, Austria.

Wesley, C., Nathaniel, G., & Trieu, M., (2021). Exploring the cost implications of increased renewable energy for the U.S. power system. *The Electricity Journal, 34*(5), 106957, ISSN 1040-6190, https://doi.org/10.1016/j.tej.2021.106957.

WNA, (2019). *Uranium Production Figures, 2009-2018*. https://www.world-nuclear.org/information-library/facts-and-figures/uranium-production-figures.aspx (accessed on 12 August 2023).

Yu, H., Yingying, S., & Junjie, W., (2021). A low-cost and efficient solar/coal hybrid power generation mode: Integration of non-concentrating solar energy and air preheating process. *Energy, 235*, 121367, ISSN 0360-5442, https://doi.org/10.1016/j.energy.2021.121367.

CHAPTER 6

The Energy Market: Transformation, Privatization, and Diversification

ROHIT BANSAL, RAKESH VERMA, and SHWETA GUPTA

Department of Management Studies, Rajiv Gandhi Institute of Petroleum Technology, Amethi (An Institution of National Importance), India

ABSTRACT

The usage of fossil fuels comes at the cost of energy security. Fossil fuels or conventional forms of energy are limited in nature; excessive use of these resources may lead to fast exhaustion and result in increased dependence on other nations for the import of oil and fuels. To fulfill this image, nations worldwide are inclined towards reducing their carbon footprints. Diversification of energy sources may reduce the risk of dependency on any one energy source. The governments of these countries may allow private players to increase the efficiency in the power sector, initially by providing attractive incentives to them. An independent institution may be formed to regulate and monitor the agreement made among these countries. This will remove the probability of political interference from the government. Diversification of energy sources may reduce the risk of dependency on any one energy source. The governments of these countries may allow private players to increase the efficiency in the power sector, initially by providing attractive incentives to them. An independent institution may be formed to regulate and monitor the agreement made among these countries. This will remove the probability of political interference from the government.

The South Asian Energy Corridor: Toward Sustainability and Security. Amritanshu Shukla, PhD, Atul Sharma, PhD & Saurabh Mishra, PhD (Eds.)

6.1 INTRODUCTION AND HISTORY OF SOUTH ASIAN ENERGY MARKET

The South Asian region is home to a total of 21% of the world's population. An important fact to state here is that a large part of this is living with very limited resources to sustain life. Most of the countries which form part of the South Asian Energy corridor are either underdeveloped or developing countries. There is an expectation of a 5% increase in the demand for energy in these countries owing to development in the household and manufacturing sectors.

While the South Asian region has diverse energy sources due to the diverse topographical setup of each country, the different forms of energy have not been put to adequate use due to the absence of proper channels for energy trade. Most of the countries of the South Asian region, namely India, Pakistan, Bangladesh, Nepal, Bhutan, and Sri Lanka, have been going through a struggle to ensure sufficient energy sources. These nations need sufficient energy sources to ensure the overall growth of the standard of living of their citizens.

All the countries of this region lack energy security. Also, the gap between the commercial energy resource endowments and the energy demand is considerable. Despite this disparity in the resources and demand for commercial energy in this region, the energy trade is very limited. The only energy trade is that of electricity between India–Bhutan and India–Nepal. Table 6.1 shows the total electricity production from different sources for South Asian countries. It states that most of the countries are dependent on a single source to produce electricity, including Bangladesh (natural gas – 75.61%), India (coal – 77.66%), Nepal (hydropower – 99.69%), and Sri Lanka (hydropower – 41.75). All the energy-related issues of this region can easily be solved by resorting to sharing resources, which can maintain an optimum balance of resources in this region and help in the development of these countries.

The SAARC Nations, in the 2014 SAARC summit in Kathmandu, had framed an agreement for energy cooperation (Electricity), which was targeted towards the free flow of electricity between SAARC nations, stating that the SAARC nations shall provide access to their transmission grids, which shall be helpful in sharing the resources through cross border trading of electricity.

In addition to SAARC (South Asian Association for Regional Cooperation), SASEC (South Asia Sub-Regional Economic Cooperation), BIMSTEC (Bangladesh, India, Myanmar, Sri Lanka, Thailand Economic Cooperation) are two other Platforms that have been incorporated to facilitate energy trade between the SACs (South Asian Countries).

TABLE 6.1 Electricity Production (As of 2018)

Countries	Electricity Production (in GWh)	Coal (Percentage of Total)	Crude Oil (Percentage of Total)	Natural Gas (Percentage of Total)	Nuclear (Percentage of Total)	Hydro (Percentage of Total)	Wind, Solar, etc. (Percentage of Total)
Bangladesh	78,612	1.91	20.72	75.61	0	1.38	0.38
India	1,498,109	77.66	0.52	4.91	2.52	10.09	4.29
Nepal	4,913	0	0	0	0	99.69	0.31
Pakistan	1,48,139	7.79	21.08	36.11	7.12	25.23	2.73
Sri Lanka	15,332	31.07	24.30	0	0	41.75	2.88
Maldives	–	–	–	–	–	–	–
Bhutan	–	–	–	–	–	–	–
Afghanistan	–	–	–	–	–	–	–

Source: Authors calculation – International Energy Agency (IEA).

On analyzing the current situation of energy trade among the SACs, we understand that in order to become a self-reliant region with respect to energy, these countries need to identify some solutions for their energy-related problems. Some such measures have been identified as follows:

- A functional structure should be developed for inter-region power exchange.
- Focusing on attracting private investment in the energy sector for both power generation and transmission.
- Harmonize legal and regulatory framework.

6.2 ENERGY SOURCES AND ENERGY CONSUMPTION PATTERN

Developing nations are characterized by a majority of the rural population, due to which one of the most easily accessible and popular kinds of energy is biomass for domestic energy consumption. The major concern for the developing nations of the South Asian region is the dependency on other parts of the world for energy sources while their RE sources are left unused due to a lack of technological expertise. Conventional energy sources form 95% of the energy consumed in these nations, and renewable energy households only have a marginal share. This becomes a cause of concern as the overdependency on conventional sources results in the fast depletion of these resources. Also, the import of these resources from other parts of the world puts a huge amount of pressure on the economy of these countries. The energy gap for this region is nearly 239.9 MTOE (Iftikhar et al., 2015).

6.2.1 TRANSFORMATION – SHIFT OF CONSUMPTION PATTERN FROM CONVENTIONAL ENERGY SOURCES TO SUSTAINABLE SOURCES

Transformation in terms of energy usage means shifting from one form of energy to another and, more importantly, shifting from conventional energy sources to sustainable energy forms. The energy demand is ever-increasing in every facet of modern-day lives. To meet this demand, new sources of energy must be identified. Over the period, with the growing necessity for the conservation of non-RE forms, various RE sources have been identified. A brief discussion about some of them is as follows:

1. **Liquid Biofuels:** These provide an alternative to engine fuels. It is a clean fuel produced from the renewable parts of plants and animals. Fossil fuels used for powering engines both for vehicles and electricity generation form a very huge part of conventional energy being used worldwide. Liquid biofuels have a promising future in the transportation sector because of the growing demand in the sector.
2. **Hydropower:** It is the energy harnessed from falling or flowing energy. The Kinetic energy of the falling or flowing water is converted and utilized for various purposes. It is a very cost-effective way to produce electricity. However, the cost-effectiveness depends upon the scale of the project.
3. **Solar Energy:** Solar photovoltaic energy is the energy harnessed from the sunlight. It is a very versatile type of energy form as it can be used to harness energy at a large scale also and can be used for power generation at an individual level for household consumption. It is a pollution-free, environment-friendly energy source.
4. **Wind Energy:** Energy harnessed from the kinetic power of wind is known as wind energy. The Kinetic energy is used to turn turbines, which creates the mechanical power that runs the generators that are used to generate electricity.
5. **Solid Biofuels:** These are the fuels prepared from biomass. It is an environment-friendly energy source as it is generated from plant-based waste.
6. **Geothermal Energy:** Heat energy generated by the subterranean radioactive decay of the materials is called geothermal energy. This energy is stored inside the earth's crust and is generated on a continuous basis as it is present in abundant amounts, but extraction or exploitation of this energy form is not easily possible as it entails huge capital costs to set up such power plants but once installed, the per unit becomes low day by day.

These are some forms of RE sources that can become part of the energy mix to make it more viable for sustainable development.

6.3 REASONS FOR TRANSFORMATION

The usage of fossil fuels comes at the cost of energy security. Fossil fuels or conventional forms of energy are limited in nature, excessive use of these resources may lead to fast exhaustion and result in increased dependence on other nations for the import of oil and fuels. Moreover, excessive exploitation of these natural resources also damages the environment. Overdependence on these limited resources may also create a hindrance in the path of growth. Most of the South Asian Countries are developing nations that require adequate energy sources both to support the growing household demand and industrial demand.

Environmental concerns are also one of the major reasons for the transformation of energy forms. In today's world, if a nation wants to make a global impact, it needs to have an all-round good image in terms of a responsible nation that is responsible towards its citizens by providing them with a decent standard of living together with being responsible towards the environment thus practicing sustainable development. To fulfill this image, nations worldwide are inclined towards reducing their carbon footprints. The carbon footprint of a nation can be defined as the measure of carbon dioxide (CO_2) emitted into the environment by the citizens of a country as a result of their day-to-day activities.

Countries can reduce their carbon footprint by following some simple steps:

1. **Stop Extracting Fossil Fuels:** Fresh extraction licenses for the extraction of fossil fuels should be discouraged. Non-issuance of fresh licenses would preserve the natural resources that are left and encourage countries to use other forms of fuels that are cleaner.
2. **Provide Subsidies for Promoting Usage of Cleaner Fuels Instead of Fossil Fuels:** To encourage industries and households to use cleaner fuels in place of carbon-emitting fossil fuels, governments should shift the subsidies available on fossil fuels to the usage of cleaner renewable fuels.
3. **Phase-Out Coal-Fired Power Plants:** One of the major carbon-emitting activities is power generation by coal-fired power plants. Countries should plan to phase out coal-fired power plants and concentrate on power generation using other forms of energy.
4. **Strengthen the Public Transport System and Promote the Use of Electric Vehicles:** The use of private vehicles for transportation damages the environment by emitting carbon dioxide (CO_2) on a large scale. It can be avoided if the public transport system is improved, people would prefer traveling by public transport, which would largely restrict carbon emissions by eliminating unnecessary usage of fossil-fueled vehicles. Promoting electric vehicles can also contribute to reducing carbon emissions. Countries should work on

making these vehicles efficient and cost-effective to make them a preferred choice for the people.

6.4 DIVERSIFICATION OF ENERGY SOURCES

Diversification, in general, is meant to mitigate the risk involved in overdependence on any one thing. As in the case of investing in the share market, it is wise to diversify one's portfolio by making many small investments rather than making one big investment, similarly, it is wise to explore and depend upon various forms of energy sources available both conventional and sustainable sources to safeguard oneself from sudden shock in the market caused due to imbalance between supply and demand. Overexploitation and overdependence on any single type of energy source will result in exhaustion of the same. Such a situation may create havoc in an economy that will have to switch to a new energy source. To escape from such a situation, it is a wise decision to diversify. Diversification in energy sources means creating a viable energy mix that consists of all the available sources in such proportions that are best suited for both socio-economic developments and do not have any negative impact on the environment by overexploitation of any one kind of resource.

A country having a diverse energy mix may consider itself safe from sudden disruptions in the supply of one single type of energy source, political disruptions may also create hindrance in the smooth flow of energy. Countries that are energy-independent can be worry-free about energy security from the point of view of geopolitical disruptions.

6.4.1 BENEFITS OF ENERGY DIVERSIFICATION

1. **Political Independence:** An importer country exposes itself to manipulation by the supplier country related to price, quantity, etc. A country with a diversified energy mix safeguards itself from any such coercion and is politically independent on the world platform.
2. **Economic Growth:** With energy security, a country guarantees its industry an undisrupted supply of energy sources and thus aids economic growth in the country. A diverse energy mix safeguards a country from disruptions in energy supply due to disruption in the supply chain of one particular energy form. An uninterrupted energy

supply is very important for any country so that its industries can work worry-free.

3. **Eco-Friendly Development:** Diversifying and adopting more and more sustainable sources of energy ensures environmental protection by less reliance on sources that pollute or harm the environment.

6.5 CHANGING SCENARIO OF THE ENERGY MARKET – PRIVATIZATION

There has been a wave of economic deregulation and privatization in developing nations in all facets of the economy. The energy market is also not untouched by this wave. Technological advancement and capacity addition in the energy industry demand huge capital investments. To grow at the pace of the increasing demand for power by developing nations, private sector investors must be attracted towards investing in the energy market.

A highly regulated energy market in developing nations may pose a hindrance in advancement and capacity building in the sector due to some factors, such as poor financial condition, pricing policy, government control over pricing, political pressure, and poor technical performance due to capacity constraints. Exploiting their potential to full extent is of utmost importance Capacity building is the need of the hour for the South Asian country in terms of power generation to ensure continuous growth of the economy. In 2018, the Ministry of Power (MoP) (India) issued guidelines on how the export and import of power and electricity will be facilitated with the neighboring countries. This came to overcome the shortcomings of 2016 guidelines that were more restrictive and defensive in terms of trade. The guidelines prevented other companies, except for Indian generators in the neighboring countries, from selling the power to India. This step was taken to restrict the dominance of China in the Indian subcontinent as China was pumping the investment in power sectors in these regions. China has a huge investment in India and its neighboring countries as well in terms of electricity generation (Figure 6.1). India's 2016 policy not only created many issues for countries like Nepal and Bangladesh but also caused economic losses to these countries as they were not able to export the power to India.

To remove all these restrictions and facilitate and promote cross-border trade, India revised the policy and brought new guidelines in 2018. These guidelines stated that power generating or distribution companies of India can export the electricity to neighboring power companies directly or through trading licenses of India after taking the approval of the government. It removes the requirement of the bilateral agreement between two countries. These were the positive steps to lead a better and more stable relationship with South Asian countries.

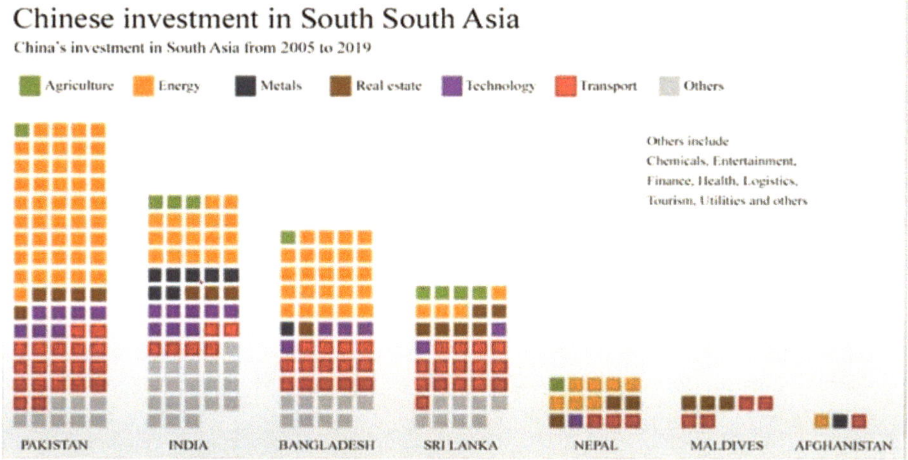

FIGURE 6.1 Chinese investment in South Asia.
Source: AEI's China Global Investment Tracker.

6.6 CONCLUSIONS

South Asian countries need more attention towards building a stable and positive relationship. These regions require robust policies to share excess energy with the deficit region. The focus should be on making a huge investment in research and development and improving the technology to generate more renewable energy. Such investment would also enhance economic growth. Diversification of energy sources may reduce the risk of dependency on any one energy source. The governments of these countries may allow private players to increase the efficiency in the power sector, initially by providing attractive incentives to them. An independent institution may be formed to regulate and monitor the agreement made among these countries. This will remove the probability of political interference from the government.

KEYWORDS

- **consumption pattern**
- **conventional energy sources**
- **energy diversification**
- **energy market**
- **energy sources**
- **privatization**
- **South Asian energy market**
- **sustainable sources**

REFERENCES

Bhattacharyya, S. C., (1995). Power sector privatization in developing countries: Will it solve all problems? *Energy Sources*, *17*(3), 373–389. https://doi.org/10.1080/00908319508946088.

Borozan, D., (2013). Exploring the relationship between energy consumption and GDP: Evidence from Croatia. *Energy Policy*, *59*, 373–381. https://doi.org/10.1016/j.enpol.2013.03.061.

Campo, J., & Sarmiento, V., (2013). The relationship between energy consumption and GDP: Evidence from a panel of 10 Latin American countries. *Latin American Journal of Economics*, *50*(2), 233–255. https://doi.org/10.7764/LAJE.50.2.233.

Editors, A., Jadhav, N., Singh, B., Patra, M. D., Ranjan, R., Kapur, M., & Bose, D., (2016). Editorial committee editorial committee. *Missiology*, *1*(2), 101.

George, J., Hammill, M., & Williamson, M. (2018). *Integrating South Asia's Power Grid for a Sustainable and Low Carbon Future*. 1–46. Asia-Pacific Economic Cooperation (APEC), 2018, Integrating South Asia's Power Grid for a Sustainable and Low Carbon Future. Asian and Pacific Energy Forum, pp. 1–46. URI: https://hdl.handle.net/20.500.12870/621 (accessed on 18 October 2023).

Ramakrishna, G., & Rena, R. (2013). An empirical analysis of energy consumption and economic growth in India: are they casually related?. *Studia Oeconomica, 58*(2), 22–40. URL: https://core.ac.uk/download/pdf/62634208.pdf (accessed on 18 October 2023).

Hou, Y., Iqbal, W., Muhammad, S. G., Iqbal, N., Ahmad, S. Y., & Fatima, A., (2019). Measuring energy efficiency and environmental performance: A case of south Asia. *Processes*, *7*(6), 325. MDPI AG. Retrieved from: http://dx.doi.org/10.3390/pr7060325.

https://data.worldbank.org/indicator/EG.ELC.FOSL.ZS?locations=8S (accessed on 12 August 2023).

https://energyeducation.ca/encyclopedia/Energy_diversification#:~:text=Diversification%20allows%20for%20a%20society,to%20disruptions%20in%20that%20sector (accessed on 12 August 2023).

https://openknowledge.worldbank.org/handle/10986/20254 (accessed on 12 August 2023).

https://reader.elsevier.com/reader/sd/pii/S2405653716302299?token=8A09CED70B09B314FACE976B74F84A9D60F8599427767C398504A1A28919A591505ADD6989261AB967030DD473A31D2B (accessed on 12 August 2023).

https://share.america.gov/diversifying-energy-sources-boosts-security/ (accessed on 12 August 2023).
https://www.adb.org/publications/overview-energy-cooperation-south-asia (accessed on 12 August 2023).
https://www.downtoearth.org.in/news/how-is-south-asia-faring-on-energy-8591 (accessed on 12 August 2023).
https://www.financialexpress.com/opinion/powering-south-asias-energy-trade-links/2088638/ (accessed on 12 August 2023).
https://www.iea.org/reports/southeast-asia-energy-outlook-2019 (accessed on 12 August 2023).
https://www.irena.org/transport/Liquid-Biofuels (accessed on 12 August 2023).
https://www.liquidbiofuels.org.nz/liquid-biofuels-opportunity (accessed on 12 August 2023).
https://www.mordorintelligence.com/industry-reports/south-asia-renewable-energy-market (accessed on 12 August 2023).
https://www.nytimes.com/guides/year-of-living-better/how-to-reduce-your-carbon-footprint (accessed on 12 August 2023).
https://www.renewableenergyworld.com/baseload/the-failure-of-privatization-in-the-energy-sector-and-why-todays-consumers-are-reclaiming-power/#gref (accessed on 12 August 2023).
https://www.researchgate.net/publication/283331050_Sustainable_Energy_for_All_in_South_Asia_Potential_Challenges_and_Solutions (accessed on 12 August 2023).
https://www.researchgate.net/publication/319341806_Energy_Interdependency_in_South_Asia (accessed on 12 August 2023).
https://www.sciencedirect.com/science/article/abs/pii/S0960148114002638 (accessed on 12 August 2023).
https://www.thehindu.com/opinion/lead/the-surge-of-geopolitics-in-south-asias-power-trade/article34144319.ece (accessed on 12 August 2023).
International Renewable Energy Agency, (2019). *Renewable Energy Market Analysis: GCC 2019* (pp. 1–110). *Irena.* https://www.irena.org/publications/2019/Jan/Renewable-Energy-Market-Analysis-GCC-2019 (accessed on 12 August 2023).
Kabir, M. H., Maknun, S. S., & Jakariya, M., (2017). *Energy Interdependency in South Asia*, 249–263. https://doi.org/10.1007/978-3-319-56747-1_14.
Lahiri-Dutt, K., (2006). Energy Resources: Will They be the Last Frontier in South Asia? Australia South Asia Research Center (ASARC) Working Paper 2006/10, pp. 1–27.
Rahman, S. H., Wijayatunga, P. D. C., Gunatilake, H., & Fernando, P. N., (2012). *Energy Trade in South Asia.*
Shukla, A. K., Sudhakar, K., & Baredar, P., (2017). Renewable energy resources in South Asian countries: Challenges, policy, and recommendations. *Resource-Efficient Technologies*, *3*(3), 342–346. https://doi.org/10.1016/j.reffit.2016.12.003.
Wijayatunga, P., & Fernando, P. N., (2013). *An Overview of Energy Cooperation in South Asia* (p. 19). South Asia Working Paper Series – Asian Development Bank. https://www.adb.org/publications/overview-energy-cooperation-south-asia (accessed on 12 August 2023).
Zeb, R., Salar, L., Awan, U., Zaman, K., & Shahbaz, M., (2014). Causal links between renewable energy, environmental degradation, and economic growth in selected SAARC countries: Progress towards green economy. *Renewable Energy*, *71*, 123–132. https://doi.org/10.1016/j.renene.2014.05.012.

CHAPTER 7

New Fossil Fuel Reserves Exploration in South Asia: Curse or Blessing?

YOGENDRA KUMAR, RAMAN SHUKLA, POOJA JAISWAL, and KOUSHIK GUHA BISWAS

Rajiv Gandhi Institute of Petroleum Technology, An Institute of National Importance, Jais, Uttar Pradesh, India

ABSTRACT

South Asia is a region of geographically stable, climatically diverse, and developing nations. This region has very high agricultural output, and forestry reserves are now looking for inexhaustible and renewable energy (RE) alternatives for fossil fuels. The current geopolitical, environmental, and technological barriers make the current energy scenario nonsustainable in this region. The bioenergy sources could be the best possible alternative to fossils due to the easy availability of biomass in this humid tropical region. Various alternative Waste to Energy conversion strategies like biochemical, thermochemical, and chemical could be effective for different types of biomasses available in this region. The different bioenergy conversion routes, operability conditions, and biomass types have been discussed in this chapter. This chapter also gives a brief description of the fossil fuel energy scenario, renewable capacity potential, and challenges in terms of South Asian countries. This chapter offers an insight into the current energy scenario, RE evolution, and highlights major hurdles and key challenges in energy tapping from biomass.

The South Asian Energy Corridor: Toward Sustainability and Security. Amritanshu Shukla, PhD, Atul Sharma, PhD & Saurabh Mishra, PhD (Eds.)

7.1 INTRODUCTION

Energy reserves fuel the development and affecting social, political, and economic interests of a country. The fossil fuel resources like coal and crude petroleum are extensively utilized in sectors like transport, power generation, industries, and irrigation. The consumption of liquid fuels in transportation and power generation accounts for 90% and 32%, respectively. The ~ 45% of world energy consumption, 35% population, and ~ 20% of world GDP will be accounted only by China and India. The estimated demand outlook for fossil resources in global transportation sectors is expected to increase by 1.8% per year from 2005 to 2035 (International Energy Outlook, 2010). Coal, natural gas, petroleum oil, and electricity are the major commercial sources of energy across the world and are widely used as an energy resource in transportation and industries.

South Asia is a region of emerging economies that consists of 8 countries, namely India, Pakistan, Bangladesh, Bhutan, Nepal, Sri Lanka, Afghanistan, and Maldives. The majority of these countries are considered to be countries of the rising middle class and the world's youngest population. In 2019, the South Asian Association of Regional Cooperation (SAARC) consists of 9 aforementioned nation that houses approximately 1.83 billion people, with a cumulative GDP of ~$3.55 trillion and accounted for 4% of the world's collective GDP (USD 87.55 trillion) (Haq et al., 2020).

The world proven crude oil reserves are approximately 1,666,300 million barrels of which major reserves are in countries like US, Venezuela, Russia, Canada, Iraq, Saudi Arabia, and Iran. The organization of major oil-exporting countries is called OPEC (Organization of the Petroleum Exporting Countries) contributes 79.4% of the world's proven oil reserve and has approximately 49% share in global oil export share (Alam et al., 2019). Higher dependencies over imported crudes make rising south Asian economies vulnerable to energy crunch as demand is rising day by day with increasing population. Moreover, series of other risks like demand, supply, price, geopolitical situation, trade openness, geographical distances, and infrastructure also contribute to making energy supply vulnerable. Most of the South Asian countries fulfilled their energy demand from the oil-rich Gulf region (Saudi Arabia, UAE, Qatar, Bahrain, Oman, Iran, and Iraq). However, the unstable geopolitical situation and US sanctions make energy supplies difficult for south Asian countries. However, each nation in south Asia has different geopolitical situations, energy usage, and energy accessibility. South Asia has a high potential of exploring new crude reserves and one of the major contributors to coal production. Among south Asian countries, India has 326.49 billion tons of proven coal reserves in 2020 and ranks fifth in the world. Statistics suggest in 2018, India produced 730 million tons and imported over 240 million tons of coal (IEA, 2020). More than 70% of electricity generation is dependent upon coal-based thermal power plants. A large share of coal reserves in India are of Gondwana descent formed 250 million years ago which has encouraged the government of India to launch 14 new coal mining projects in 2020 which is projected to increase our coal production to 750 million tons in the next fiscal year. India has total estimated 618.95 million tons of crude and 1380.63 billion cubic meters of natural gas reserves in 2019. Bangladesh is having nearly 322 million tons of coal and 28 billion barrel of crude reserves (REN 21, Worldometer). Similarly, Pakistan has 3.37 G tons of coal reserves and 335-million-barrel crude reserves. Despite these fossil reserves energy tapping from fossils is not environment friendly and contributes highly to global warming. Countries have now decided to cut their carbon emission to reduce global emission levels. Bioenergy resources will play a vital role for south Asia in creating self-reliance over imported fuels and reducing overall emission levels. Various bioenergy conversion strategies have been adapted over the years for sustainable development. Biofuels could easily be blended with conventional fuels to make environmentally sustainable substitutes for transportation, i.e., ethanol in gasoline and biodiesel in diesel. Likewise, biogas and syngas may potentially replace natural gas and liquefied petroleum gas (LPG) used for cooking.

7.2 ENERGY SCENARIO IN SOUTH ASIAN COUNTRIES

Most of the petroleum reserves are either in the middle east or Northern America which do not house the major population. Southern Asia with more habitable conditions does not have much fuel reserve. The global market scenario ensures that the oil prices almost never go down with the exception of the global pandemic scenario. To be independent from big corporate countries dictating the terms of business energy sector needs a reformation and renovation with abundantly available biomass. South Asia is home to 1.83 billion people and has very high growth potential in the future but lacking in quality infrastructure and insufficient investments in physical and social buildings are major hurdles in sustainable energy resource development. All South Asian countries are facing issues of energy deficiency and oversized population therein have an adverse effect on public welfare. The shortage of energy resources and low economic activities during the COVID-19 pandemic force to shut down of nearly half of medium and small-scale businesses resulted in unemployment and social unrest in this region. Most South Asian economies have a high trade deficit as they rely upon imported crudes. Most exportable commodities need a continuous energy supply to shorten the gap between imports and export.

The energy scenario in south Asian countries is non-sustainable due to geopolitical, economic, and environmental issues. The negligible territorial energy exchange between south Asian economies has a facing major challenge in enhancing energy cooperation between countries. The introduced research discoveries recommend that to accomplish the goal-oriented objective to decrease emanation release by up to 30% constantly 2030 under Intended Nationally Decided Contributions (INDCs), the Governments of the three nations should take preemptive measures. It incorporates the stage-wise decrease of endowments on petroleum derivatives, market mix inside the area, and quick acknowledgment of the current activities through solid political will, great administration, reception of the most recent advancements, and energy collaboration across the locale (Yu et al., 2018; NITI Aayog, 2017).

India in this region outpaced other countries and 120 GW power generation capacity added of which 73.3 GW is contributed by thermal power. Coal is contributed 45% in India total thermal power generation. The total demand of POL has increase with 5% CAGR and sell of passenger vehicles rise 10% during fiscal year 2013–2018. Similarly, energy demand in Pakistan have risen 5.2% CAGR from the fiscal year 2013–2018. Pakistan reported 7.9% increase in petroleum consumption driven by power and transportation sectors. Bangladesh have reported 6.6% CAGR GDP growth and energy demand of country rise 7.4% in 2012–2018 (Shakeel et al., 2021). The demand of petroleum product in Bangladesh is rise 4.7% CAGR in the fiscal years 2012–2018. The power generation capacities of India, Pakistan and Bangladesh is estimated to have 416, 43.775 and 24.991 GW respectively as of 2023. The electrical energy demand of Bangladesh, Pakistan, and India in 2018–2019 have reached 71,102, 1,22,710, 1,415,000 GWH, respectively. The energy statistics shows India dominates over overall fuel consumption and power generation among these countries (Figures 7.1 and 7.2). The diesel is major power source in transportation among these countries and contribute nearly half of the overall petroleum derivative demand (Figure 7.3) (Haq et al., 2019).

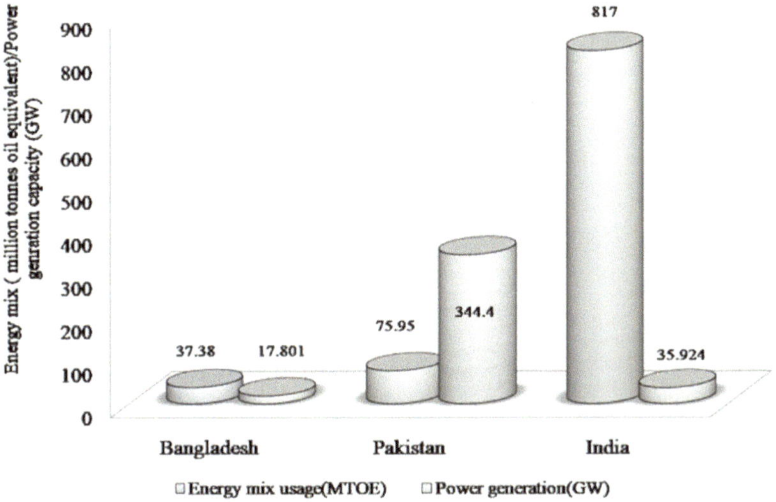

FIGURE 7.1 Energy statistics of South Asian Economies (2018).

Source: Haq et al., 2020. https://creativecommons.org/licenses/by/4.0/

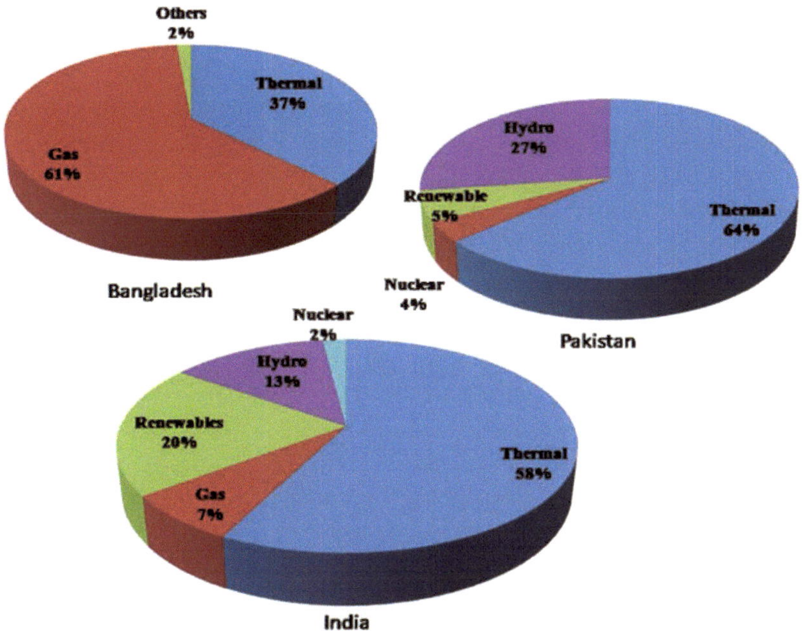

FIGURE 7.2 Power generation statistics in India, Pakistan, and Bangladesh source wise (2018–2019).

Source: Haq et al., 2020. https://creativecommons.org/licenses/by/4.0/

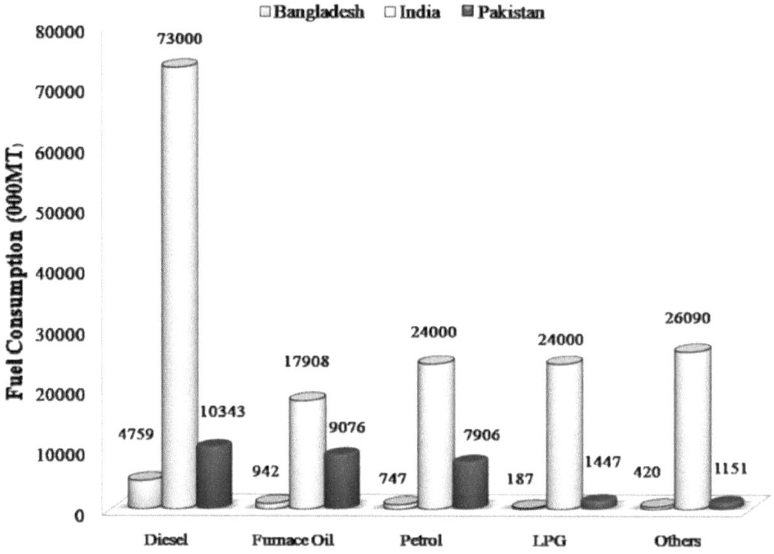

FIGURE 7.3 Fuel consumption among south Asian countries based on petroleum derivatives.
Source: Haq et al., 2020. https://creativecommons.org/licenses/by/4.0/

The primary reason behind use of biofuels in place of fossils is that crude oil and coal take millions of years to form and with the current extraction rates they will be exhausted soon. The energy security of South Asian countries is at risk as South Asian economies still highly dependent on imported fuels. Although some efforts have been made by these countries to shift some energy demand over renewable sources of energy to reduce greenhouse emissions and create energy independence (Table 7.1).

TABLE 7.1 Renewable Energy Share and Per Capita Usage among South Asian Countries (MNRE 2019)

Country	Total Energy Consumption (TPES) (2018)	Renewable Energy Share (2018)	Net Energy Import (%) (2014)	Energy Use/Capita (2014)	Electrification Rate (%) (2017)
India	8,62,483	24%	34	637	93
Bangladesh	39,555	24	17	222	88
Pakistan	95,701	39	24	484	71
Sri Lanka	11,702	43	50	516	98

The total share of renewable energy (RE) in south Asian countries is 24, 24, 39 and 43% respectively in India, Bangladesh, Pakistan, and Sri Lanka (Bioenergy Status MNRE, 2019). Although, the full potential of RE tapping is still not achieved by these countries. India itself has a very high potential for energy generation through RE resources. The total potential of renewable power generation 1,097,465 MW that includes solar power (7,48,990 MW), wind power (3,02,251 MW), small hydro (21,134 MW), biomass power (17,536 MW), bagasse power cogeneration (5,000 MW), and waste to energy (2,554 MW) (Bioenergy Status MNRE, 2019; Mospi, 2019).

South Asia has immense workforce and nearly 24.8% of the world population but the growth is not satisfactory compared to the world. Government must push for sustainable growth of the region through concerted efforts through intergovernmental actions and cross border energy trade. The improvements in energy sector could unleash forces for faster economic growth for this region that eventually improves living standard and social security in this region.

7.3 ENCUMBRANCE WITH ENERGY TAPPING FROM FOSSIL FUEL

The increasing globalization leads to elevated status of living with comes with a cost of increased energy demand for industrial production. India's estimated oil demand is projected to 1,516 and gas production will touch 90 billion cubic meters by the year 2040. To create self-reliance and reduce our dependence over imported crude south Asian economies needs to invest billions of dollars in oil exploration. Although, south Asia has sufficient coal reserve act as an armor to create energy independence over imported crude. Large exploration costs and less return are one of the major hurdles in oil and natural gas exploration. The huge energy and water consumption and environmental risk made oil exploration difficult in south Asia (Kayoma et al., 2017; Martins et al., 2019). Cost-effective oil and natural gas exploration in most the burning of fossils not just meets our energy needs but also produces a large amount of CO_2 that drives the current global warming crisis as well. Fossil fuels produce toxic emissions even before they burn. Benzene (blood disorder and leukemia) and formaldehyde (Carcinogenic) fumes are having serious health repercussions. Fossil-driven cars, trucks, vehicles are major contributors to toxic gasses NO_x, SO_x, CO_x, and soot. Similarly, coal fire power plants are a major contributor to toxic mercury emissions. South Asia is home to large populations that might threaten millions of lives (Lelieveld et al., 2019; Perera et al., 2018). Exploring new fossil reserves is a costly and troublesome process in south Asia. The techno-economic feasibility of oil and natural gas reserve below 2–3 km depth is very low and require huge investment to explore oil from reserves. Although, Biofuels (biodiesel and bioethanol) could help to reduce dependency on imported crude.

7.4 SOLUTIONS THROUGH TOTAL PRODUCTIVE UTILIZATION OF BIOMASS TO GENERATE FUEL/BIOENERGY

Asia being biomass rich with plant and animal waste most of it gets unexplored due to lack of technology but recently a lot of the research is directed towards solving this problem. Production of bioenergy from biomass resources is an inexhaustible, sustainable, and promising source to combat rising economic, technological, and environmental issues with conventional biofuels. These biowastes are generated in many forms from agriculture to industry at a different stage and their reckless disposal has a detrimental effect on human health as well environment.

Biomass is organic matter derived mainly from plants (both terrestrial and aquatic) that includes forest, crops residues, abattoir waste, animal manures, fishery waste, and cow dung. Biomass is considered a renewable source of energy as its energy comes from the sun and it can regrow in a relatively shorter period. Biomass is produced through the photosynthesis route when trees absorb CO_2 from atmosphere and produce monosaccharides. These monosaccharides are further converted into polysaccharides (Lignocellulose) within plant cells over a period. When plant leaves and stems die, they released back CO_2 into the atmosphere (Anwar et al., 2014). Hence biomass to biofuel conversion strategies is considered as an environment-friendly approach towards sustainable development. At present, utilizing biomass for energy generation has gained momentum because of the scarcity of conventional energy resources (Demirbas, 2009). In India, biomass-based energy resources dominate the rural energy consumption patterns, accounting for over 80% of total energy consumed (Rabindranath et al., 1995). In rural areas, livestock dung, crop residues, fuelwood are used (Rabindranath et al., 2005).

Bhattacharya et al. (2006) studied total biomass-based energy consumption in India and reported fuel dependency of firewood and crop residue are 64%, 36% respectively. They also reported 30% of rural household are dependent over traditional energy resources (Figure 7.4).

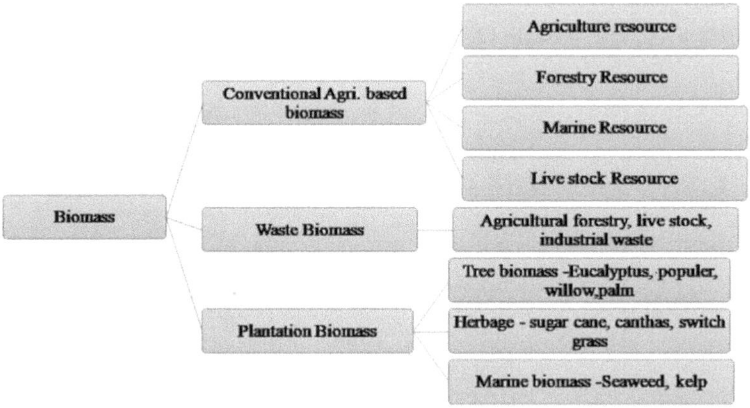

FIGURE 7.4 Various conventional biomass resources.

7.4.1 PRETREATMENT

Pretreatment of biomass is primarily done to disrupt compact biomass structure and expose cellulosic fibers. Various chemical and physical changes have been done to overcome recalcitrance. The pretreatment process has been done to increase surface area, hydrolysis yield, and to reduce cost of operation, inhibitor accumulations (Figure 7.5). The fibrous component of cellulosic, hemicellulose biomass be hydrolyzed, But the structural features or properties of biomass such as degree of crystallinity, degree of polymerization, degree of water swelling, and hardness of the wood, these properties make biomass less amenable to hydrolysis. The pretreatment process is used to increase the monosaccharides in raw biomass.

The pretreatment process generally decomposes the highly ordered cellulosic structure and convert this to digestible polysaccharides. The other advantages of pretreatment removal of lignin and increase the surface area available to enzymes or acid (Lu et al., 2008).

It also increases the rate and extent of cellulosic saccharification. The physical treatment process involves its reduction to small mesh size can lead to a high surface area for the hydrolysis process. The shearing and compressive forces involve during milling reduce crystallinity and mean degree of polymerization and increase the bulk density of the digestive feed. The milled feed reduces in volume hence it requires low reactor volume (Low initial installation cost). The power requirement for milling makes these pretreatment processes are so expensive. Chemical treatment with strong acid and bases, such as sulfuric acid and sodium hydroxide or other cellulosic swelling agents or dissolving agents (Gupta et al., 2015).

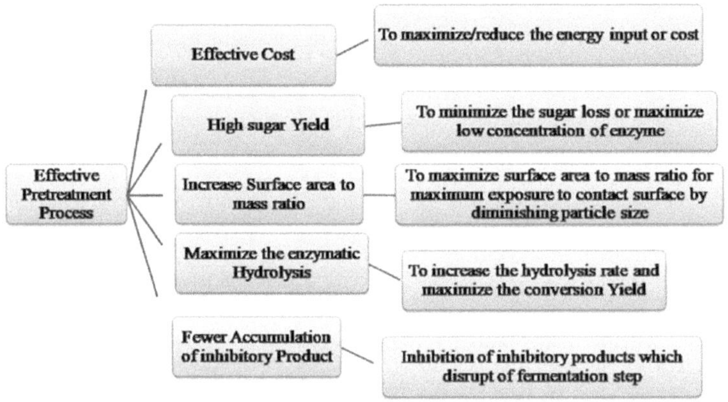

FIGURE 7.5 Pretreatment process, and its benefits.

Various physical and chemical pretreatment methodologies are used like shredding, mechanical milling, steam treatment, hot water treatment, acid, and alkaline treatment. These processes are grouped depending on nature of process into three categories, Physical, Chemical, Biological treatment. These pretreatments are listed below in info graphics (Figure 7.6).

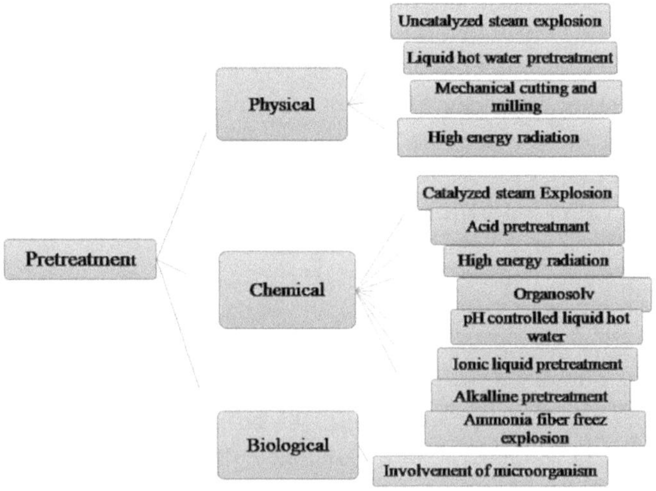

FIGURE 7.6 Different pretreatment methodologies.

Another pretreatment process is the steam explosive process carried out in 2,000–3,400 KPa and elevated temperature is about 215–260°C. In this Process when water is released at elevated temperature the cellulosic fibers are separated.

7.4.2 BIOMASS (WASTE) TO ENERGY CONVERSION ROUTES

A plethora of options available to convert biomass to produce biofuel through total productive resource utilization strategies. Biochemical conversion routes are simple (digestion and fermentation). Moreover, thermochemical routes (combustion, Incineration) are a bit cumbersome and require a huge initial investment. These thermochemical routes have been tested in large-scale commercialized plants. These thermochemical routes used biomass that usually consists of agriculture refuse, Industrial, and lignocellulose biomass. Some of the different biomass to biofuel conversion routes are listed below (Kumar & Samadder, 2017):

1. **Thermo-Chemical:**
   i. Combustion;
   ii. Pyrolysis;
   iii. Gasification;
   iv. Incineration.
2. **Biochemical Conversion:**
   i. Fermentation (bioethanol);
   ii. Anaerobic digestion (biogas).
3. **Chemical Conversion:**
   i. Transesterification;
   ii. Chemical hydrogen breakdown.

Thermochemical routes are usually used to convert biomass to bioenergy/biofuel at larger scales. All thermochemical routes are endothermic high-temperature destructive processes. The required temperature in these processes is < 800°C. The biomass is treated before using as feedstock using pretreatment steps.

These aforementioned routes are energy densification strategies and different high-value products are obtained. The efficacy of these conversion routes depends on raw material quality and physical properties. Moreover, the large-scale utilization of these technologies is underway. The cogeneration of chemicals using end products/byproducts of transesterification, pyrolysis, and gasification are underway at a research scale. Although separation of byproducts is still challenging. The different conversion routes, features, end products, and applications are listed in Table 7.2.

TABLE 7.2 Different Biomass Resources and Conversion Routes (Kumar & Samadder, 2017; Moya et al., 2017)

Biomass Resource	Conversion Process/ Features	Energy Products	Applications
Dry biomass Wood Plant residue Plant leaves Paper Plastics	• Combustion Presence of air Complete oxidation	Heat Steam Electricity	Industry Domestic Electricity
	• Pyrolysis Absence of air Thermal breaking	Oil Char Gas	Industry Transport Cogeneration of chemical
	• Gasification Absence of air Presence of high-temperature steam Partial oxidation of biomass	Synthesis gas ($CO + H_2$) Electricity Chemicals	Industry Cooking Heating
	• Hydrolysis and Fermentation Chemical conversion of lignocellulose.	Ethanol	Transport Industry
Wet biomass	• Anaerobic Digestion	Methane	Industry Domestic
Lignocellulose Sugar from fruit molasses	• Fermentation	Ethanol	Transport Chemical
Water Sewage	• Photo Catalysis Photochemical catalyzed Photobiological catalyzed	Hydrogen Hydrogen	Industry Chemicals

Biomass-to-energy conversion technologies are diverse and can be classified into several categories based on the processes used to convert biomass into usable energy. Here are some different biomass-to-energy conversion technologies listed in Table 7.3 with their advantages and disadvantages. The choice of biomass-to-energy conversion technology depends on factors such as the type and availability of biomass feedstock, energy needs, environmental considerations, and economic viability. The choice biomass to biofuel conversion strategy is determined by several factors, i.e., resource location, raw material physical properties, biomass availability, the market for the products, and economics of competing processes. Additionally, ongoing research and development efforts continue to improve these technologies and explore new pathways for biomass utilization in the quest for sustainable and renewable energy sources.

TABLE 7.3 Comparative Analysis of Different Conversion Processes (Kumar & Samadder, 2017; Moya et al., 2017)

Technology	Advantage	Disadvantage
Incineration	• Suitable for high calorific waste. • The process is hygienic. • Low space requirement. • Reduce transportation cost.	• Not suitable for high moisture low calorific value, and chlorinated waste. • Low overall efficiency. • Skilled labor requirement. • High capital requirement. • Higher particulate material, and toxic metal in ash and emission.
Pyrolysis/gasification	• Fuel gas/fuel oil generation. • Better control of pollution as compared to Incineration.	• Low energy recovery due to excessive moisture content, burning. • Transportation are problem generating due to high viscosity.
Anaerobic digestion	• Generation of fuel gas with high-grade soil manure. • Compact design requires low space. • Generate high-grade fuel gas for cooking, lightening. • No power requirement, odorless free from rodent and fly menace.	• Require waste segregation to improve digestion efficiency. • Not suitable for waste containing low organic substance.

TABLE 7.3 *(Continued)*

Technology	Advantage	Disadvantage
Fermentation	• High octane bioethanol fuel. • Low-cost process. • Can be used for medium and high scale plant.	• Require specific condition of temperature and reactor. • Microbe, yield depends on nature of microbes.
Transesterification	• Generate high-quality and low-cost renewable fuel. • High lubricity and cetane number than diesel fuel.	• Generate highly black smoke and exhaust high NO_x smoke.
Combustion	• Highly used for refuse treatment and waste management processes.	• Generate high level of pollutant.

These aforementioned technologies are used to convert low energy density biomass into high-density bioenergy/biofuel. Several commercially viable secondary energy conversion strategies have been developed in the last few decades and many others at research and development (R&D) stage.

7.4.2.1 THERMOCHEMICAL CONVERSION

Thermochemical conversion is a waste-to-energy (WtE) conversion route in which heat is required to catalyze chemical transformations into energy and other chemicals. Three different thermochemical routes are adopted in the last decade namely combustion, pyrolysis, and gasification. The chemical transformations are taken place in excess air in combustion process and heat produced. The heat is utilized in a waste heat recovery system to convert this high-pressure steam. The high-pressure steam was eventually used to drive the motor or alternator to produce electricity. Conversely, gasification and pyrolysis are taken place in absence of oxygen. The pyrolysis process destructive chemical transformation at elevated temperatures. The liquid fuel is formed due to the thermal breaking of higher hydrocarbons into smaller hydrocarbons during this chemical transformation. The gasification process is also used at elevated temperatures. The partial oxidation of biomass in presence of high-temperature steam by thermal breaking (Moya et al., 2017). Fuel gas is one of the byproducts obtained at elevated temperatures by the thermal decomposition of biomass. Its main component carbon monoxide and hydrogen, but its calorific value and composition are depending on the way that they produce and raw. The thermochemical conversion is a sequential process of a set of intermediate stages given below:

1. **Feed Preparation:** Feedstock's used for the thermochemical conversion process is pretreated with series of sequential steps like shredding, magnetic separation, and milling to improve physical properties of biomass. The waste is segregated and separated by a magnetic separation unit, where ferrous materials are separated with the help of a magnetic separator. These pre-treatment operations can either be carried out in the plant operation or externally before arriving at the plant. Currently on source segregation of waste is trending in south Asia than onsite segregation. The low associated cost with 'on source' segregation makes overall operation competitive. The feed bunkers or silos are used to store pretreated waste. The feed hopper is used to feed pretreated biomass to the thermochemical chamber (Figure 7.7).

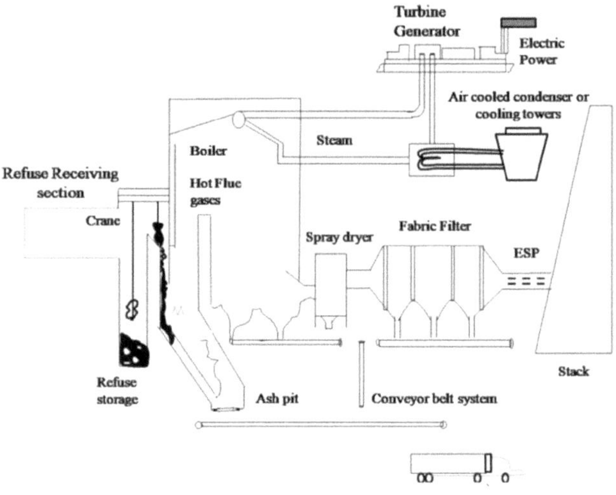

FIGURE 7.7 Schematic of a waste to energy (WtE) plant.

Source: Reprinted with permission from Siddiqi et al., 2019. https://creativecommons.org/licenses/by/4.0/

2. **Thermal Conversion or Oxidation Step:** Thermal conversion takes place in different types of thermal converters, i.e., fluidized, fixed combustors, and moving bed converters. Each thermochemical

transformations have different set of conditions required, i.e., pyrolysis requires inert condition for thermal breaking, whereas gasification takes place in the limited presence of oxidizers and combustion require excess air supply for complete thermochemical transformations. The different sets of requirements are mentioned in these thermochemical strategies are mentioned in Table 7.4.

TABLE 7.4 Comparative Analysis of Different Thermal Conversion Process (Islam et al., 2019; Motghare et al., 2015; Tadeu de et al., 2018)

Process	Reaction Temperature (°C)	Reaction Time	Reaction Atmosphere	Feedstock Moisture Content
Pyrolysis	300–500	Seconds	Inert	Dry
Gasification	500–800	Seconds	Partial oxidation in air or steam	Dry
Incineration	850–1,200	Seconds/minutes	Excess air/O_2	Dry
HTC	180–250	Hours-days	Inert/lim.O_2 sat	Wet

Product Distribution							
Process	**Char**			**Liquid**		**Gas**	
	Dist. (wt.%)	**Carbon (wt.%)**	**Energy (MJ/Kg)**	**Dist. (wt.%)**	**Energy**	**Dist. (wt.%)**	**Energy**
Pyrolysis	12–35	24–95	13–35	30–75	10–35	13–35	5–30
Gasification	10	4–46	NA	5	NA	85	2–20
Incineration	15–20	2–10	NA	NA	NA	80–90	12–16
HTC	50–80	58–83	18–36	5–20 (as TOC)	NA	2–5	NA

Note: NA: Not available.

3. **Heat Recovery or Steam Generation Plant:** The excess heat from the thermochemical step is to recover in the heat recovery section and utilized excess heat to steam generation. The heat recovery units are essential for effective utilization of energy and maximizing profits. The heat recovery unit ensures complete rejection of heat from the flue gas stream and utilizes it for steam generation. The unit used for energy recovery is known as heat recovery steam generator (HRSG)

section or waste heat recovery Unit. HRSG consists of a water-tube boiler or smoke-tube boiler and steam economizer system. The shot ball system is used for cleaning the heat transfer surfaces to reduce fouling overheat exchanger tubing. The continuous supply of water is maintained using a feed water overhead tank (Singh et al., 2010). The boiler system is designed to produce saturated or superheated steam. Although, that depends on the number of passes and water coil.

4. **Air Pollution Control and Flue Gas Cleaning:** Thermochemical WtE conversion routes are equipped with simple, dry, and wet fuel gas cleaning systems that consist of lime and activated carbon-based absorbers. The absorber unit traps fly ash and other solid content in the exhaust flue gas. The cyclone separator system is also used to traps heavy particulate deposits just before the adsorption unit. The gasses stream and toxic gasses from flue gasses neutralized in wet scrubbers' system. The lime and carbon are supplied from a storage tank or silo and the filter residue is collected and stored in a filter dust silo.

   The lime adsorbs acidic components from the flue-gas stream, while the activated carbon adsorbs dioxins, furans, and heavy metals from the waste flue gas stream. Sometimes water wet gas venturi scrubber or cyclone wet scrubber is also used for this purpose. An electrostatic precipitator system is most common in the waste conversion process containing high carbon content.

5. **Process Control and Monitoring:** An automatic logic control system is utilized to optimize both environmental and operational performance. Emissions are regulated by respective countries and monitored continuously, and data is collected and made available to the government regulatory body. The logic systems are operated in automatic mode that minimizes human efforts. The control system also features a separate and independent automatic shutdown system based on various control logic.

7.4.2.1.1 PROCESS OF THERMAL CONVERSION

1. **Combustion:** It is a thermochemical change measures acted in presence of overabundance air. The combustion process is utilized to dispose of an enormous amount of waste and to extricate power from

it. Direct combustion required biomass with a moisture content of under 15%, so drying of the waste is likewise utilized as an upstream interaction. It produces steam both for mechanical use and for delivering power. The cycle of combustion is assumed an indispensable part in paper plants, sawmills, Municipal organizations, and different enterprises. Most WtE combustors are having the accessibility of conventional fuel however they utilized a mix of waste and fuels for disposal credit and to economize measure. The process has become highly competitive with existing fuels by hybrid combustors. Ordinary combustion technologies produce steam through energy created by the ignition of biomass. This steam may then be expanded by a conventional turbo-alternator and produce power. Few combustion advancements have been created in couple of years. Underfeed stokers are appropriate for small-scale boilers up to 6 MWth (Islam et al., 2019). In present days Grate type boilers are broadly conveyed. They have moderately low speculation costs, low working expenses, and great activity adaptability at partial loads. Yet, they can have higher NO_x emissions and diminished efficiencies because of the necessity of abundance air, and they have lower operational efficiencies (Table 7.5).

TABLE 7.5 Heat of Combustion of Different Biomasses

Biomass Type	Heat of Combustion (MJ/kg)
Wood	10–11
Household rubbish	5–8
Farm produce, farm waste:	10–11
• Rice husk, wheat	11–12.2
• Rice straw, wheat	12–13.8
• Wood chips	16–17.8
• Bagasse of sugar cane	7–8.8
• Saw dust	24–27
• Ground nutshell	16–17.6
Secondary energy forms:	
• Butane C_4H_{10}	50
• Methane CH_4	56
• Propane C_3H_8	50

Fluidized bed combustors (FBC), which utilize a bed of hot inert material like sand and Zeolite, are a later turn of events. Bubbling FBCs are for the most part utilized at the range between 10 and 30 MWth capacities while coursing FBC are more relevant at bigger production units. Benefits of FBCs are that they can endure a wide scope of low-quality fuel while producing low NO_x levels and other contamination levels.

2. **Pyrolysis:** It is thermochemical change at raised temperatures in a latent environment. Inert air works with the thermal disintegration of cellulosic, hemicellulose, and lignocellulosic compounds into fluid hydrocarbon. Pyrolysis may yield solid, fluid, or vaporous fuel after thermal breaking of biomass. Without oxygen, the energy parts to break the substance obligations of high atomic weighted biomass into more modest particles. The response is more unpredictable. The pyrolysis cycle is broadly utilized for waste management processes. In significant south Asian urban areas (Delhi, Dhaka, Karachi, Colombo, and Calcutta) huge loads of wastes are pyrolyzed to deliver esteem added items comprise of straight-run hydrocarbon. Pyrolysis of cow excrement, wood sawdust, and domestic wastes are created H_2, N_2, CO, CO_2, C_2H_4, C_2H_6, and C_7H_8. Hydrogen and carbon monoxide further changed over into methanol, gas, diesel, and Ammonia for composts, medications, building, and bonding material, and synthetic textiles. The smallness, low-pressure activity, irrelevant waste, and high conversion efficiency of the order 83%. Pyrolysis is the term given to the thermal degradation of wood without oxygen (The measure which is completed in absence of oxygen can create fuel gas).

Biomass is to be changed over to a blend of solid singe material, Fuel gas, and fluid bio-oil. Pyrolysis advancements are for the most part classified as "fast pyrolysis," or "slow Pyrolysis" rely upon the time taken for handling the feed into pyrolysis items. These items are created in generally equivalent proportions with slow pyrolysis. Utilizing a quick pyrolysis measure, bio-oil yield can be just about as high as 80% of the item on a dry fuel premise. Bio-oil produce in this cycle can go about as a liquid fuel or as a feedstock for chemical production (Goyal, 2009). A scope of bio-oil creation measures is under ongoing improvement advancements, including liquid bed reactors, entrained stream reactors, ablative pyrolysis, and vacuum pyrolysis and turning cone reactors (Table 7.6).

TABLE 7.6 Comparative Analysis of Different Pyrolysis Process (CPHEEO, 2000; Mohan, 2006; Katyal, 2007; Demirbas, 2009; Singh et al., 2011)

Parameters	Conventional	Fast pyrolysis	Flash Pyrolysis
Temperature (K)	500–900	850–1,250	1,050–1,350
Heating rate (K/S)	0.1–1	10–100	>1,000
Particle size (mm)	5–50	<1	<0.2
Residence time (S)	300–3,600	0.5–10	<0.5

3. **Gasification:** It is a thermochemical change measure that requires a restricted oxygen source to gasify biomass into synthetic gas. The gasification changes over solid or fluid into a vaporous fuel without leaving any solid carbonaceous residue. The fractional oxidation of biomass or dried municipal waste is occurred in presence of high-temperature steam/restricted oxygen. The gasifier is a compound reactor where different unpredictable and actual cycles are occurred. Solid fuel is changed over by a progression of thermochemical measures like drying, pyrolysis, oxidation, and reduction to vaporous fuel. The air is ordinarily utilized as a gasification specialist, the producer gas comprises of carbon monoxide, oxygen, and hydrogen. Gasification of biomass happens in a limited stock of oxygen and happens through starting volatilization of the biomass, burning of unstable material and scorch is first occurred, and further reduction happens to deliver a fuel gas which is rich in carbon monoxide and hydrogen. This flammable gas produce in the gasification cycle has a lower calorific worth than natural gas however It can in any case be utilized as fuel for boilers, motors, and as a likely hotspot for burning turbines in the wake of cleaning the gas stream of tars and particulate matter.

   Assuming gasifiers are 'air blown,' air nitrogen weakens the fuel gas to a degree of 10–14% that of the calorific estimation of oxygen blown fuel. The gasifiers that utilization oxygen and steam has blown will deliver gas with a to some degree higher calorific worth than the air blown gasifiers. Compressed gasifiers are a work in progress to decrease the actual size of significant equipment items which can expand the gasifier effectiveness and design (Lu et al., 2008). A wide assortment of gasifiers has been created over the last few decades. These incorporate the limited scale fixed bed updraft reactor, downdraft Reactor, and cross-stream gasifiers, just as fluidized bed gasifiers, which are additionally utilized for bigger applications. For limited scope units, downdraft gasifiers are reasonable for their generally low tar creation, however they are not appropriate for powers with low melting point (like straw, these can effortlessly volatilize at low temperature). They additionally require fuel moisture levels to be controlled inside narrow levels.

The advancement program in the field of gasifier innovation begins during the 1980s. The Indian Government has dispatched its improvement in 1987 and at first spotlight on the little wood gasifier to run diesel motors for the mostly water pumping and other homegrown activities. The second period of the gasifier is begun during the 1990s and focused on power generation and planning bigger gasifiers just as market direction through a reduction in subsidies. Diesel Engines are ordinarily combined with gasifiers for power generation and working in a double fuel mode. The Engines worked on 100% producer gas are presently accessible in India. Three significant plants of this innovation are authorized in India in Coimbatore, Tamil Nadu of 250 MW limit each and having 750 MW total capacities in 2004. The first biggest undertaking in India depends on 100% producer gas.

i. **Impact of Biomass Gasification Development Program:** South Asia is one of the biggest regions in terms of population and suffers from a huge energy crisis. There is also a huge amount of biowaste and MSW are generated from our metropolitan cities. Both of these problems can be solved by using gasification technology. So, for biomass gasification process solved many hurdles of energy crisis and solid waste management. Tons of waste and agricultural biomass disposed of using the gasification process. The initial technical problem of tar decomposition is effectively solved by using high temperature steam gasification process. Better gasifier design and introducing a gas cleaning system to reduce the environmental issues. The problems of pollutant water disposed of gas cleaning remain. The large-scale involvement of private enterprise and national programs has gained large growth in the development of gasification technology in recent years, playing a key role in the electrification of rural areas at reasonable costs. Apart from electricity, biomass gasification also has potential to generates employment in rural areas and utilization of local skills. Rice husk and rice straw can be used to generate electricity in capacitive power plants for power or steam generation. Bagasse and other agricultural residue are also used to generate electricity in sugar mills and other food or grain processing industries (Balaji et al., 2014).

ii. **Application of Gasification Process:** The gasification process has tremendous potential in terms of its process output (in its electrical, mechanical, and Heat potential) and its application in various industrial process plants.

The small size gasifiers (up to 10 kW) find their application in a rural area to generate shaft power for agriculture pump, processing machinery, and other agricultural machinery like Thrashers, Straw choppers, etc.

- The Medium size of gasifier used in shaft line work in various extensive rural industries like sawmills, mechanical fabrication shops, and carpentry workshops as well as small and medium-sized rice mills. They also find their extensive application as a centralized source of energy in the milk chilling process and play an appreciable role in the rural electrification process.

- The large-scale gasifiers (up to 50 MW) find extensive application in rural as well urban industries, besides being a source of electrification. This type of gasifier is used in shaft work requiring jobs like and power generation directly or indirectly for various industries like Mineral processing, oil mills, Dairy, brick manufacturing, and ceramic and pottery industries. These are also used in Oil mining process forest-based wood processing industries (Figure 7.8).

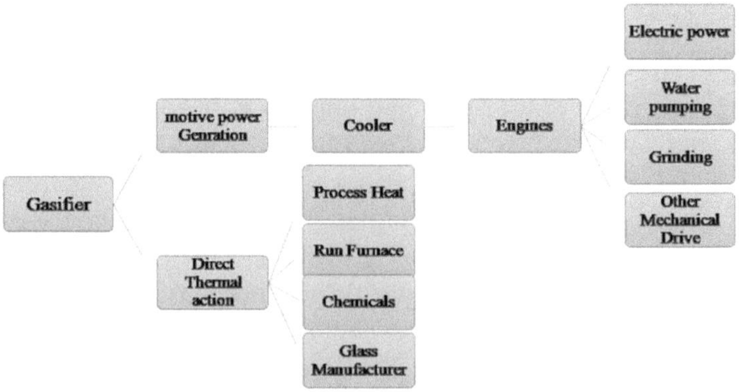

FIGURE 7.8 Application of modern gasification process.

Direct thermal application of gasifiers is useful where coal/lignite is used to generate process heat or to run furnace/kiln. Industries that commonly use gasification process in place of fuel oil or coal/lignite are:

- o Cement manufacturer;
- o Glassmaking;
- o Brick making;
- o Ceramic and pottery processing;
- o Rubber manufacture, food processing;
- o Brewing, crops drying, chemicals production.

7.4.2.2 BIOCHEMICAL CONVERSION

Biochemical conversion is the slow taking place by the microorganism converting biomass to biofuel at very low temperatures. Production of bioethanol from lignocellulosic biomass and biogas is the example of biochemical conversion. The hydrolysis + fermentation process is used to produce 2nd generation bioethanol from lignocellulosic biomass whereas anaerobic digestion (AD) is used to produce biogas from organic refuse.

7.4.2.2.1 Bioethanol

Ethanol will play crucial role in energy independency of south Asian countries. Although India is one of the major producers of ethanol, but majority of its share is coming from 1st generation ethanol (Molasses to ethanol). The conversion of lignocellulosic biomass into bioethanol will gear up ethanol production in these countries. The ethanol could replace petrol in transportation sector. Ethanol has very high-octane number and it is now used as blending agent in gasoline. In 2022, the worldwide ethanol market reached a valuation of USD 99.06 billion. It is anticipated that by the year 2032, during the forecast period spanning from 2023 to 2032, this market will exceed approximately USD 162.12 billion, experiencing a compound annual growth rate (CAGR) of 5.1% (Precedence Research, 2023).
Second generation bioethanol production from lignocellulosic biomass is carried out in three steps:

i. Pretreatment (to enhance hydrolysis process and reduce loss of sugars especially pentose);

ii. Enzymatic or acid hydrolysis (convert polysaccharides to monosaccharides (fermentable sugar)); and
iii. Fermentation (convert monosaccharide to ethanol).

Cellulosic biomass, derived from non-food sources, like trees and grasses or from biomass, is also being developed and used as a feedstock for ethanol production. Lignocellulose consists of 35–40% cellulose, 25–40% hemicelluloses and 10–20% lignin. The pretreatment process makes cellulose and hemicellulose readily available for hydrolysis process. The aim of pretreatment is removed or decompose lignin in prehydrolysis steps, increase amount of fermentable solids, increase fermentable sugars in prehydrolysis mixture, avoid sugar loss, reduce formation of inhibitors and reduction of total cost. In hydrolysis process nonfermentable sugars are converted to fermentable sugars (pentose, glucose, xylose) through acid hydrolysis. The enzymatic hydrolysis is used to decompose cellulose and hemicelluloses. It has been observed that untreated biomass is difficult to digest and require higher hydrolytic enzymes that increases overall operational cost. The presence of nonbiodegradable lignin, low digestibility of crystalline cellulose and hemicelluloses are main obstacle in this process (Ghoshima et al., 2013). Various enzymes *Aspergillus nidulans*, *Aspergilus niger*, *Trichoderma reesei* has overserved to have ability to produce endoglucanase, beta-glucosidase, and exo-cellobiohydrolase. The amorphous cellulose regions are attacked by endoglucanase causing depolymerization of indigestible cellulose. In fermentation process, the main aim to convert fermentable sugars (Pentose, Hexose) into ethanol in presence of microorganism (Sarkar et al., 2012; Tong et al., 2012). The yield of bioethanol will depend on incubation time, incubation temperature, pH, and nitrogen source. Globally ethanol fuel production reached over 108.54 billion liters (29.03 billion gallons) in 2019, with the United States and Brazil as the world's top producers, accounting together for 84% of global production. The USA and Brazil are dominating over bioethanol production and contributes 54.35 and 29.52% share in global ethanol production (AFDC, 2019). Ethanol can be used in petrol engines as a replacement for gasoline; it can be mixed or blended to with gasoline to any percentage. Existing car petrol engines can run on blends of up to 15% Bioethanol with petroleum/gasoline. E85 is most used alternative fuel in USA and about 90% RE transportation id depends on it. E85 fuel consist of 85% ethanol blended with gasoline. Ethanol has advantage over fossils that it increases engines compression ratio that eventually increases thermal efficiency. Bioethanol has advantage that it comes from renewable biomass and can be used in winters/cold weathers.

- **Application of Bioethanol and its Future Expansion:** Bioethanol obtain from Biochemical conversion is generally used as a fuel additive to cut down vehicles and engines. The cut-down vehicles are generally run on 80–85% ethanol. Bioethanol reduces greenhouse emissions as well as air pollution and it plays a vital role in global climate change and reduces carbon discharge to the atmosphere. Ordinary Fossil fuel leads to emitting various pollutant gasses like CO_2, N_2O, CH_4, and SF_6, and chlorofluorocarbons whereas bioethanol generates less amount toxic substances and produce carbon-free smoke (Figure 7.9).

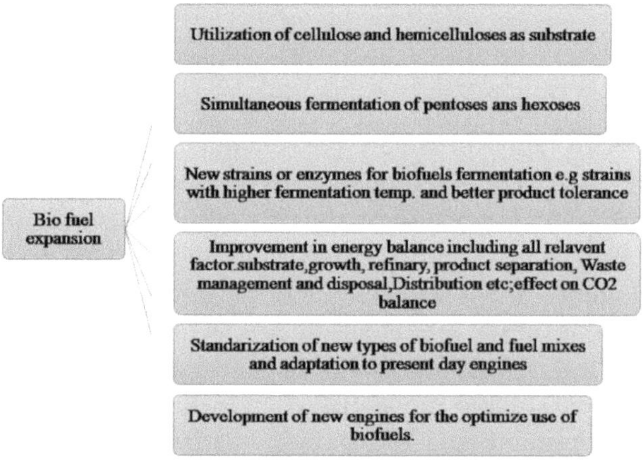

FIGURE 7.9 Biofuel future expansion strategies for South Asia.

The liquefaction process involving certain steps to convert these wastes into methanol/ethanol or any other high heating value gas. Alcohols have a high-octane rating. So, alcohol can improve engine performance, and these are cleaner than petrol and diesel and burns with higher efficiency and at low temperature. Although low calorific value means it has less energy density that fossils. Methanol contains 25% less energy/gallon than ethanol, and 50% less than petrol.

7.4.2.2.2 Biogas

Biogas is the mixture of gasses produced after AD of organic matter and lignocellulosic biomass. Biogas mostly consists of methane (40–60%) and carbon dioxide (CO_2) which finds vivid applications in domestic heating, cooking purposes. The production of biogas is more cost-effective than other thermochemical WtE routes. AD of biomass occurs in presence of anaerobic bacteria that eat or digest organic matter and produce methane-rich biogas. These bacteria naturally occur in water bodies, swamps, lakes, and even in animal and human digestive systems. The focus of new research is based on increasing methane fraction in biogas. Various alkaline, acid, and steam-based pretreatment processes adapted to enhance overall biogas yield and methane fraction. Although the increase in cost, risk of producing inhibitors, and corrosion are major roadblocks associated with pretreatment processes. Multistage anaerobic digesters and high-pressure digestion approaches are adapted to enhance overall process yield (Achinas et al., 2017; Plugge, 2017). In two-stage digester hydrolysis/acidogenesis and acetogenesis/methanation process are carried out separately in two reactors. The hydrolysis process is the liquid phase acidification process whereas acetogenesis is the process in which gas and liquid phase are participating simultaneously and acetate and methane form. The researcher observed that 95% methane yield in a single-stage digester is possible at higher working pressure (100 bar). High-pressure single-stage AD aims to produce biogas $< 5\%$ CO_2 is dissolved in water at higher pressures. The biogas genesis is completed in four steps hydrolysis, acidogenesis, acetogenesis, and methanogenesis.

During hydrolysis, extracellular enzymes decompose organic matter into simple water-soluble molecules. Extracellular enzymes are produced by the hydrolytic microbes. The complex carbohydrates, proteins, fats are hydrolyzed into sugars, fatty acids, and amino acids in hydrolysis process. The acidogenic bacteria (acidogens) convert these soluble compounds into small carbon chains that consist of volatile fatty acids (VFA) and alcohols. In the next step acetogenic bacteria (acetogens) further convert these fatty acids into acetic acids, CO_2, and hydrogen. In the last step methanogens produce biogas by the process of methanogenesis. In methanogenesis acetates, hydrogen, and acetic acids are converted to methane that is the primary constituent of biogas (Achinas et al., 2017; Scarlet et al., 2018). The microbial dynamics of different enzymes are complex and interactive. The quantities of these microbial groups are disproportionate that influences overall reaction dynamics. Moreover, methanogens have the slowest growth rate among others and methanogenesis is considered as a rate-limiting step in the whole AD process. That's why methanogenesis is most sensitive to change in pH, redox, temperature, and inhibitors.

7.4.2.3 CHEMICAL CONVERSION

7.4.2.3.1 Biodiesel (FAME)

Biodiesels are fatty acid methyl/ethyl esters (FAME/FAEE) derived from the transesterification of natural fatty acids. The transesterification process is taking place in presence of acid/alkaline catalysts that derive an overall transesterification mechanism (Kumar et al., 2022a). Different types of batch and continuous reactor systems are used to increase overall FAME yield. The use of microreactor technologies and process intensification strategies like pulsation, vibration, and microwave for higher FAME yield are tested extensively in the last decade. Although, saponification and higher product separation cost are limiting the application of homogeneous catalysts (Sun et al., 2008, 2010). The use of the heterogeneous catalyst for biodiesel production tested extensively to reduce overall product separation cost and improve product purity. Various acid-catalyzed and base-catalyzed solid catalysts are tested for biodiesel production (Kumar, Das, & Biswas, 2022). Biodiesel is produced via a chemical transesterification conversion route. Natural triglycerides (Oils) are converted into fatty acid alkyl esters and glycerol in the presence of alcohols and acid/alkaline catalysts. Biodiesels have inherent lubricity, low sulfur content, high flash point, high biodegradability, reduced emission, and high solubility with conventional diesel (Ogunkunle & Ahmed, 2019; Sharma & Singh, 2008).

7.4.2.3.2 Hydrotreated Vegetable Oil/Hydrotreated Esterified Fatty Acids (HVO/HEFA)

Biodiesel is the leading biofuel used in Europe as it holds 79.7% of total biofuel consumption. Although, low energy density and increased NO_x exhaust in low duty diesel engines limit its operability both in transient and steady-state conditions. Biodiesel is associated with the problem of cold flow properties due to its high viscosity that may affect fuel injection and cold start performance of the engine (Huber et al., 2007; Bjorgen et al., 2020). Despite several advantages, biodiesels fail to meet the expectation as a green substitute for diesel due to low oxidation stability, inferior cold flow properties, high feedstock cost, and higher NO_x emission. Hydrotreated vegetable oils are used to improve the flow and burning properties of biodiesel/fatty acids. The calorific value and cold flow properties of biodiesel/fatty acid improve after hydrotreating. Catalytic hydroprocessing of fatty acids is an alternate for biodiesel as biofuel. The HVO are straight-chain paraffinic hydrocarbons that are free from aromatics and sulfur. The HVO has a higher calorific value, improved cold flow properties, oxidation stability, and a high cetane number. Catalytic hydrotreating is done in series of steps, in the first step triglycerides, are converted into several intermediates (monoglycerides, diglycerides, and carboxylic acid). These intermediates are converted into alkanes via several pathways namely – (i) hydrogenation; (ii) hydrodeoxygenation (HDO); and (iii) hydrodecarboxylation (HDC) (Vasquez et al., 2017). The HVO has a higher heating value due to improved hydrogen content, low density due to paraffinic nature, and low lubricity due to the absence of sulfur and oxygen. These properties should be adjusted before use in existing diesel engines by additives (Dimitriadis et al., 2018).

7.5 BIOENERGY STATISTICS IN SOUTH ASIA

The reported urban waste generation is reported 0.45 kg/capita in 2012 which is estimated to grow 0.77 kg/capita in 2025. Although, it is well below global per capita usage estimated 1.19 kg/capita and 1.42 kg/capita (estimated) in 2012 and 2025, respectively. In India, 40% of the country's primary energy demand fulfilled using direct biomass burning resulting in a harmful level of air pollution. India with installed nearly 5 million biogas digesters only 0.4% of rural households are using biogas as the cooking medium. India has 90.3 GW total renewable power generation capacity out which 11.4% is comes from bioenergy. The total renewable power generation capacity of India is in Figure 7.10.

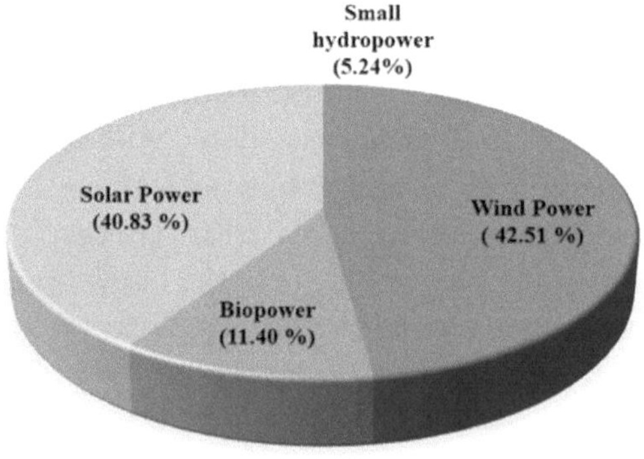

FIGURE 7.10 Total percentage renewable power generation capacity distribution of India.

India has added 4.5 GW biopower capacity since 2015 producing a total 10.314 GW biopower capacity as of 2020 (CEA, 2020). The power from combustion and biomass+ bagasse co-generation conversion reached upwards to 10.14 GW and an additional 168.64 MW from waste to energy in 2019. Indian is now producing 4.5 billion and 670 million liters of bioethanol and biodiesel in 2019. In Bangladesh state-run Infrastructure Development Company Limited (IDCOL) supports the construction of biogas plants for cooking gas and organic manure. Bangladesh has 0.4 MW of biomass and 0.68 MW biogas power biopower capacity and state rum IDCOL now operates 9 biogas powerplant across the country in 2019. Nearly 48,500 brick-based community biodigesters are supported by IDCOL for biogas generation. Similarly, biomass is used as fuel for 72% of industrial activities in Sri Lanka. Pakistan has a total of 432 MW of total installed bioenergy capacity and utilized 80% of its installed potential in 2019.

7.6 HURDLE AND CHALLENGES WITH BIOENERGY SOURCES

Bioenergy resources certainly play a vital role to create energy independence, reduce import bills, and decentralized toxic emissions. Although bioenergy sectors in south Asian countries are still facing technological, economic, and logistic, and supply chain-related barriers. Most of the challenges in WtE production not technical, alternatively they are related to biomass supply chain and product marketability. The public in South Asia is not aware of WtE technologies and they burn or dump precious biomass instead. The product marketability and lack of a small electricity grid are some infrastructural roadblocks. Government policies to promote WtE startups are lacking in most South Asian countries. The low energy density of biomass has also created an issue of storage as large silos or bins are required. South Asian countries lack the technical capability, appropriate business model, financial resources, and conducive environment. The waste management and source segregation of municipal kinds of waste is still challenging due to a lack of appropriate awareness and mechanism. On-site, segregation increases the overall cost of the process.

7.7 CONCLUSION

This is a comprehensive outlook of various bioenergy resources available for South Asian countries to substitute existing fossil fuels. Various techno economic concerns with conventional fossil fuel are discussed. The various WtE conversion routes are also discussed in line with South Asia along with hurdles and challenges. The above discussion shows South Asia is heavily relying upon the highly geopolitically unstable Middle East for energy supplies. Over few next decades, the south Asian economies will show surge in energy demand due to rapid urbanization, industrialization, population growth. The geopolycally unstable Middle East will threaten South Asia energy supplies. Another issue with fossils that it is finite and will last in next 50 years. The bioenergy resources could be helpful for south Asian countries to create self-reliance over energy demand. The regional cooperation groups and inter-governmental organizations like SAARC and BIMSTEC technology collaboration programs can improve technology and knowledge exchange for sustainable development of bioenergy between South Asian economies. There is significant potential for economic and technological cooperation between south Asian countries to utilized waste biomass for sustainable fuel for the future. The effective utilization of resources not only transition South Asian region to low carbon economy but also fuel economic development in this region.

KEYWORDS

- **bioenergy**
- **biomass**
- **energy conversion routes**
- **energy tapping**
- **fossil**
- **South Asia**
- **statistics**
- **waste to energy**

REFERENCES

Achinas, S., Achinas, V., & Euverink, G. J. W., (2017). A technological overview of biogas production from biowaste. *Engineering, 3*(3), 299–307.

AFDC, (2019). *Alternative Fuel Data Center*. www.afdc.energy.gov/data/ (accessed on 12 August 2023).

Alam, F., Saleque, K., Alam, Q., Mustary, I., Chowdhury, H., & Jazar, R., (2019). Dependence on energy in South Asia and the need for a regional solution. *Energy Procedia, 160*, 26–33.

Anwar, Z., Gulfraz, M., & Irshad, M., (2014). Agro-industrial lignocellulosic biomass a key to unlock the future bioenergy: A brief review. *Journal of Radiation Research and Applied Sciences, 7*(2), 163–173.

Balaji, R., Govindaraj, A., Karthikeyan, C., & Raghupathy, A., (2014). Simulation of biomass gasification. *Applied Mechanics and Materials, 592–594*, 1771–1775.

Baruah, J., Nath, B. K., Sharma, R., Kumar, S., Deka, R. C., Baruah, D. C., & Kalita, E., (2018). Recent trends in the pretreatment of lignocellulosic biomass for value-added products. *Frontiers in Energy, 6*, 141.

Bhattacharyya, S. C., (2006). Energy access problem of the poor in India: Isrural electrification a remedy? *Energy Policy, 34*, 3387–3397.

Bjorgen, K. O. P., Emberson, D. R., & Lovas, T., (2020). Combustion and soot characteristics of hydrotreated vegetable oil compression-ignited spray flames. *Fuel, 266*, 116942.

CPHEEO, (2000). *Manual on MSW*. Ministry of urban development central public health and Environment Engineering Organization, New Delhi.

Demirbasm, A., (2009). Biofuels securing the planet's future energy needs. *Energy Conversion Manage, 50*, 2239–2249.

Demirbras, A., (2009). Biorefineries: Current activities and future development. *Energy Convers Manage, 30*, 38–44.

Dimitriadis, A., Natsios, I., Dimaratos, A., Katsaounis, D., et al., (2018). Evaluation of a hydrotreated vegetable oil (HVO) and effects on emissions of a passenger car diesel engine. *Frontiers in Mechanical Engineering, 4*, 7.

Goshima, T., Tsuji, M., Inoue, H., Yano, S., Hoshino, T., & Matsushika, A., (2013). Bioethanol production from lignocellulosic biomass by a novel *Kluyveromyces marxianus* strain. *Bioscience, Biotechnology, and Biochemistry, 77*(7), 1505–1510.

Goyal, H. B., (2009). Biomass based energy fuel through bio chemical routes. *Renew Sust. Energy Rev., 13*, 167–178.

Gupta, A., & Verma, J. P., (2015). Sustainable bio-ethanol production from agro-residues: A review. *Renewable and Sustainable Energy Reviews, 41*(C), 550–567.

Haq, A. U., Jalal, M., Sindi, H. F., & Ahmad, S., (2020). Energy scenario in South Asia: Analytical assessment and policy implications. *IEEE Access, 8*, 156190–156207.

Huber, G. W., O'Connor, P, & Corma, A. (2007). Processing biomass in conventional oil refineries: Production of high-quality diesel by hydrotreating vegetable oils in heavy vacuum oil mixtures. *Applied Catalysis A*, 120–129.

IEA, (2010). *World Energy Outlook*. International Energy Agency.

Islam, M. R., Rafi, J. A., & Hossain, M. S., (2019). A Comparative study on waste-to-energy (WTE) technologies for municipal solid waste (MSW) management in Chittagong, Bangladesh. *IEEE Int. Con. Adv. Sci., Eng. Robo. Tech., 1–6.*

Katyal, S., (2007). Effect of carbonization temperature on combustion reactivity of bagasse char. *Energy Resource Part A, 29*, 1477–1485.

Kayoma, K., (2017). The role and future of fossil fuel. *IEEE Energy Journal*, 80–84.

Kumar, A., & Samadder, S. R., (2017). A review on technological options of waste to energy for effective management of municipal solid waste. *Waste Management, 69*, 407–422.

Kumar, Y., Das, L., & Biswas, K. G., (2022), Biodiesel: Features, Potential Hurdles, and Future Direction, In *Clean Energy Production Technologies*. Springer Nature, Singapore.

Kumar, Y., Jaiswal, P., Nigam, K. D. P., Panda, D., & Biswas, K. G., (2022). A critical review on nanoparticle-assisted mass transfer and kinetic study of biphasic systems in millimeter-sized conduits. *Chemical Engineering and Processing – Process Intensification 170*, 10867.

Lelieveld, J., Klingmüller, K., Pozzer, A., Burnett, R. T., Haines, A., & Ramanathan, V., (2019). Effects of fossil fuel and total anthropogenic emission removal on public health and climate. *PNAS, 116*(15), 7192–7197.

Lu, Y., & Mosier, N. S., (2008). Current technologies for fuel ethanol production from lignocellulose plant biomass. *Genetic Improvement of Bioenergy Crops*, 161–182.

Martins, F., Felgueiras, C., Smitkova, M., & Caetano, N., (2019). Analysis of fossil fuel energy consumption and environmental impacts in European countries. *Energies*, *12*, 964.

MNRE, (2019). *Bioenergy Status in India, Ministry of New and Renewable Energy*. Govt. of India.

MNRE, (2022), https://mnre.gov.in/bio-energy/current-status (accessed on 12 August 2023).

Mohan, D., Pittman, C. J., & Steele, P., (2006). Pyrolysis of wood/biomass for bio-oil: A critical review. *Energy Fuels, 20*, 848–889.

MOSPI, (2019). Energy Statistics., https://www.mospi.gov.in/sites/default/files/publication_reports/Energy%20Statistics%202019-finall.pdf (accessed on 12 August 2023).

Motghare, K. A., Rathod, A. P., Wasewar, K. L., & Labhsetwar, N. K., (2015). Comparative study of different waste biomass for energy application. *Waste Manag, 47*(A), 40–45.

Moya, D., Aldas, C., Jaramillo, D., Jativa, E., & Kaparaju, P., (2017). Waste-to-energy technologies: An opportunity of energy recovery from municipal solid waste, using Quito - Ecuador as case study. *Energy Procedia, 134*, 327–336.

NITI Aayog, (2017). Draft National Energy Policy. National Institution for Transforming India. Government of India, New Delhi, India. [Online]. Available: https://www.udyami.org.in/storage/schemes/file/Y0lZk0ox9ekHl6nO3Eq3NC4lhNqGsrO8D3ynOTxh.pdf (accessed on 12 August 2023).

Ogunkunle, O., & Ahmed, N. A., (2019). A review of global current scenario of biodiesel adoption and combustion in vehicular diesel engines. *Ene. Rep., 5*, 1560–1579.

Perera, F., (2018). Pollution from fossil-fuel combustion is leading environmental threat to global pediatric health and equity: Solutions exist. *Int. J. Environ. Res. Public Health, 15*(1).

Plugge, C. M., (2017). The contribution of microbial biotechnology to sustainable development goals. *Microbial Biotechnology 10*(5), 1128–1130.

Precedence research, (2023), Ethanol market, https://www.precedenceresearch.com/ethanol-market#:~:text=The%20global%20ethanol%20market%20size,forecast%20period%202023%20to%202032 (accessed on 12 August 2023).

Ravindranath, N. H., & Hall, D. O., (1995). *Biomass Energy and Environment: A Developing Country Prospective from India*. Oxford: Oxford University Press.

Ravindranath, N. H., Somashekar, H. I., Nagaraja, M. S., Sudha, P., Sangeetha, G., Bhattacharya, S. C., et al., (2005). Assessment of non-plantation biomass resources potential for energy in India. *Biomass Bioenergy, 29*, 178–190.

Renewable Energy Status Report, (2019). REN 21, Renewable Now, Asian Development Bank.

Sarkar, N., Ghosh, S. K., Bannerjee, S., & Aikat, K., (2012). Bioethanol production from agriculture waste an overview. *Renewable Energy, 37*(1), 19–27.

Scarlat, N., Dallemand, J. F., & Fahl, F., (2018). Biogas: Developments and perspectives in Europe. *Renewable Energy, 129*(A), 457–472.

Shakeel, M., (2021). Economic output, export, fossil fuels, non-fossil fuels and energy conservation: Evidence from structural break models with VECMs in South Asia. *Environ. Sci. Pollut. Res. Int., 28*(3), 3162–3171.

Sharma, Y. C., & Singh, B., (2008). Development of biodiesel from Karanja, a tree found in rural India. *Fuel, 67*, 1740–1742.

Singh, J., & Gu, S., (2010). Biomass conversion to energy in India—A critique. *Renewable and Sustainable Energy Reviews, 14*(5), 1367–1378.

Singh, R. P., Tyagi, V. V., Allen, T., Ibrahim, M. H., & Kothari, R., (2011). An overview for exploring the possibilities of energy generation from municipal solid waste (MSW) in Indian scenario. *Renewable and Sustainable Energy Reviews, 15*(9), 4797–4808.

Sun, J., Ju, J., Ji, L., Zhang, L., & Xu, N., (2008). Synthesis of biodiesel in capillary microreactors. *Ind. Eng. Chem. Res., 47*(5)398–1403.

Sun, P. Y., Sun, J., Yao, J. F., & Zhang, L. X., (2010). Continuous production of biodiesel from high acid value oils in microstructured reactor by acid-catalyzed reactions. *Chem Eng J., 162*(1), 364–370.

Tadeu De, E., Ferreira, F., & Balestieri, J. A. P., (2018). Comparative analysis of waste-to-energy alternatives for a low-capacity power plant in Brazil. *Waste Management & Research, 36*(3), 247–258.

Tong, Z., Pullummanappallil, P., & Teixeira, A. A., (2012). How ethanol is made from cellulosic biomass constituents of cellulosic biomass agriculture and biological engineering? University of Florida IFAS extension, AE 493.

Vasquez, M. C., Silva, E. E., & Castillo, E. F., (2017). Hydrotreatment of vegetable oils: A review of the technologies and its developments for jet biofuel production. *Biomass and Bioenergy, 105*, 197–206.

World Energy Statistics, www.worldometer.info (accessed on 12 August 2023).

Yu, S., Evans, M., Kyle, P., Vu, L., Tan, Q., Gupta, A., & Patel, P., (2018). Implementing nationally determined contributions: Building energy policies in India's mitigation strategy. *Environ. Res. Lett., 13*(3), 034034.

CHAPTER 8

Wind Energy Status in South Asian Countries

ANKIT SRIVASTAVA,[1,2] ABHISHEK ANAND,[1] AMRITANSHU SHUKLA,[3] and ATUL SHARMA[1]

[1]*Department of Basic Sciences, Rajiv Gandhi Institute of Petroleum Technology, Jais, Amethi, Uttar Pradesh, India*

[2]*Department of Mechanical Engineering, Feroze Gandhi Institute of Engineering and Technology, Raebareli, Uttar Pradesh, India*

[3]*Department of Physics, University of Lucknow, Lucknow, U.P., India*

ABSTRACT

In the economic growth of a region, energy plays an important role. Because of large populations, the disparity between energy demand and supply has widened in most South Asian countries, limiting the region's economic development. Fossil fuels, which are not only finite but also costly, are the primary source of energy in these countries. Renewable energy (RE) sources, especially wind energy, are abundant in the South Asian region. This natural asset can be used to satisfy the energy gap and prevent the energy dilemma of potential demand in the country with proper preparation. This section of the book seeks to address opportunities for the production of wind energy by reviewing the status, capacity, and policy of the region. The current status of wind power has been given in those countries. Finally, the chapter outlined guidelines of policy in South Asian countries for the production of wind power.

The South Asian Energy Corridor: Toward Sustainability and Security. Amritanshu Shukla, PhD, Atul Sharma, PhD & Saurabh Mishra, PhD (Eds.)

8.1 INTRODUCTION

South Asian Countries include countries like India, Bangladesh, Pakistan, Sri Lanka, Bhutan, Afghanistan, and Maldives. Due to huge population growth in these countries' energy requirement is increasing continuously, and fossil energy sources are continuously depleting day by day. Also, these fossil sources produce carbon emissions. To meet the energy requirement in these countries, these countries are exploring the possibilities of renewable sources. Geographically, these countries are tropical and humid, so wind power is available in adequately large amounts. As a sustainable renewable energy (RE) source, it is advancing rapidly all over the world. It does not pollute the atmosphere in the same way as conventional fossil fuels do, and it is cost-effective in multiscale applications. These countries are constantly working to convert wind power into usable form.

The demand for wind energy in South Asia is enormous. India has a 1,02,778-megawatt (MW) wind power potential, with a total installed wind power capability of 32,878 MW (Iftikhar et al., 2015). Pakistan has a 1,000-kilometer coastline with consistent wind speeds of 7–8 m/s. The country is expected to have a major wind power resource of about 3,46,000 MW, with a total installed wind power capability of 792 MW (Shami et al., 2016). Bangladesh has a 574-kilometer coastline with an average wind speed of 7 meters per second. The total installed wind power capacity is 3 MW, with a potential of more than 20,000 MW (Saifullah, Karim, & Karim, 2016).

The objective of this chapter is to study low carbon growth through the use of wind energy in South Asian countries, such as Pakistan, India, Sri Lanka, Bhutan, and Bangladesh. Technically speaking, this study contributes a great deal to the current literature. Firstly, we studied the status and diverse sources of energy that these countries use to fulfill their energy requirements in these countries. Secondly, in those countries, we have compared the demand for wind energy and installed wind power capability. We also drew up critical wind energy growth policies for the country. By choosing a strategy better suited to their interests, investors, stakeholders, and citizens will benefit from this study. Finally, the conclusions of the report reveal important policy recommendations.

8.2 WIND ENERGY STATUS IN SOUTH ASIAN COUNTRIES

8.2.1 INDIA

India's wind energy market is dominated by its own wind energy industry, which has made steady progress. The wind industry has grown to a strong setting, plant operations, and production base of about 10,000 MW per annum.

The wind industry has been growing. Presently, with an installed capacity of 35.5 GW (as of 31 March 2019), the fourth highest wind installed capacity in the world, the country produced about 52.66 billion units in 2017–2018 (Ministry of New and Renewable Energy, 2022). By offering numerous fiscal and financial incentives, including accelerated depreciation benefits, the government encourages wind energy projects in the whole country through private investment and exemption from the custom duty of certain components of wind turbines. In addition, the wind project scheme established before 31 March 2017 was given for the generation-based incentive (GBI) Program. Wind is a site-specific and intermittent energy resource, so an extensive assessment of wind resources is crucial in the identification of possible locations. The government established over 800 wind weather stations in the region through the National Institute of Wind Energy (NIWE) (Wind Data Sharing Policy, 2020). The wind capacity maps of 50 meters, 80 meters, and 100 meters above ground level have been released. The latest evaluation shows 302 GW of gross wind power in the country at 100 meters above ground level. In seven windy States, as seen in Table 8.1, the greatest potential exists.

TABLE 8.1 Wind Potential in States of India (Ministry of New and Renewable Energy, 2022)

SL. No.	State	Wind Potential (MW)
1.	Gujarat	84431.33
2.	Rajasthan	18770.49
3.	Maharashtra	45394.34
4.	Tamil Nadu	33799.65
5.	Madhya Pradesh	10483.88
6.	Karnataka	55857.36
7.	Andhra Pradesh	44228.6
8.	Kerala	1699.56
9.	Telangana	4244.29
10.	Odisha	3093.47
11.	Chhattisgarh	76.59
12.	West Bengal	2.08
13.	Puducherry	152.83
14.	Lakshadweep	7.67
15.	Goa	0.84

TABLE 8.1 *(Continued)*

SL. No.	State	Wind Potential (MW)
16.	Andaman & Nicobar	8.43
	Total in MW	302251.49
	Total in GW	302

The development of wind power in India is impressive and demonstrates the progress of Union and State governments' policies and regulations over a period of time. The wind industry's activities have brought the nation to the peak of 37+GW. State-wise and Year-wise Wind Power Installed Capacity (in MW) (as on 31-03-2020) of India is listed in Table 8.2.

TABLE 8.2 India's State-Wise and Year-Wise Wind Power Installed Capacity (in MW) (as on 31-03-2020) (Adapted from Indian Wind Energy Association, 2021)

State	Up to 2019	2019–2020
Tamil Nadu	8968.89	335.44
Karnataka	4694.9	95.7
Maharashtra	4794.18	206.2
Rajasthan	4299.65	0
Andhra Pradesh	4090.5	2
Madhya Pradesh	2519.9	0
Kerala	52.9	10
Gujarat	6073.05	1468.45
Telangana	128.1	0
Others	4.3	0
Total	35626.37	2117.79
Cumulative capacity	35626.37	37744.16

Because of the country's 7,500-kilometer-long coastline, the Indian government has agreed to promote the production of offshore wind after the success of its onshore wind power scheme. The MNRE has announced an offshore wind energy supplement of 5 GW by 2022 and 30 GW by 2030 for medium to long-term targets. To take advantage of offshore wind power capacity in the Indian exclusive economic zone (EEZ) along the Indian coastline, the ministry issued the National Offshore Wind Policy in October 2015. The NIWE is conducting offshore wind potential studies, and preliminary findings show that the coasts of Tamil Nadu and Gujarat have a lot of potential. One LiDAR has been deployed along the Gujarat coast for accurate wind quality measurements, and it has been producing data about the quality of offshore wind since November 2017. The NIWE has proposed the first-ever open bidding process for the development of the First Offshore Wind Project for India (FOWPI), which will have a capacity of around 1 GW and be located in Gujarat's Gulf of Khambhat (Indian Wind Energy Association, 2021).

8.2.2 BANGLADESH

Bangladesh is a South Asian country that is predominantly rural and agricultural. Bangladesh has a large sea to the south and a long coastline, making wind energy a viable option. For decades, all wind energy experiments were focused on data from the Meteorological Department, which seemed to be insufficient due to the incorrect height for wind energy production. Advanced technologies have been used in recent years to locate significant wind areas in Bangladesh, especially for the study of electric power generation. Via direct participation in wind velocity estimation and a host of national workshops and international conferences, various government and non-government programs have enhanced the goal. The updated wind speed data not only indicate that wind energy has a promising future in Bangladesh, but it also has a significant impact on rural electrification and irrigation programs, two critical sectors of the country's socioeconomic growth. The annual average wind speed at 30 meters is over 5 meters per second. The wind speed in northeastern Bangladesh is over 4.5 m/s, although it is about 3.5 m/s in the rest of the world. Patenga, Feni, Kuakata, Kutubdia, Magnaghat, and Munshigonj all have a lot of capacity for wind energy production. These sites have average wind speeds of more than 6.2 m/s, which, according to NREL (USA), is suitable for wind energy harvesting (Khan, Iqbal, & Mahboob, 2004). Bangladesh's first-ever generation of electricity from the wind at a 900-kilowatt plant has sparked fresh hope for low-cost power generation in the region. The wind turbines at the power plant, which is located near the Muhuri Dam in the southeastern Feni district, each have a capacity of 225 KW (11 small wind turbines in different coastal sites), and Grameen Shakti (two wind generators of 300 W and 1 KW at its Chakoria Shrimp Farm) have also installed small wind generators. The cost of producing 1 kilowatt hour of electricity (kWh) from wind energy is just around half that of generating an equal unit of power from solar energy, according to the

Board. The extension of wind energy capacity will be critical if Bangladesh is to meet its national goal of supplying electricity to all of its citizens by 2020 (Foreign Affairs, 2019).

Bangladesh's electricity demand is expected to reach 10,283 MW by the end of 2015. While having a long coastline and a vast land area, only 100 MW of the massive demand is expected to be met by wind power. The majority of this generation will be stationed along the coast and on nearby islands. The research takes into account wind speed data from the years 2002 to 2011. The annual average wind speed of 6 m/s is expected for grid-connected wind turbines to be financially viable. On Kutubdia Island in the Cox's Bazaar district, the Bangladesh Power Development Board (BPDB) installed a 1,000 KW wind battery hybrid power plant. As a result, wind-generated electricity in coastal areas can be transferred to other parts of the world through high-voltage transmission lines. Wind turbines will need very little operation and maintenance throughout their entire lifetime, and no fuel will be used to generate energy from the wind (Golam et al., 2004).

The cost of implementing the technology, as well as a shortage of knowledge about wind energy and difficulties obtaining historical data from accurate and credible sources, are the main concerns. The formation of policies to support this sector's growth in this country is hampered by a lack of formal supervision and assessment, as well as insufficient expertise. Weak regulatory incentives have failed to entice the private sector to invest in clean energy (Shaikh et al., 2017) (Figure 8.1).

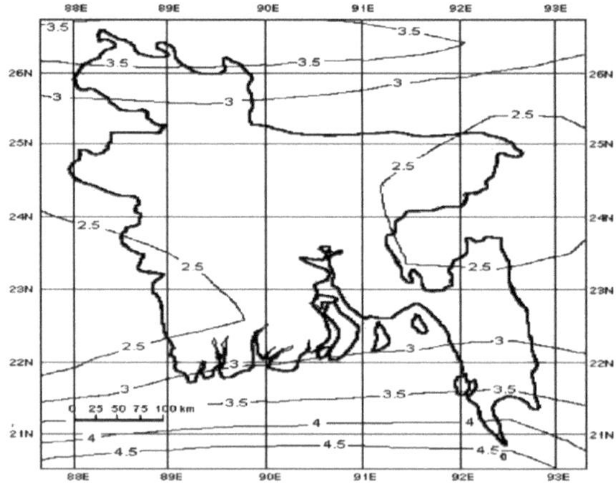

FIGURE 8.1 Wind map of Bangladesh using NASA–SSE-data set at 50 m height (m/s)

Source: Reprinted with permission from Khan, Iqbal, & Mahboob, 2004. Copyright 2004. Elsevier

8.2.3 PAKISTAN

In rural areas, one-third of Pakistan's population is without power, and load shedding of 10–12 hours is normal in urban areas. While Pakistan has always had a deficit in traditional resources, it has made little progress in renewable resources such as wind and solar energy. As a result, it is preferable to make use of these natural resources in order to meet the country's energy needs. Pakistan has a coastal line of around 1,100 kilometers (km) with wind energy capacity. Wind energy is a long-term energy supply with possible advantages such as a carbon source, minimized toxic gas emissions, and an environmentally safe operating protocol. Due to various political and financial challenges, Pakistan is among the top countries experiencing the worst energy crisis. Pakistan has a big wind energy opportunity because it meets all of the basic conditions for harnessing energy, such as windy areas and high wind speeds. Pakistan should take advantage of wind energy's ability to alleviate the country's energy crisis and transition to a green economy from a traditional fuel economy.

Wind is a major source of RE, and Pakistan is fortunate to have a large opportunity for this fuel. The Pakistan Meteorological Department (PMD), the Alternative Energy Development Board (AEDB), and the National Renewable Energy Laboratory (NREL) (a US-based company) are the key contributors to the data on Pakistan's wind energy capacity. In Pakistan, there is a projected wind power capacity of 3,46,000 MW, of which only 1,20,000 MW is viable (Saulat et al., 2020). Sindh province's south-eastern region, Khyber Pakhtunkhwa (KPK) province's northwestern region, and Balochistan province's south-western region all have significant wind energy capacity. The coastal areas of Balochistan and Sindh provinces are ideal locations for wind turbine construction. Wind speeds in the coastal areas of Sindh and Balochistan provinces range from 4 to 9 m/s at a height of 10 m to 12.5 m/s at a height of 50 m. Sindh has an 88,460 MW wind power generation capability, with possible sites such as Hyderabad, Kotri, Jerruck, Gharo, Jhimpir, Lakha, Khuttikun, and Bhambore. According to a report conducted by PMD, Sindh's coastal areas have a gross projected wind power capacity of 43,871 MW, of which 11,000 MW is viable (Mirza et al., 2007). Table 8.3 shows the different capacities of wind energy projects installed in Pakistan.

TABLE 8.3 Wind Power Plant Projects in Pakistan (Baloch, Kaloi, & Memon, 2016)

Name of City	Name of State	Name of Company	Installation Agreement (MW)
Bhambore	Sindh	Zephyr Power Ltd.	50
Jhimpir	Sindh	Yunus Energy Limited	50
Jhimpir	Sindh	Wind Eagle (Pvt.) Limited	50
Jhimpir	Sindh	UEPL Wind Power Pakistan (Pvt.) Ltd.	100
Jhimpir	Sindh	Trident Energy (Pvt.) Ltd.	50
Jhimpir	Sindh	Tricon Boston Corporation	50
Jhimpir	Sindh	Titan Energy Pakistan (Pvt.) Ltd.	50
Jhimpir	Sindh	Three Gorges First Wind Farm Pakistan Ltd.	49.5
Thatta	Sindh	Thatta Power Plant	150
Khutti Kun, Mirpur Sakro	Sindh	Tenaga Generasi Limited	49.5
Jhimpir, Thatta	Sindh	Tapal Wind Energy Ltd.	30
Sindh	Sindh	Sapphire Wind Power (Pvt) Limited	49.5
Jhimpir	Sindh	Sachal Engineering Works (Pvt.) Limited	50
Jhimpir	Sindh	Metro Wind Power Co Limited	50
Jhimpir	Sindh	Master Wind Energy Limited	49.5
Jhimpir	Sindh	Jhimpir Wind Power Plant	56.4
Jhimpir	Sindh	Jhimpir Wind Energy Project (FFCEL)	49.6
Bhambore	Sindh	HydroChina Dawood Power Ltd.	50
Jhimpir	Sindh	Hawa Energy Ltd.	50
Jhimpir	Sindh	Hartford Alternate Energy	50
Jhimpir	Sindh	Gul Ahmed Wind Power Limited	50
Gharo	Sindh	Gharo Wind Power Plant	50
Khutti Kun, Gharo	Sindh	Foundation Wind Energy – I & II Pvt. Ltd.	100
Jhimpir	Sindh	Finerji (Pvt.) Ltd.	50
Jhimpir	Sindh	Dewan Energy (Pvt.) Ltd.	50
Bhambore, Gharo	Sindh	Dawood Power (Pvt.) Limited	50
Jhimpir	Sindh	China Sunec Energy (Pvt.) Ltd.	50
Gharo	Sindh	Bhambore Wind Project	50
Punjab	Punjab	AM Pak Energy	50

The Pakistan Meteorological Department conducted its own wind energy capacity survey of the country's major regions. PMD has collected wind data for 42 separate locations in Pakistan's northern regions, including the Federally Administered Tribal Areas (FATA). At two suitable heights, 10-minute average wind direction and speed data were obtained (10 m and 30 m). PMD also gathered wind speed and direction data at 20 separate locations in Sindh's coastal area. In Balochistan's coastal regions, similar measurements were taken. Figure 8.2 depicts wind masts that have been built in numerous locations throughout various provinces.

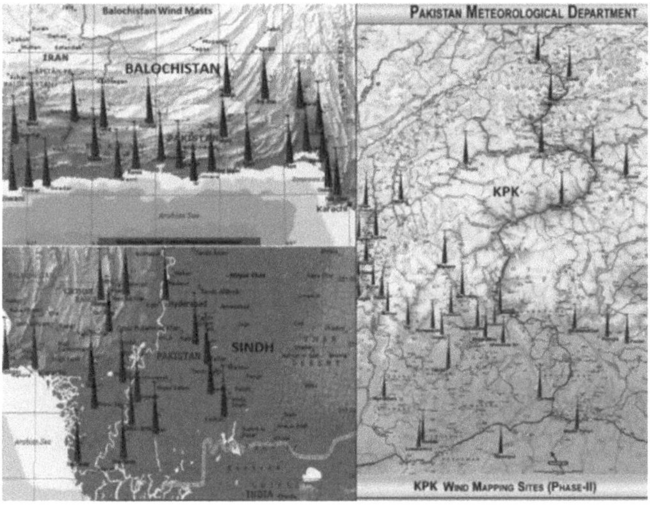

FIGURE 8.2 Installed wind masts in three provinces of Pakistan (Shami et al., 2016).

Source: Reprinted with permission from Shami et.al., 2016. Copyright 2016. Elsevier.

8.2.4 SRI LANKA

In Sri Lanka, wind energy has been described as one of the most promising, commercially feasible, and environmentally sustainable energy sources. Sri Lanka's geographical location makes it an ideal location for wind energy growth (Young & Vilhauer, 2003). However, since wind potential is based on many geographical variables, the best locations for installing wind power plants are specific. Geospatial techniques were used to identify possible wind power plant locations. Five decision factors were derived from established Geodatabases, and their relative weights were calculated using Spatial Multi-Criteria Analysis methods in a GIS environment: elevation, wind density, land usage, restricted area, and proximity to road network. According to the findings, Sri Lanka has enough wind power capacity in 13,130 km^2 (20%)

of its land area to increase RE contribution to the national power supply. Furthermore, the majority of the possible areas were located in the country's sparsely inhabited coastal belt. Spatial distribution of wind power potential under the influence of individual factors is obtained (Amarasinghe & Perera, 2021). Potential sites in Sri Lanka based on wind power density shown in Figures 8.3 and potential sites in Sri Lanka based on protected areas shown in Figures 8.4. Potential sites in Sri Lanka based on elevation and land used are shown in Figures 8.5 and 8.6, respectively.

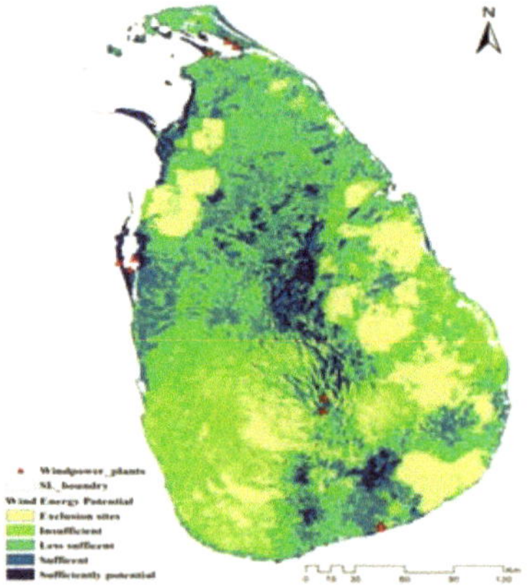

FIGURE 8.3 Potential sites based on wind power density.

Source: Reprinted with permission from Amarasinghe & Perera, 2021. Copyright 2021. Springer Nature

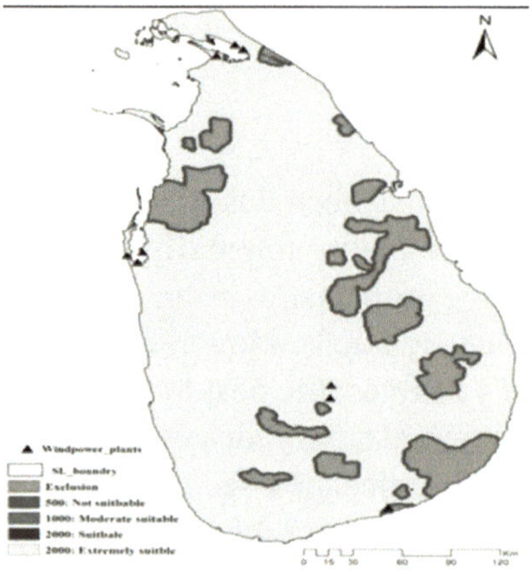

FIGURE 8.4 Potential sites based on protected areas.

Source: Reprinted with permission from Amarasinghe & Perera, 2021. Copyright 2021. Springer Nature

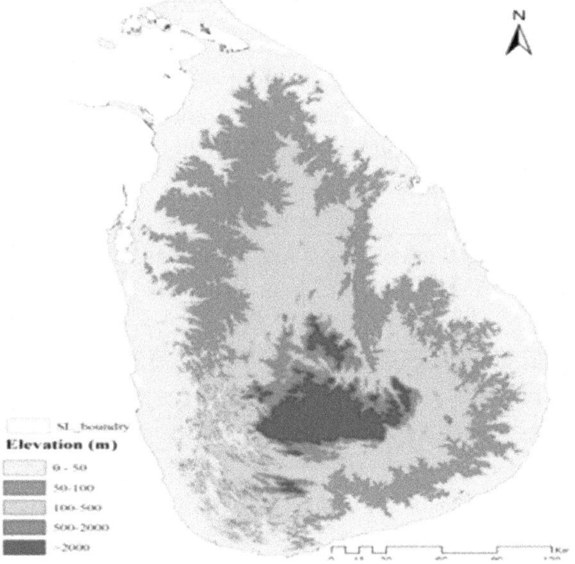

FIGURE 8.5 Potential sites based on elevation.

Source: Reprinted with permission from Amarasinghe & Perera, 2021. Copyright 2021. Springer Nature

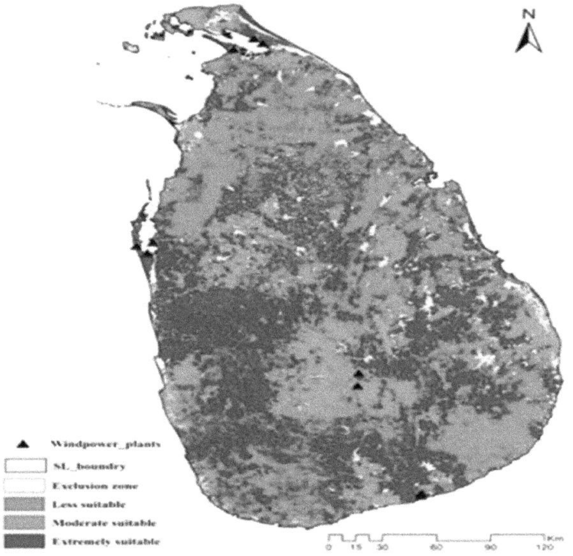

FIGURE 8.6 Potential sites based on land use.

Source: Reprinted with permission from Amarasinghe & Perera, 2021. Copyright 2021. Springer Nature

In Sri Lanka, there are approximately 5,000 km^2 of windy areas with decent to excellent wind resource capacity. Land makes up around 4,100 km^2 of the overall windy area. Tables 8.4 and 8.5 show the proportion of windy land and lagoon possible wind energy in each wind power segment. This windy land accounts for around 6% of Sri Lanka's overall land area (65,600 km^2) (Sayanthan & Kannan, 2017). This windy land could support more than 20,000 MW of potential installed capacity, based on a conservative assumption of 5 MW per km^2. The windy lagoon areas are estimated to cover 700 km^2 and have a 3,500 MW installed capacity. Sri Lanka's combined capacity is projected to be over 24,000 MW. Figure 8.7 reveals that there are five provinces with a wind capacity of at least 1,000 MW. Additional research is needed to reliably measure the wind energy capacity, which must take into account considerations like the existing transmission system and usability (Ratnasiri, 2008).

When new areas with modest wind resource capacity are considered (which will possibly become possible in the near future with more advances in wind technology), the proposed total windy land and lagoon areas (as seen in Table 8.4) rise to over 11,000 km^2, or around 15% of Sri Lanka's total land area (Figure 8.7). This volume of windy land could accommodate more than 51,000 MW of installed power (55,000 MW if lagoons are included). Figure 8.7 reveals that there are seven provinces with wind capacity above 2,000 MW (Elliot et al., 2003).

TABLE 8.4 Sri Lanka – Wind Electric Potential (Good-to-Excellent Wind Resource at 50 m) (Elliot et al., 2003)

Wind Resource Utility Scale	Wind Power at 50 m (W/m^2)	Wind Speed at 50 m (m/s*)	Land Area (km^2)	Lagoon Area (km^2)	Total Area (km^2)	Percent Windy Land	Total Capacity Installed (MW)
Good	400–500	7.0–7.5	2,341	664	3,005	3.6	15,000
Excellent	500–600	7.5–8.0	788	41	829	1.2	4,150
Excellent	600–800	8.0–8.8	517	0	517	0.8	2,600
Excellent	> 800	> 8.8	501	0	501	0.8	2,500
Total			4,147	705	4,852	6.4	24,250

TABLE 8.5 Sri Lanka – Wind Electric Potential (Moderate-to-Excellent Wind Resource at 50 m) (Elliot et al., 2003)

Wind Resource Utility Scale	Wind Power at 50 m (W/m²)	Wind Speed at 50 m (m/s*)	Land Area (km²)	Lagoon Area (km²)	Total Area (km²)	Percent Windy Land	Total Capacity Installed (MW)
Moderate	300–400	6.4–7.0	6,119	196	6,315	9.5	31,600
Good	400–500	7.0–7.5	2,341	664	3,005	3.6	15,000
Excellent	500–600	7.5–8.0	788	41	829	1.2	4,150
Excellent	600–800	8.0–8.8	517	0	517	0.8	2,600
Excellent	> 800	> 8.8	501	0	501	0.8	2,500
Total			10,266	901	11,167	15.9	55,850

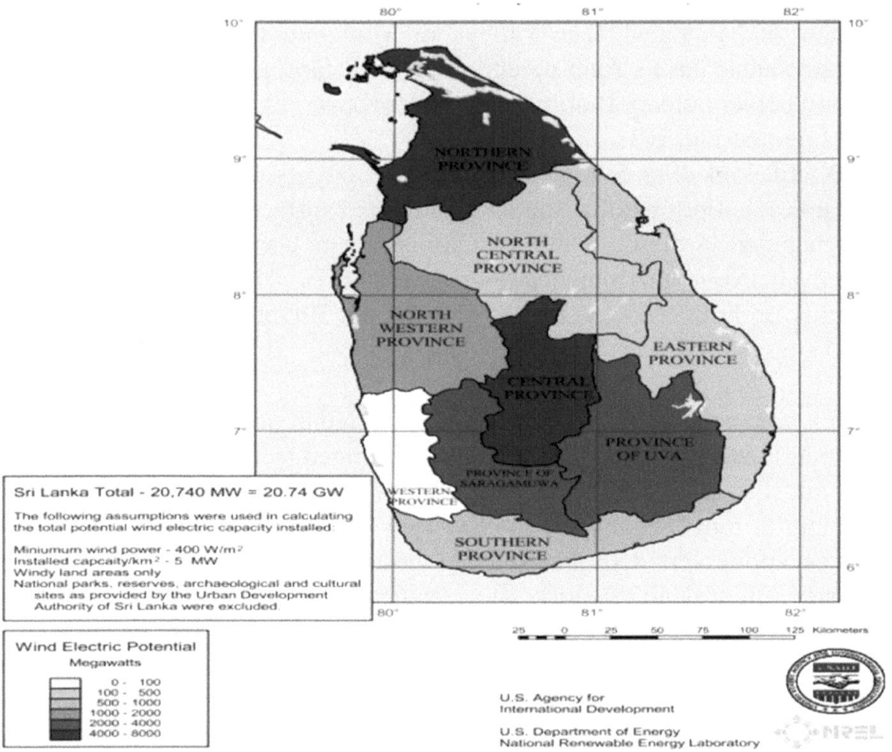

FIGURE 8.7 Wind energy potential of Sri Lanka.

Source: Reprinted from Elliot et al., 2003

8.2.5 BHUTAN

Bhutan is a landlocked nation with a population of around 6,90,000 people situated between India and China. The climate ranges from tropical in the southern plains to cold winters and hot summers in the central valleys to harsh winters and cool summers in the northern Himalayas. Bhutan's population is split between rural and urban areas, with 70% living in rural areas and 30% in urban areas. Bhutan generates green energy by using its rivers in a run-off river hydropower scheme that is distributed across the world. Apart from hydropower, small-scale wind, and standalone solar PV systems are currently operational; however, there are only a few studies on the capacity of solar and wind energy in Bhutan (Irfan et al., 2019). Using assessed data from local weather stations as well as NREL climate data, measuring the wind and solar resource capacity of some of Bhutan's selected locations. The wind data were used to determine the density and capacity of wind power. In addition, after extrapolating wind speeds, the wind power capacity at various heights is calculated. The valleys of Wangdue Phodrang, Tsirang, and Trashiyangtse have strong potential for wind farm production, according to a wind power potential estimation report (Karma Chophel School of Energy and Engineering, 2011).

NREL looked at data from a variety of outlets to assess Bhutan's wind resources. Unfortunately, Bhutan's integrated surface hourly (ISH) data was not included in the US National Climatic Data Center (NCDC), so NREL relied on ISH data from neighboring countries. These data were used to identify preliminary wind resource signs near Bhutan (Uddin, Taplin, & Yu, 2007).

Then, using meteorological data from 12 stations around the region, NREL created a model. The data was collected in one-hour increments. The period of data collection for such stations ranged from one to four years. But for four stations at a height of 20 m, the majority of the data was collected at a height of 5 m. The annual and monthly averages of wind speed and wind power output were determined using these data by NREL. It also looked at the strength of wind direction and speed combined.

A color-coded wind power map of Bhutan is the product of the NREL's wind resource study of Bhutan. The wind power density at a height of 50 meters is used to classify wind power. The wind power map of Bhutan, as seen in Figure 8.8, is the final contribution of the entire study (Dorji, Urmee, & Jennings, 2012).

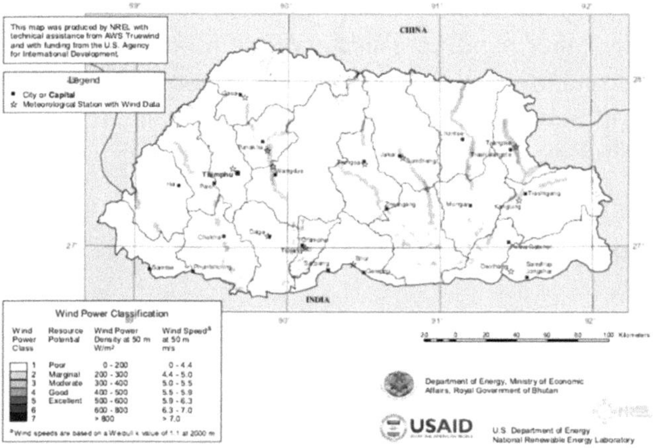

FIGURE 8.8 The wind potential of Bhutan at 50 m.

Source: Reprinted with permission from Dorji, Urmee, & Jennings, 2012. Copyright 2012. Elsevier

According to the NREL report, wind power density would be 50 meters above ground. The study was carried out by vertical extrapolation to a height of 50 m, referred to as synthesized 50, in order to allow proper comparisons (Tables 8.6 and 8.7). Table 8.6 shows the similarities between two different outcomes.

TABLE 8.6 Comparison of Results (National Renewable Energy Laboratory, (n.d.)

Name of Station	Results from NREL (Wind Power Class)	Wind Power Density Range at 50 m (W/m^2)	Wind Power Density at 50 m (kW/m^2)	Mean Wind Speed at 50 m (m/s)
Gelephu	1 (poor)	0–200	131	4.71
Kanglung	1 (poor)	0–200	60	3.76
Simtokha	1 (poor)	0–200	48	2.59
Tsirang	2 (marginal)	200–300	261	4.74
Wangdi	2 (marginal)	200–300	266	5.19

TABLE 8.7 Wind Power Classification (National Renewable Energy Laboratory, (n.d.)

Wind Power Class	Resource Potential	Wind Power Density at 50 m (W/m^2)	Wind Speed at 50 m (m/s)
1	Poor	0–200	0.0–4.4
2	Marginal	200–300	4.4–5.0
3	Fair	300–400	5.0–5.5
4	Good	400–500	5.5–5.9
5	Excellent	500–600	5.9–6.3

For the most part, Bhutan's wind resource is insignificant. Wind power density, on the other hand, is usually strong along the valleys of Bhutan's main rivers. In addition, the majority of these areas are connected by road and power grid networks. As a result, these areas could be suitable for the production of utility-scale wind power in the future.

8.2.6 AFGHANISTAN

Afghanistan is a country in South Asia with a total area of 6,47,800 km^2. Its population is mostly made up of young adults and children who live mostly in rural areas. Afghanistan is a war-torn country with a volatile political and social situation. Since Afghanistan is a mountainous country with snow-covered mountains, it is not affected by runoff water for several months of the year. In addition, the nation has the ability to use solar and wind energy (Jahangiri et al., 2019). In Afghanistan, there are vast areas with attractive wind energy prospects. The energy potential of wind is estimated to be about 67,000 MW in a report for MEW's Renewable Energy Department, while NREL estimates it to be 1,58,000 MW. The latter survey, which used sophisticated modeling and simulation techniques to create high-resolution wind resource charts, classifies 5% (31,600 km^2) of Afghanistan's total land area (6,50,000 km^2) as "class 4+," with "good-to-excellent" potential for utility-scale applications (Ershad, Brecha, & Hallinan, 2016). The western provinces of Nimroz, Farah, and Herat, as well as the northeastern provinces of Balkh and Takhar, have significant wind capital areas. In the east, wind corridors can be found near Jabalsaraj, Sarobi, and Tirgari, while in the central and southern regions, they can be found near Qalat, Gadamsar, Walakhor, Golestan, and Gorzanak. This listed two sites in Kabul, three sites in Core, and five sites in Mazar-e-Sharif that have the best current potential, taking into account a variety of reasons, including security concerns. Afghanistan has the capacity to produce 3,42,521 GWh per year, according to another report (Fahimi & Upham, 2018) (Figure 8.9).

The minimum wind speed was determined using the NREL wind speed classification for Afghanistan. We considered areas with wind power density class 3 (300–400 W/m^2 wind power density at a height of 50 m) and above, as well as wind speeds greater than 6 m/s. However, the minimum wind speed in references is 5 m/s, and the maximum wind speed is 5.5 m/s. Afghanistan's wind energy potential (Good-to-Excellent Wind Resource at 50 m) is listed in Table 8.8.

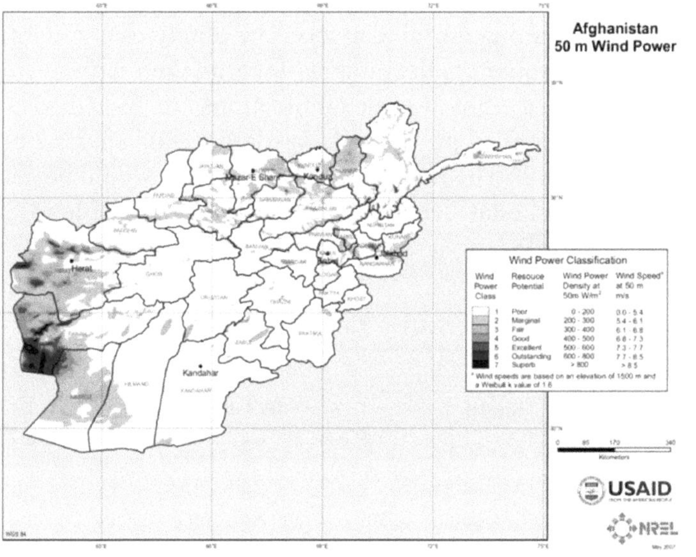

FIGURE 8.9 Afghanistan 50 m wind power potential map (Anwarzai & Nagasaka, 2017).
Source: Reprinted with permission from Anwarzai & Nagasaka, 2017. Copyright 2012. Elsevier

TABLE 8.8 Afghanistan Wind Power Potential (Good-to-Excellent Wind Resource at 50 m) (Elliott, 2011)

Wind Resource Utility Scale	Wind Class	Wind Power (W/m²)	Wind Speed (m/s)	Land Area (km²)	Percent Windy Land	Total Capacity Installed (MW)
Good	4	400–500	6.8–7.3	15,193	2.4	75,970
Excellent	5	500–600	7.3–7.7	6,633	1.0	33,160
Excellent	6	600–800	7.7–8.5	6,615	1.0	33,100
Excellent	7	> 800	> 8.5	3,169	0.5	15,800
Total				31,610	4.9	158,030

Despite the fact that the country's west is known for wind, there is significant capacity near established hydropower plants such as Kajaki and Naghlo, as well as in the provinces of Zabul, Nangarhar, Baghlan, Takhar, Sari Pul, and Faryab (Elliott, 2011).

8.2.7 MALDIVES

In the Maldives, the wind resource in exposed areas varies little, but it is important to highlight the areas of the country where the resource is stronger.

Table 8.9 shows a basic wind classification system with short wind power intervals for representing the wind assets. The Maldives is suitable for village and resort wind turbine applications as long as there are no trees or other barriers along the way. Small-scale applications are described as those with 50 kW or less of installed power, whereas large-scale applications are those with more than 50 kW. The feasibility of any project would be determined by economic and social considerations as well as the available resources (Van, Hekkert, & Van, 2008).

TABLE 8.9 Wind Power Classification (Elliot et al., 2003)

Resource Potential		**Wind Power Density (W/m^2) @ 50 m agl**	**Wind Speed (m/s) @ 50 m agl**
Large	**Small**		
Fair	Moderate	225–250	5.8–6.0
Fair	Moderate	250–275	6.0–6.2
Fair	Moderate	275–300	6.2–6.4
Moderate	Good	300–325	6.4–6.5
Moderate	Good	325–350	6.5–6.7

The greatest wind resource is located between 4.5N Lat and 6.5°N Lat. The strongest opportunity in the Maldives is projected to be in the North Maalhusmadulu, South Maalhusmadulu, North Miladhunmadulu, South Miladhunmadulu, and Faadippolhu Atolls. This amount of resource is thought to be adequate for small-scale applications and moderate for large-scale ones. The wind resource is marginally lower to the north and south of this region, but it is still considered strong for small-scale and moderate for large-scale wind applications (Wijayatunga, George, Lopez, & Aguado, 2016).

The Maldives is dominated by monsoon rains, but the annual and seasonal trends differ from north to south because monsoon flow strengths vary with proximity to the equator. The strongest west monsoon occurs in the Maldives' extreme north, while the strongest northeast monsoon occurs in the country's north-central region. The Maldives' highest wind resource is situated where the northeast monsoon is best. The west monsoon weakens in the southern Maldives but remains moderately strong down to Addu Atoll. The northeast monsoon weakens quickly south of Male' and is barely visible at Addu Atoll. In reality, since west winds prevail at Addu Atoll during the year, the monsoon flows can be said to come to a halt somewhere north of that area. The weakening of the west and northeast monsoons causes a noticeable change in the high wind resource months from the northern to the southern areas of the region, but the moderate west winds across the southern Maldives hold the difference in the overall resource in the Maldives relatively minimal (Elliot et al., 2003). Figure 8.10 shows the wind resource map and the atoll names for the Maldives.

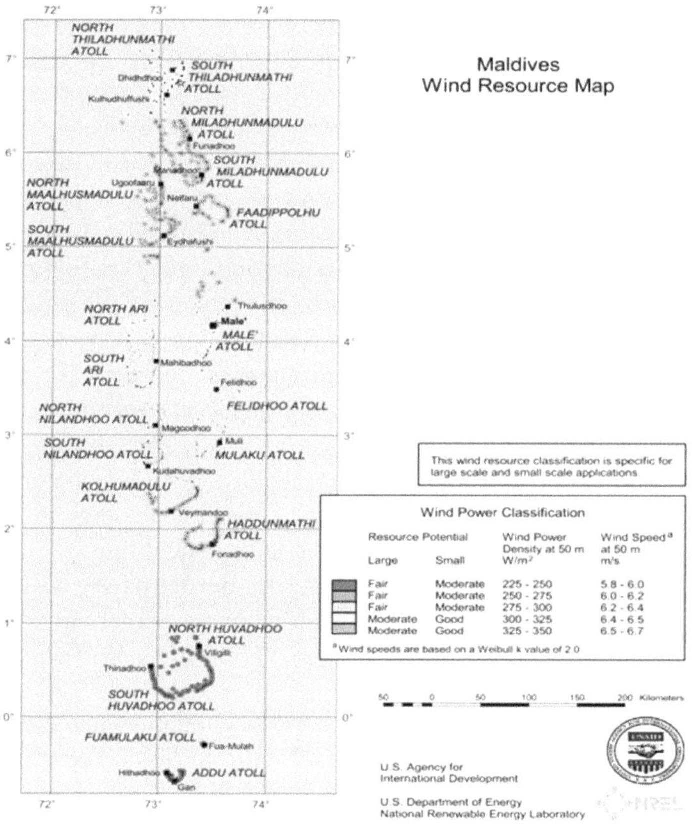

FIGURE 8.10 Wind resource map and the atoll names for the Maldives (Elliot et al., 2003).

8.3 SOUTH ASIAN COUNTRIES POLICY RECOMMENDATIONS FOR WIND ENERGY

In comparison to fossil fuel-based power generation, wind energy has been shown to be a true sustainable clean energy source around the globe. The cost of generating electricity from wind is now comparable to the cost of generating electricity from fossil fuels. Wind energy is critical for creating a world with RE. Due to large greenhouse gas emissions from nuclear power plants and rising fossil fuel prices, wind power production would be the most cost-effective option for mitigating global environmental and fossil fuel depletion issues. For the countries of South Asia, the use of wind energy for power generation is extremely important because of their dense populations, heavy power needs, and rapidly growing economies. To make an improved regional wind power market, encouraged by a regional power exchange and connected to established South Asian power exchanges, the following steps are recommended:

1. An analysis of the SAC's wind power system systems, including legal and regulatory elements, as well as the participating countries' power transmission system protection and stability requirements and their compatibility from a regional wind power trading perspective.
2. The development of a framework for a regional wind power exchange that can connect with existing South Asian power exchanges, with appropriate organizational modifications, and centrally promote the expansion of the South Asian power market to accommodate regional power trading.
3. A study of power generation scheduling and dispatch processes, energy accounting systems, and financial settlement systems for electricity transactions in each country; detailing of proposals for harmonization to allow feasible cross-border power trade; and analysis of the administrative, regulatory, and commercial conditions for cross-border power trade (George, Hammill, & Williamson, 2018).

By using domestic energy sources, including the huge unexploited hydroelectric capacity that spans the entire sub-region, South Asia will come up with a joint solution. Intra-regional resource gaps, along with variations in load curves across South Asia, provide possibilities for sub-regional energy generation approaches and their effective usage. This includes mutual energy investment and joint power delivery management across cross-border grid links and allied networks. As well as the physical network interconnection's "strong" infrastructure, it is important to develop: (i) the "warm" interconnection infrastructure – power exchange agreements and the harmonization of legislation, regulations, and standards – that are necessary to link the adjacent power systems; and (ii) to reinforce the capacities of the concerned institutions (Khan et al., 2012).

There are some relevant policies recommended following a thorough analysis of the wind energy position, capacity, and policies in South Asia:

- Policy relaxation and subsidies should be given for buyers and entrepreneurs to spend more freely.
- The governments of the South Asian countries should allow local residents and private developers to build smaller wind power stations to meet rural and remote energy needs.
- The local population should be mindful of the advantages of small community pilot programs, including proper preparation when using wind energy.
- Subsidies and tax reductions must be granted to households and industrial consumers.
- Smooth strategies should be implemented to encourage international investment and wind energy technology expertise.
- A wind energy production fund could encourage small investors to participate on simple terms and conditions.
- The transmission network should be prioritized such that power to households and factories is continuously and uninterrupted.
- Local wind energy staff and organizations should be educated by the world-leading countries that have wind power technology skills (Irfan et al., 2019).

For the future exploration of the wind energy market in those areas, further research can be carried out in other countries in South Asia.

8.4 CONCLUSION

As the best alternative to conventional fossil fuel power generation, wind energy has arisen. Wind is a RE source that can alleviate environmental issues globally. With the advancement of wind technologies, wind energy production costs are now pretty much near the expense of fossil-power generation. South Asian countries, particularly because of the massive population and rising economies, face energy demand and supply gaps. Earlier wind energy research did not take into account the political designs of South Asian countries in general and wind energy in particular. This chapter presents wind energy potential and status in South Asia. South Asia has enormous wind energy promise. If wind energy can be used properly and as a whole, it has many advantages. By reducing GHG emissions, it would enhance environmental sustainability. There will also be an increase in living standards. The country will save a great deal, and energy efficiency will be enhanced by reducing dependency on imported fossil fuels. Taking these advantages into account and announcing several initiatives in this respect, Southern Asian policymakers have launched several schemes to promote public understanding of the importance of wind power. There is a shortage of professional human and technical knowledge in this field in South Asian countries. The wind potentials of these countries are enormous. The governments of these countries must make sincere and rigorous attempts to make full use of wind power.

KEYWORDS

- **environmental sustainability**
- **GHG emissions**
- **policy recommendations**
- **policy relaxation**
- **potential**
- **South Asian countries**
- **status**
- **wind energy**

REFERENCES

Amarasinghe, A. G., & Perera, E. N. C., (2021). Modeling predictive suitability to determine potential areas for establishing wind power plants in Sri Lanka. *Model Earth Syst. Environ., 7*, 443–454. https://doi.org/10.1007/s40808-020-00868-w.

Anwarzai, M. A., & Nagasaka, K., (2017). Utility-scale implementable potential of wind and solar energies for Afghanistan using GIS multi-criteria decision analysis. *Renew. Sustain. Energy Rev., 71*, 150–160. https://doi.org/10.1016/j.rser.2016.12.048.

Baloch, M. H., Kaloi, G. S., & Memon, Z. A., (2016). Current scenario of the wind energy in Pakistan challenges and future perspectives: A case study. *Energy Reports, 2*, 201–210. https://doi.org/10.1016/j.egyr.2016.08.002.

Dorji, T., Urmee, T., & Jennings, P., (2012). Options for off-grid electrification in the Kingdom of Bhutan. *Renew. Energy, 45*, 51–58. https://doi.org/10.1016/j.renene.2012.02.012.

Elliot, D., Schwartz, M., Scott, G., Haymes, S., Heimiller, D., & George, R., (2003). *Wind Energy Resources Atlas of Sri Lanka and the Maldives*, 175.

Elliott, D., (2011). *Wind Resource Assessment and Mapping for Afghanistan and Pakistan.* Renew Energy Lab Golden, Color USA.

Ershad, A. M., Brecha, R. J., & Hallinan, K. (2016). Analysis of solar photovoltaic and wind power potential in Afghanistan. *Renew. Energy, 85*, 445–453. https://doi.org/10.1016/j.renene.2015.06.067.

Fahimi, A., & Upham, P., (2018). The renewable energy sector in Afghanistan: Policy and potential. *Wiley Interdiscip. Rev. Energy Environ., 7*, e280. https://doi.org/10.1002/wene.280.

Foreign Affairs, (2019). *Baseline Study Wind Energy Bangladesh.*

George, J., Hammill, M., & Williamson, M. (2018). *Integrating South Asia's Power Grid for a Sustainable and Low Carbon Future,* 1–46.

Golam, M., Khan, K., Rahman, T., & Alam, M. M., (2004). Wind energy in Bangladesh: Prospects and utilization initiatives. In: *3rd Int. Conf. Electr. Comput. Eng. ICECE 2004* (pp. 28–30). Dhaka, Bangladesh.

Iftikhar, M. N., Najeeb, F., Mahazzam, S., & Khan, S. A., (2015). Sustainable energy for all in South Asia: Potential, challenges, and solutions. *Sustainable Development Policy Institute 2015*, 1–33.

Indian Wind Energy Association, (2021). *Wind Energy in India Wind Power Installation.*

Irfan, M., Zhao, Z. Y., Mukeshimana, M. C., & Ahmad, M. (2019). Wind energy development in south Asia: Status, potential and policies. In: *2019 2nd Int. Conf. Comput. Math. Eng. Technol.* (pp. 1–6). IEEE. https://doi.org/10.1109/ICOMET.2019.8673484.

Jahangiri, M., Haghani, A., Mostafaeipour, A., Khosravi, A., & Raeisi, H. A. (2019). Assessment of solar-wind power plants in Afghanistan: A review. *Renew. Sustain. Energy Rev., 99*, 169–190. https://doi.org/10.1016/j.rser.2018.10.003.

Karma Chophel School of Energy and Engineering, (2011). *A Preliminary Analysis of Wind Data from Bhutan.*

Khan, I., Chowdhury, H., Rasjidin, R., Alam, F., Islam, T., & Islam, S., (2012). Review of wind energy utilization in South Asia. *Procedia Eng., 49*, 213–220. https://doi.org/10.1016/j.proeng.2012.10.130.

Khan, M. J., Iqbal, M. T., & Mahboob, S., (2004). A wind map of Bangladesh. *Renew Energy 29*, 643–660. https://doi.org/10.1016/j.renene.2003.10.002.

Ministry of New and Renewable Energy, (2022). *Wind Energy Current Status.*

Mirza, U., Ahmad, N., Majeed, T., & Harijan, K., (2007). Wind energy development in Pakistan. *Renew. Sustain. Energy Rev., 11*, 2179–2190. https://doi.org/10.1016/j.rser.2006.03.003.

National Renewable Energy Laboratory, (n.d.). *Potential for Development of Solar and Wind Resources in Bhutan.*

Ratnasiri, J., (2008). Alternative energy – prospects for Sri Lanka. *J. Natl. Sci. Found Sri Lanka 36*, 89. https://doi.org/10.4038/jnsfsr.v36i0.8049.

Saifullah, A. Z. A., Karim, A., & Karim, R., (2016). Wind energy potential in Bangladesh. *American Journal of Engineering Research (AJER): Am. J. Eng. Res., 5*, 85–94.

Saulat, H., Khan, M. M., Aslam, M., Chawla, M., Rafiq, S., Zafar, F., et al. (2020). Wind speed pattern data and wind energy potential in Pakistan: Current status, challenging platforms, and innovative prospects. *Environ. Sci. Pollut. Res.* https://doi.org/10.1007/s11356-020-10869-y.

Sayanthan, S., & Kannan, N., (2017). Renewable energy resource of Sri Lanka ! A review. *Int. J. Environ. Agric. Res., 3*, 80–85.

Shaikh, M. A., Chowdhury, K. M. A., Sen, S., & Islam, M. M. (2017). Potentiality of wind power generation along the Bangladesh coast. *AIP Conf Proc.*, 1919. https://doi.org/10.1063/1.5018553.

Shami, S. H., Ahmad, J., Zafar, R., Haris, M., & Bashir, S. (2016). Evaluating wind energy potential in Pakistan's three provinces, with proposal for integration into the national power grid. *Renew. Sustain. Energy Rev., 53*, 408–421. https://doi.org/10.1016/j.rser.2015.08.052.

Uddin, S., Taplin, R., & Yu, X., (2007). Energy, environment, and development in Bhutan. *Renew. Sustain. Energy Rev., 11*, 2083–2103. https://doi.org/10.1016/j.rser.2006.03.008.

Van, A. K., Hekkert, M. P., & Van, S. W., (2008). Renewable energy technologies in the Maldives-realizing the potential. *Renew. Sustain. Energy Rev., 12*, 162–180. https://doi.org/10.1016/j.rser.2006.07.006.

Wijayatunga, P., George, L., Lopez, A., & Aguado, J. A., (2016). Integrating clean energy in small island power systems: Maldives experience. *Energy Procedia, 103*, 274–279. https://doi.org/10.1016/j.egypro.2016.11.285.

Wind Data Sharing Policy, (2020). National Institute of Wind Energy.

Young, M., & Vilhauer, R., (2003). Sri Lanka wind farm analysis and site selection assistance. *Natl. Renew. Energy Lab*, 1–44.

CHAPTER 9

The Role of the Solar Water Heating System in South Asian Countries

SHAILENDRA SINGH,[1] ABHISHEK ANAND,[1] AMRITANSHU SHUKLA,[2] and ATUL SHARMA[1]

[1]*Non-Conventional Energy Laboratory, Rajiv Gandhi Institute of Petroleum Technology, Jais, Amethi, Uttar Pradesh, India*

[2]*Department of Physics, University of Lucknow, Lucknow, U.P., India*

ABSTRACT

The incessant increase in the level of greenhouse gas emission is the main driving force behind the development of renewable energy sources (RES). South Asian countries (SACs), i.e., India, Pakistan, Bangladesh, Nepal, Bhutan, Sri Lanka, and the Maldives, having a respectable part of the world's population, major energy consumers. During the years 1991 and 2000, the SAC's primary energy consumption started to increase and continued to grow in the same way in the upcoming decades. The governments in these countries had started many thermal power projects to meet this exorbitant energy demand. The emission from these power plants leads to some serious environmental issues like global warming, climate change, and air pollution. The scientific community all around the world is in search of technologies that are based on RES. In many parts of the world, direct solar radiation is considered one of the potential sources of energy. The status of solar energy in SAC, particularly for the water heating systems (SWHs), is presented in this chapter.

The South Asian Energy Corridor: Toward Sustainability and Security. Amritanshu Shukla, PhD, Atul Sharma, PhD & Saurabh Mishra, PhD (Eds.)

9.1 INTRODUCTION

In the present century, climate change is a crucial issue that requires the attention of scientists and common people alike. The increase in the population, together with the high rate of economic growth, leads to an increase in energy demand. The rise in energy demand contributed to the accumulation of greenhouse gases (GHGs) because the major portion of this energy comes from thermal power plants. The rise in GHGs in the atmosphere contributed to global warming. In SAC countries (i.e., Figure 9.1), the rise in energy demand is the key factor in the increase of greenhouse (GHG) emissions. The per capita electricity consumption of the region is shown in Table 9.1. Energy consumption has been increasing annually by about 3.3% since the year 1990. It was observed that between the years 1990 and 2005, the total GHG emissions from energy-using activities increased by 98.2% across the SAC region compared to an increase in global emissions of 30.8% (Sudhir & Doig, 2014). Due to global warming, the average temperature of the Earth's surface is increasing regularly, and events like the melting of glaciers and forest fires have become common. It is quite a known fact that about 78% of GHG emission results from fossil fuel combustion and industrial activities happening around the world. If this rate of emission continues, then it is expected that CO_2 emissions will double or triple by the year 2050 (Sudhir & Doig, 2014). The United Nations Framework Convention on Climate Change (UNFCCC) has recommended keeping the temperature rise below 1.5°C of the pre-industrial time.

The energy demand in SAC is continuously increasing as population and industrial activities are flourishing, and this demand cannot be fulfilled by conventional sources of energy alone. To meet this demand, the government of SAC must focus on non-conventional sources of energy like solar energy. Solar energy can be applied in various energy-intensive sectors, i.e., residential, commercial, and industrial. The SACs are primarily dependent on the Gulf countries to meet their energy demands in the form of crude oil or petroleum products. The data shows that for the financial year 2009, India imported 35% of its total energy demand, Bangladesh imported 94% of its oil demand and 45% of its coal demand, Bhutan imported 25% of its commercial demand, and the Maldives imported 100% of its energy demands. The economic growth of SAC is directly linked with energy production and consumption in their countries. The electricity consumption per capita of SAC is presented in Table 9.1. The energy demand and consumption of SAC are indicators of a growing economy, but it also has to keep in mind the climate change issue. There must be the adoption of renewable-based technologies to fulfill the growing energy demands in the various sectors.

TABLE 9.1 Per Capita Electricity Consumption (Ahmed, Mahmood, Hasan, Sidhu, & Butt, 2016)

South Asian Countries	Per Capita Electricity Consumption in (kWh)
India	644
Sri Lanka	636.3
Pakistan	457
Afghanistan	119.8
Bangladesh	278.1
Nepal	454.1

The global warming effect can be lowered to a certain level by replacing the existing technology with renewable-based technology. The hot water demand also consumes lots of electrical energy or fossil fuel indirectly. This hot water demand is generally fulfilled by an electrical heating system or by burning fossil fuels, which is a major source of GHG emissions. The emission can be controlled at large by the application of the SWH system.

Among all renewable energy (RE) resources, solar energy is a prominent source and is available globally. The world's overall solar energy potential is around 5.6 GJ per square meter per year. The highest solar resource potential is in the Red Sea area, including Egypt and Saudi Arabia (Parks et al., n.d.). Solar energy is available free of cost everywhere in the world. It only requires RES-based technologies to convert it into a usable form for different solar thermal applications. It is considered a clean and environmentally friendly source of energy that is free from all harmful effects of pollutants. Like other parts of the world, SACs are also receiving an ample amount of solar radiation throughout the year except for some cloudy and foggy days. The solar radiation received, potential, and installed capacity of SACs are presented in Table 9.2. Table 9.2 shows that a good amount of solar energy is received per square meter of the Earth's surface per day in the SAC region. It can also be observed from the table that each country has sufficient potential for solar-based energy technology, but only a small fraction has been harnessed till now. The availability of solar radiation is a good sign of the application of solar-based technologies like the SWH system.

TABLE 9.2 Solar Radiation and Power Potential of the SAC Region (Sudhir & Doig, 2014)

Country	Installed	Potential	Solar radiation (kWh/m²/day)
India	0.239 MW	58.00 GW	4.00-7.00
Bhutan	15.00 MW	NA	4.00-5.50
Pakistan	NA	1600.00 GW	4.70-6.20
Bangladesh	2600.00 MW	600000.00 GW	3.20-6.10
Nepal	64300.00 SHS	2920.00 GWh	3.60-6.20
Sri Lanka	1.38 MW	NA	2.00-6.00

Source: Low-Carbon Development in South Asia Report 2014.

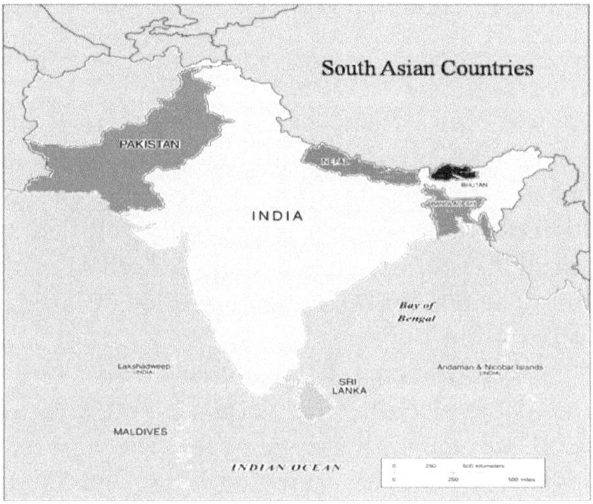

FIGURE 9.1 South Asian countries

Source: Reprinted with permission from Pranti et al., 2013. https://creativecommons.org/licenses/by/4.0/

9.2 SOLAR WATER HEATING TECHNOLOGY

Solar water heating technology is not new. It has continuously been used since ancient times. The world's first commercialized SWH system was patented in the United States in 1891. The SWH system contains mainly three components, i.e., solar collector, storage tank, and auxiliary heating system. Solar collectors are used in stockpiling solar radiation, and this heat is used to heat cold water at the desired temperature. The cold water advances from the bottom side, and hot water is left at the upper side of the collector. Three types of collectors are used in SWH systems, i.e., unglazed, glazed, and evacuated tube-type collectors, as shown in Figure 9.2.

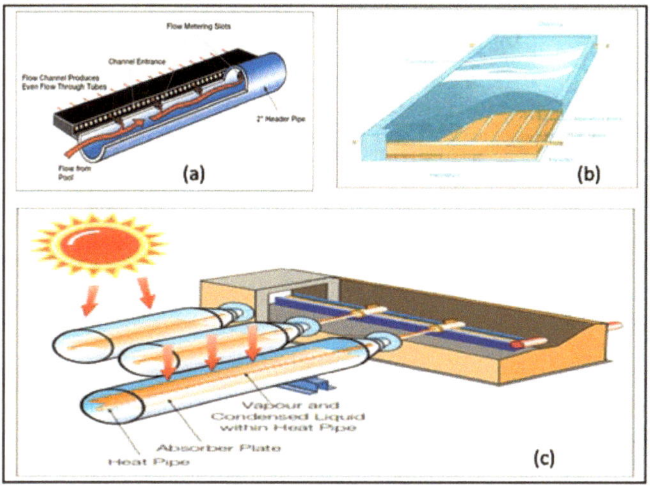

FIGURE 9.2 Types of collectors – (a) unglazed; (b) glazed; and (c) evacuated (Natural Resources Canada, 2005).

The unglazed collectors are usually made of polymer box, and it is simply laid down on wooden floors for capturing solar radiation. Generally, the desired coatings are not used in this type of collector due to the low cost associated with it. In this case, thermal losses to the environment are very high, so this type of collector is not preferred for attaining a high temperature. This type of collector is not recommended for cold climatic countries but can be recommended for the summer season only. While in the case of a glazed type of collector, a desired type of coating is applied to prevent the thermal losses in the single or double box of absorber plate. The backside of a collector is also covered with insulation materials. The glazed type of solar collectors is more efficient than the unglazed type of collectors. In the case of an evacuated tube-type solar collector, the absorber plate with the desired coating is fitted inside of glass tube. The thermal losses to the environment are very less, and this type of collector is the most efficient among all other collectors. This type of collector is recommended for getting a high temperature of water in the case of cold climatic countries, and even this is suitable for all types of weathers of any country. Presently, the cost of the evacuated type of collector is cut down, and the rates are quite similar to that of glazed type of collectors. Due to this, the evacuated types of collectors are the most preferred type for SWH applications.

There are two types of SWH systems available, i.e., natural circulation and forced circulation. In the case of the natural circulation type, the water is flowing due to the density difference of water in the circulation loop of the system. Here the pump is not needed for the circulation of water. The storage tank is installed above the top surface of the solar collector, as shown in Figure 9.3(a). This type of SWH system is not recommended for cold climatic countries generally but can be recommended for warm, humid climatic countries. In the forced circulation type of the SWH system, a pump is required for maintaining the flow of water in the circulation loop, as shown in Figure 9.3(b). This type of SWH system is most preferred for cold climatic countries. In cold climatic countries, SWH systems also suffer from freezing problems during the winter season. To avoid freezing, a heat exchanger is used in the loop with suitable heat transfer fluid, and this fluid is separated in the heat storage tank. Generally, the auxiliary heating arrangement is also fitted before the delivery of hot water to maintain the required temperature of hot water to the utility point. Some other components, like the piping arrangement and cold supply mechanism, are also needed in the circulation loop.

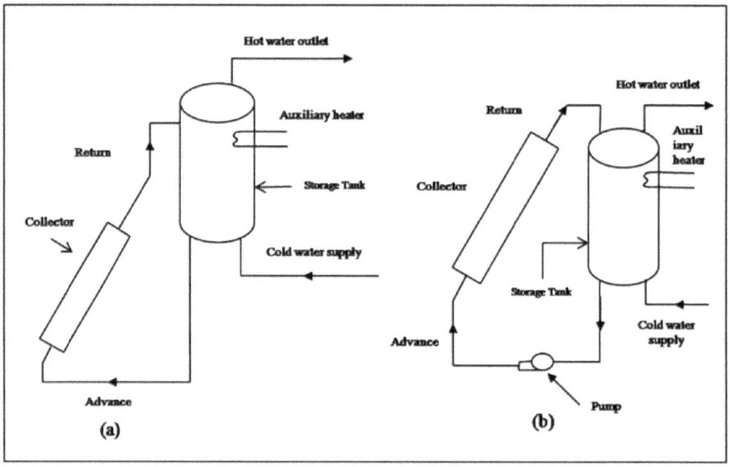

FIGURE 9.3 (a) Natural circulation SWH system; and (b) forced circulation SWH system.

The thermal performance of an SWH system is described in terms of the solar fraction (SF). It is the ratio of energy provided by the SWH system to the total energy required for heating water up to the desired temperature. The SF is given in Eqn. (1).

$$SF = \frac{Q_{saved,solar}}{Q_{aux,no\ solar}} \tag{1}$$

where;

$$Q_{saved,solar} = Q_{aux,\ no\ solar} - \left(Q_{aux,\ solar} + Q_{parasitic}\right) \tag{2}$$

In Eqns. (1) and (2), $Q_{saved,solar}$ is the amount of energy saved by using the SWH system instead of a conventional water heating system on an annual basis, $Q_{aux,no\ solar}$ is the amount of annual energy needed to meet the water heating load without the use of the SWH system, $Q_{aux,solar}$ is the amount of annual energy needed to meet the water heating load with the use of the SWH system, and $Q_{parasitic}$ is the energy consumed by other components like pumps and auxiliary heating systems. Another common performance metric is the annual system efficiency (η_{swh}), which is the ratio of energy saved by using an SWH system ($Q_{saved,solar}$), and solar energy reaches the surface of collectors annually ($Q_{incident}$).

$$\eta_{swh} = \frac{Q_{saved,solar}}{Q_{incident}} \tag{3}$$

9.3 CLIMATIC RESOURCE ASSESSMENT OF SOUTH ASIAN COUNTRIES

The climatic condition plays an important role in assessing the feasibility and performance of the SWH system. The studies have revealed that the application of the SWH systems is advantageous for any type of climatic condition like cold, hot, and humid. But one thing should be noted here: the performance of the SWH system may differ at different locations of the world and may vary according to the climatic conditions. So, the study of climatic resources for the SWH system is also important before getting approval from the governments of the SAC region. The climatic resource includes some environmental parameters for assessing the suitable site of SWH systems like air temperature, relative humidity, daily solar radiation, wind speed, earth temperature, heating degree days, and cooling degree days at selected sites of facility locations. Among all the parameters, the daily solar radiation is a significant parameter for assessing the appropriate location of SWH systems. The available solar radiation must be greater than 3.5 kWh/m^2/day for all months in a year, excluding some cloudy and foggy days, for generating the required heat for heating water up to the desired temperature. Here, the case study of residential SWH for 7 occupants has been presented for each SAC. The climate parameter for each country is given in Tables 9.3(a)–9.3(g).

TABLE 9.3(A) India (Facility Location – Delhi: Latitude–28.6°N and Longitude 77.2°E)

Month	Air Temperature (°C)	Relative Humidity (%)	Daily Solar Radiation-Horizontal ($KWh/m^2/d$)	Wind Speed (m/s)	Earth Temperature (°C)	Heating Degree-Days 18°C (°C-d)	Cooling Degree-Days 10°C (°C-d)
January	14.3	70.9	3.74	2.2	13.2	115	133
February	16.8	63.0	4.76	2.5	16.9	34	190
March	22.3	53.9	5.89	2.8	23.7	0	381
April	28.3	39.8	6.79	3.1	31.2	0	564
May	32.5	39.7	6.96	3.1	36.8	0	698
June	33.4	52.3	6.30	3.1	37.9	0	702
July	30.8	70.9	5.27	2.5	34.1	0	645
August	30.0	75.7	5.10	2.2	30.8	0	620
September	29.5	70.1	5.46	2.2	28.3	0	585
October	26.3	62.5	5.04	1.7	24.1	0	505
November	20.8	63.7	4.13	1.4	18.9	0	324
December	15.7	70.1	3.55	1.9	14.4	71	177
Annual	25.1	61.1	5.25	2.4	25.9	220	5,524

Source: NASA Database.

TABLE 9.3(B) Sri Lanka (Facility Location – Anuradhapura: Latitude–8.3°N and Longitude–80.4°E)

Month	Air Temperature (°C)	Relative Humidity (%)	Daily Solar Radiation-Horizontal (KWh/m²/d)	Wind Speed (m/s)	Earth Temperature (°C)	Heating Degree-Days 18°C (°C-d)	Cooling Degree-Days 10°C (°C-d)
January	24.9	80.0	4.54	4.9	25.4	0	462
February	26.0	73.6	5.51	4.2	26.9	0	448
March	27.9	67.9	6.12	3.1	29.3	0	555
April	28.6	72.9	5.75	3.0	29.7	0	558
May	28.2	78.0	5.54	5.9	28.8	0	564
June	27.6	77.4	5.21	7.4	28.2	0	528
July	27.3	76.1	5.24	7.1	28.1	0	536
August	27.3	75.7	5.39	6.9	28.2	0	536
September	27.4	75.6	5.33	6.4	28.3	0	522
October	26.8	79.7	4.72	4.5	27.5	0	521
November	25.9	83.8	4.07	3.8	26.3	0	477
December	25.2	83.1	3.98	4.7	25.5	0	471
Annual	26.9	77.0	5.11	5.2	27.7	0	6,179

Source: NASA Database.

TABLE 9.3(C) Bangladesh (Facility Location – Dhaka: Latitude–23.70°N and Longitude–90.39°E)

Month	Air Temperature (°C)	Relative Humidity (%)	Daily Solar Radiation-Horizontal (KWh/ m^2/d)	Wind Speed (m/s)	Earth Temperature (°C)	Heating Degree-Days 18°C (°C-d)	Cooling Degree-Days 10C (°C-d)
January	18.4	51.6	4.36	2.5	18.1	0	260
February	22.2	42.6	4.92	2.5	22.2	0	342
March	27.4	40.0	5.59	2.8	28.0	0	539
April	30.6	51.9	5.76	3.6	31.8	0	618
May	30.6	65.3	5.30	3.8	31.6	0	639
June	29.5	79.4	4.53	4.3	30.0	0	585
July	28.7	85.1	4.23	4.3	29.0	0	580
August	28.5	85.6	4.29	3.7	28.8	0	574
September	27.9	85.3	4.02	3.0	28.1	0	537
October	26.5	77.9	4.32	2.2	26.6	0	512
November	23.1	68.4	4.28	2.1	22.9	0	393
December	19.4	61.8	4.21	2.2	19.0	0	291
Annual	26.1	66.4	4.65	3.1	26.4	0	5,869

Source: NASA Database.

TABLE 9.3(D) Pakistan (Facility Location – Islamabad: Latitude–33.62°N and Longitude–73.10°E)

Month	Air Temperature (°C)	Relative Humidity (%)	Daily Solar Radiation-Horizontal (KWh/ m^2/d)	Wind Speed (m/s)	Earth Temperature (°C)	Heating Degree-Days 18°C (°C-d)	Cooling Degree-Days 10°C (°C-d)
January	10.0	69.7	2.13	1.6	7.0	248	0
February	13.1	64.0	3.22	2.1	9.6	137	87
March	17.9	59.4	3.77	2.7	15.2	3	245
April	23.4	50.1	4.56	3.2	21.5	0	402
May	28.8	42.0	5.26	3.4	28.6	0	583
June	31.3	47.2	5.67	3.8	32.8	0	639
July	29.6	69.3	4.79	3.1	31.0	0	608
August	28.6	75.3	4.63	2.5	28.2	0	577
September	26.8	69.6	4.36	1.9	25.1	0	504
October	22.1	62.4	4.33	1.6	19.5	0	375
November	15.9	64.8	3.13	1.2	13.5	63	177
December	11.3	69.6	2.32	1.3	8.8	208	40
Annual	21.6	62.0	4.02	2.4	20.1	659	4,236

Source: NASA Database.

TABLE 9.3(E) Nepal (Facility Location – Kathmandu: Latitude–27.71°N and Longitude–85.31°E)

Month	Air Temperature (°C)	Relative Humidity (%)	Daily Solar Radiation-Horizontal (KWh/m^2/d)	Wind Speed (m/s)	Earth Temperature (°C)	Heating Degree-Days 18°C (°C-d)	Cooling Degree-Days 10C (°C-d)
January	8.7	38.3	4.26	1.9	6.2	288	0
February	10.9	35.7	5.15	2.2	9.4	199	25
March	15.5	29.7	6.18	2.6	14.9	78	171
April	19.8	28.0	6.76	2.8	20.3	0	294
May	22.2	41.6	6.68	2.6	23.6	0	378
June	23.0	60.9	5.75	2.5	24.7	0	390
July	21.6	81.4	4.79	2.3	22.8	0	360
August	21.0	85.3	4.80	2.0	21.7	0	341
September	19.7	83.9	4.56	1.8	20.0	0	291
October	16.9	67.4	5.13	1.8	16.2	34	214
November	13.2	50.6	4.72	1.8	11.1	144	96
December	10.1	42.1	4.15	1.7	7.2	245	3
Annual	16.9	53.9	5.24	2.2	16.5	988	2,563

Source: NASA Database.

TABLE 9.3(F) Bhutan (Facility Location – Samtse: Latitude–27.3°N and Longitude 90.6°E)

Month	Air Temperature (°C)	Relative Humidity (%)	Daily Solar Radiation-Horizontal (KWh/ m^2/d)	Wind Speed (m/s)	Earth Temperature (°C)	Heating Degree-Days 18°C (°C-d)	Cooling Degree-Days 10°C (°C-d)
January	5.2	68.0	3.91	1.3	2.6	397	0
February	6.6	64.1	4.37	1.6	4.2	319	0
March	9.7	60.2	4.79	1.8	7.7	257	0
April	13.0	65.3	5.43	1.9	11.6	150	90
May	15.9	75.9	5.47	1.8	15.2	65	183
June	18.1	82.7	5.19	1.9	17.9	0	243
July	18.5	86.0	4.92	1.8	18.3	0	264
August	18.4	85.1	4.66	1.6	18.0	0	260
September	17.0	85.3	4.38	1.5	16.4	30	210
October	13.8	79.8	4.53	1.4	12.6	130	118
November	9.9	74.0	4.14	1.3	7.8	243	0
December	6.9	70.3	3.82	1.2	4.3	344	0
Annual	12.8	74.8	4.63	1.6	11.4	1,936	1,368

Source: NASA Database.

TABLE 9.3(G) Maldives (Facility Location – Male: Latitude–4.20°N and Longitude–73.53°E)

Month	Air Temperature (°C)	Relative Humidity (%)	Daily Solar Radiation-Horizontal (KWh/m²/d)	Wind Speed (m/s)	Earth Temperature (°C)	Heating Degree-Days 18°C (°C-d)	Cooling Degree-Days 10°C (°C-d)
January	28.1	78.5	5.84	5.9	28.0	0	561
February	28.6	77.3	6.63	5.2	28.3	0	521
March	29.1	76.4	6.96	4.0	29.1	0	592
April	29.5	78.5	6.39	3.7	29.7	0	585
May	29.2	80.6	5.49	5.7	29.4	0	595
June	28.8	80.8	5.20	5.4	28.9	0	564
July	28.5	80.1	5.37	5.1	28.8	0	574
August	28.3	79.6	5.63	4.7	28.7	0	567
September	28.1	80.8	5.75	5.1	28.6	0	543
October	28.2	80.8	5.96	5.3	28.4	0	564
November	27.9	82.3	5.39	4.2	28.5	0	537
December	27.9	80.2	5.31	5.9	28.2	0	555
Annual	28.5	79.7	5.82	5.0	28.7	0	6,758

Source: NASA Database.

The annual values of daily solar radiation of all SA countries, i.e., India (Delhi), Sri Lanka (Anuradhapura), Bangladesh (Dhaka), Pakistan (Islamabad), Nepal (Kathmandu), Bhutan (Samtse), and Maldives (Male) at the facility location are shown in Figure 9.4. It can be evaluated that the minimum value of daily solar radiation-horizontal is 4.02 kWh/m^2/d for Pakistan, and the maximum value is 5.85 kWh/m^2/d for the Maldives at the selected facility location. These values are far away from the required minimum value of 3.5 kWh/m^2/d. This figure also demonstrates that solar radiation is available in enriched quantity at all mentioned locations, which is a good indication of the feasibility of SWH systems in the countries.

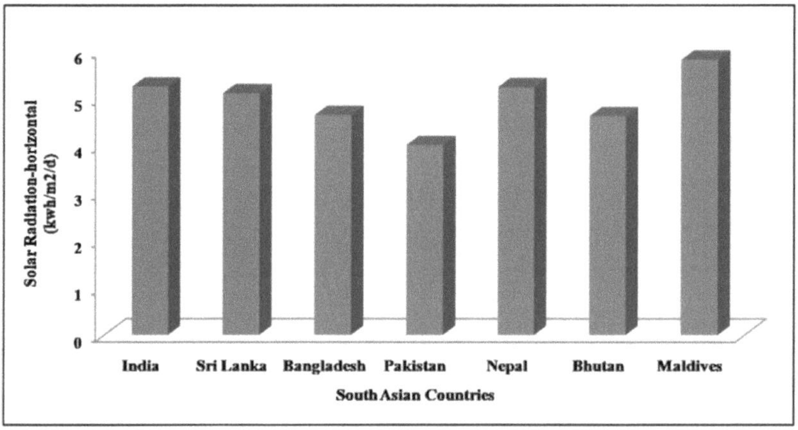

FIGURE 9.4 Annual solar radiation in SAC.

9.4 THE ENVIRONMENTAL BENEFIT OF SWH SYSTEMS

The SWH technology has a large potential for GHG emissions mitigation if it is used in place of the conventional heating system, i.e., electric geysers. In the case of conventional heating, the heat is delivered using electricity produced by the burning of coal. The burning of coal is the main cause of the generation of GHGs. The increasing concentration of GHGs in the atmosphere is acting as a trap that absorbs the heat radiated from the Earth's surface, thus raising the surface temperature of Earth. A schematic diagram of the greenhouse effect is presented in Figure 9.5. Carbon dioxide (CO_2) is the most important human-generated GHG. Its annual emissions grew by about 80% between 1970 and 2004. During 1870, the atmospheric concentration of CO_2 was between 280 and 300 ppm. In 1990, the value increased by more than 340 ppm (WHO, 2008). The amount of CO_2 reached 385 ppm in 2006 at an unprecedented level in the past 6,50,000 years. The top three countries contributing the highest per capita input are the USA, Saudi Arabia, and Australia. In 2002, these countries have contributed CO_2 ranging from 18–20 tons per capita. On the contrary, Bangladesh has only contributed 0.2 tons per capita in the same year (WHO, 2008). Scientists estimate that the CO_2 level will rise another 30% during the next 50 years (Bhuiyan, n.d.).

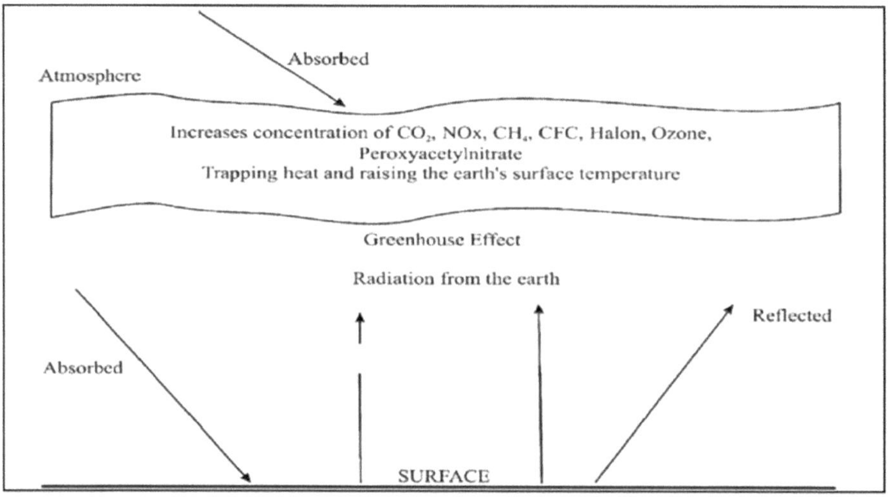

FIGURE 9.5 Greenhouse effect.

Source: Reprinted with permission from Panwar, Kaushik, & Kothari, 2011. Copyright © 2010 Elsevier Ltd.

These gases are greatly responsible for increasing the temperature of the atmosphere. To look into the environmental benefit of utilizing solar energy instead of conventional sources of energy, the different emissions resulting from the solar system operation are estimated and compared to those of a conventional electricity or fuel system and presented in Table 9.4.

TABLE 9.4 Environmental Impact of Thermosiphon SWH System with Electricity Backup (Kalogirou, 2009)

Emissions	Units	Conventional	Solar System	Savings (%)
Carbon dioxide (CO_2)	tons/year	1.546	0.449	70.1
Carbon monoxide (CO)	g/year	374.6	109.7	70.7
Nitrogen oxides (NO_x)	g/year	56.3	16.3	71.1
Nitrous oxide (N_2O)	g/year	6.3	2.1	66.7

TABLE 9.4 *(Continued)*

Emissions	Units	Conventional	Solar System	Savings (%)
Methane (CH_4)	g/year	9.3	2.7	71.0
Hydrocarbons	g/year	37.7	11.0	70.8
Sulfur dioxide (SO_2)	g/year	562.7	164.5	70.8
Dust	g/year	188.1	54.8	70.9
Savings in GHG	%	–	–	70.3

These gases could be decreased up to a certain level by using the SWH technology for heating water in various sectors like residential, commercial, and industrial. Here the potential of GHG mitigation for residential SWH systems for each case study of SAC is presented in Figure 9.6. The annual GHG emission reduction potential is calculated with the help of RETScreen Clean Energy Management Software. From the figure, it can be evaluated that the maximum and minimum values are 4.3tCO_2 and 2.2tCO_2, corresponding to India and Pakistan, respectively. The second highest value is 2.5tCO_2 for Sri Lanka, followed by Bangladesh having 2.4tCO_2. The GHG emission reduction is not available for countries Nepal, Bhutan, and the Maldives by the RETScreen software.

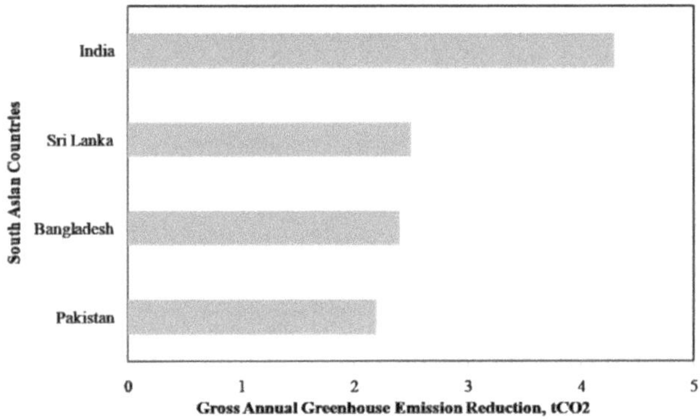

FIGURE 9.6 Gross annual GHG emission reduction potential.

9.5 ENERGY-SAVING POTENTIAL OF SWH TECHNOLOGY

Energy generation and consumption are the backbones of any economy. It shows how fast the economy is growing in comparison to other countries of the world. Reducing the dependence on non-renewable resources is not only environmentally friendly but also economical and sustainable. Dependence on fossil fuels makes countries more susceptible to oil price hikes and can lead to macroeconomic unsteadiness. RE will not only avoid such unsteadiness but also lead to substantial savings and an improvement in the balance of payments due to a reduction of high-cost imports (such as oil). At a household level too, RE will limit the use of expensive fossil fuels for motorized equipment and machinery and augment savings (Iftikhar, Najeeb, Mahazzam, & Khan, 2015). In this sequence, energy savings are also an important phenomenon of a growing economy like the SAC region of India, Bangladesh, Sri Lanka, Pakistan, Bhutan, Nepal, and the Maldives. The energy savings could be achieved by the application of green energy technologies such as SWH systems. The SWH does not use electrical energy for heating water but only uses solar radiation for heating water and qualifies as a clean energy technology that is free from all pollutants. The application of the SWH system demonstrates the energy-saving potential. The saved energy could be utilized in various other sectors. The saved energy is an attractive parameter for the governments to promote the new installation of SWH systems in their respective countries. Even it is well known that SACs are suffering from the rise in energy demand, which could be only solved by energy-efficient technologies such as SWH systems. Countries like Nepal, Bhutan, and the Maldives have lesser capabilities of power generation compared to India, Pakistan, Bangladesh, and Sri Lanka. Therefore, for these countries, SWH systems should be more popular for the heating of water. Even the remaining countries should also offer an attractive mechanism for the application of the SWH system to decrease the energy crises. Here, the energy-saving potential of India, Sri Lanka, Bangladesh, Pakistan, Nepal, Bhutan, and Maldives for residential SWH systems for seven occupants has been calculated with the help of RETScreen software and presented in Figure 9.7. It can be evaluated from the figure that the highest value of energy-saving is 6.2 MWh for Nepal, followed by Bhutan, which has a value of 5.7 MWh. The minimum value of energy-saving was found for Maldives, i.e., 3.6 MWh, and other values are 4.2 MWh for India, 4.5 MWh for Sri Lanka, 3.7 MWh for Bangladesh, and 4.7 MWh for Pakistan. These annual values of energy savings of SACs are an attractive parameter for the application of the SWH system in these countries. The government should also take necessary and essential steps toward the installation of SWH systems in these countries.

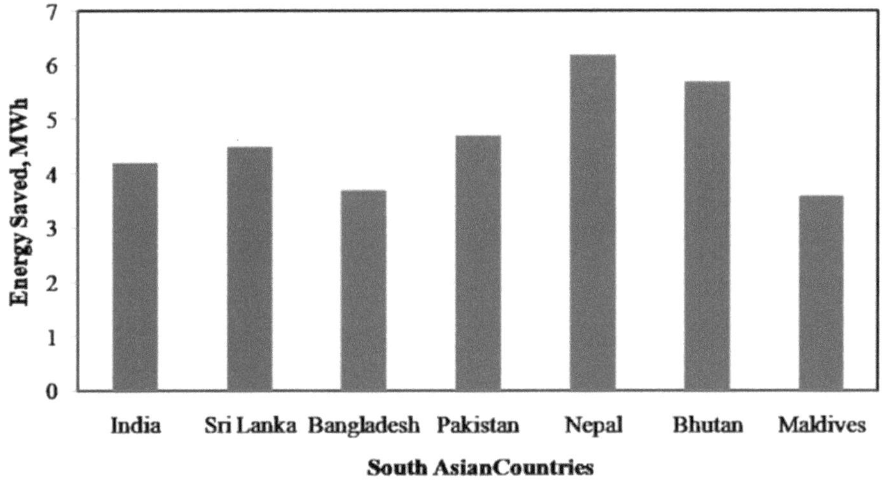

FIGURE 9.7 Energy saving potential in SACs.

9.6 ECONOMIC ANALYSIS OF SWH SYSTEM

The economic analysis of the SWH system in terms of a simple payback period (SPP) is also an important parameter for decision-makers for implementing SWH systems across the SACs. The SPP is the time in which the initial cost of the SWH system can be recovered. It is a well-known fact that the economy of SACs is at the developing stage. These countries have to be assured whether the SWH system is a profitable project or not. It will thus help in assuring whether the technology is viable or not. Suppose if SPP is large, then there is a need for some kind of financial support by the government of that country. In this regard, some developed countries must come forward to support financially the SACs in terms of loans for this type of sustainable energy project so that these projects can be implemented. Here, the SPP of each SAC for residential SWH for seven occupants is calculated with the help of RETScreen clean energy management software as calculated by Singh et al. (2020). The SPP in terms of the cumulative cash flow diagram for each country is presented in Figure 9.8(a)–(g).

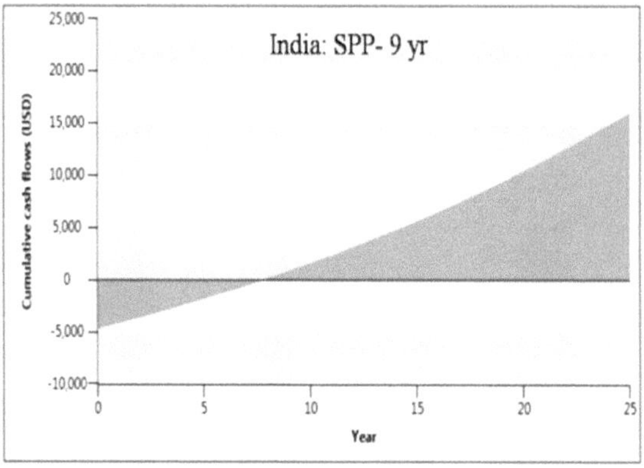

FIGURE 9.8(A) Cumulative cash flow for India.

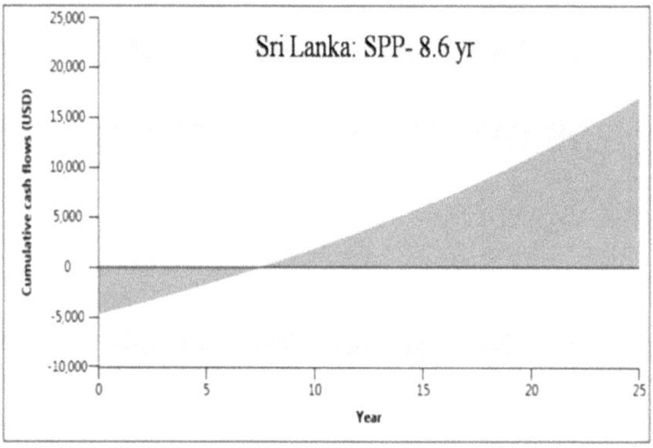

FIGURE 9.8(B) Cumulative cash flow for Sri Lanka.

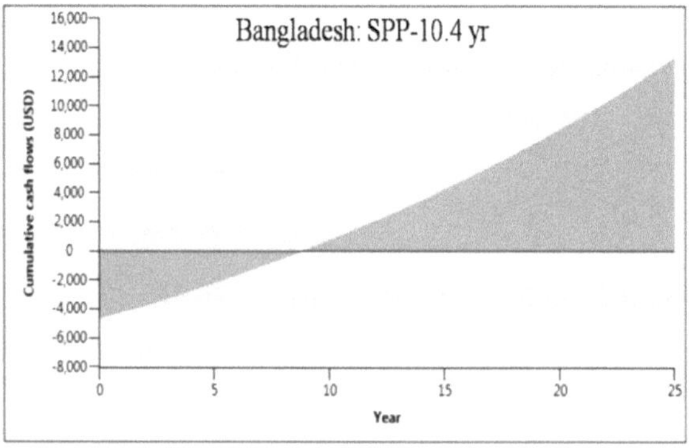

FIGURE 9.8(C) Cumulative cash flow for Bangladesh.

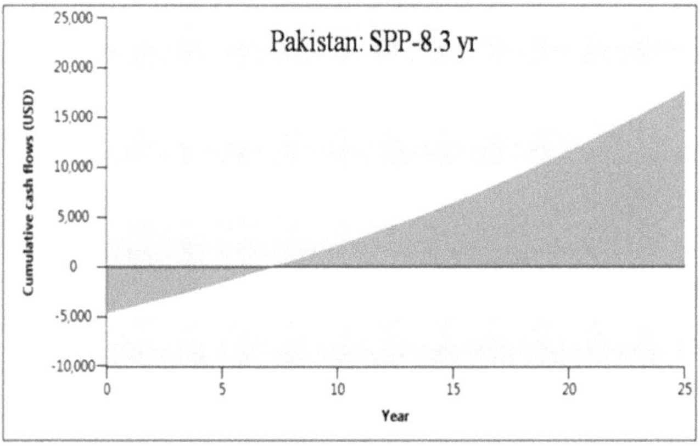

FIGURE 9.8(D) Cumulative cash flow for Pakistan.

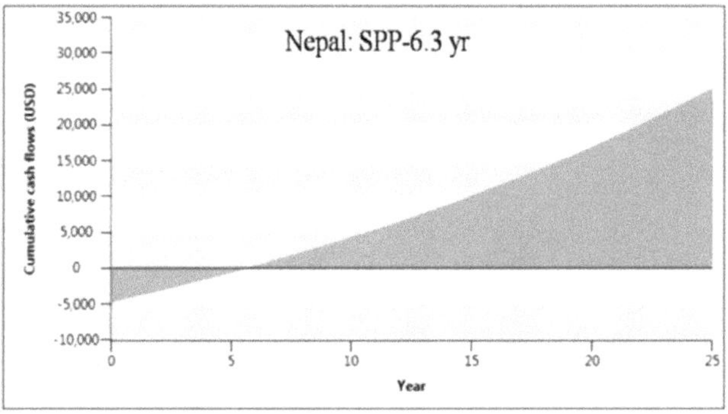

FIGURE 9.8(E) Cumulative cash flow for Nepal.

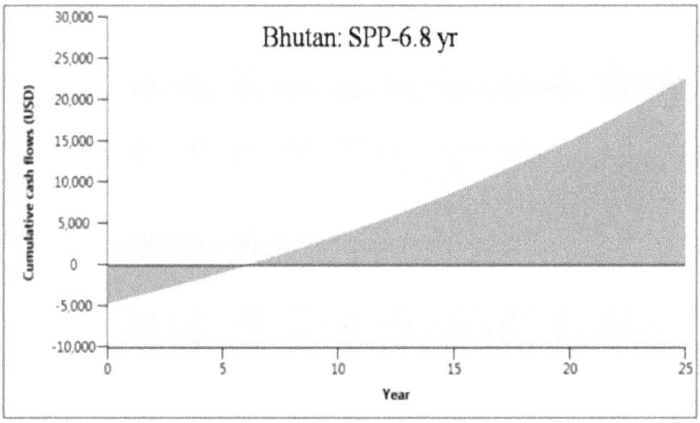

FIGURE 9.8(F) Cumulative cash flow for Bhutan.

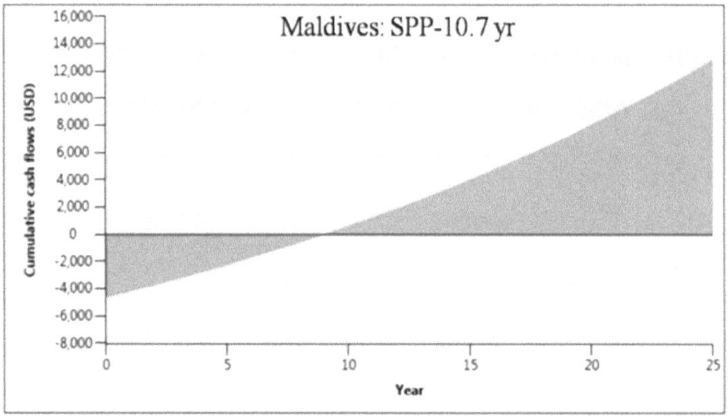

FIGURE 9.8(G) Cumulative cash flow for the Maldives.

The comparative values of SPP of all SACs are presented in Figure 9.9.

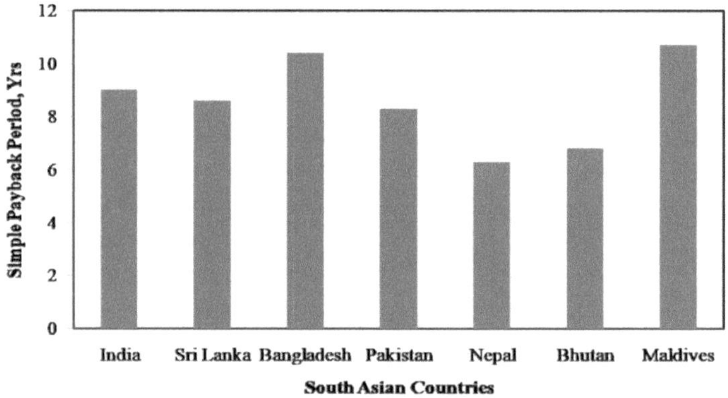

FIGURE 9.9 The SPP of SACs.

It can be evaluated from the figure that the maximum value of SPP is 10.7 years for the Maldives, and the minimum value is 6.3 years for Nepal. The second highest value is 10.4 years for Bangladesh. The other values of SPP are 9 years for India, 8.6 years for Sri Lanka, 8.3 years for Pakistan, and 6.8 years for Bhutan. The analysis indicates that the incentive and grants needed for at least three successively highest values of SPP for the countries of Maldives, Bangladesh, and India. The incentives and grants should be provided by the governments of their respective countries. The incentive and grant are a powerful mechanism for increasing the number of installations in the SACs. Even there should be some provision for developed countries to provide financial support to developing countries to foster such types of green energy technologies.

9.7 BARRIERS OF SWH SYSTEMS IN SACS

The SWH systems are admirable RE technology. However, this technology faces several barriers to execution in the SACs, as shown in Figure 9.10. The major hurdle to the implementation of this is the technology itself. There are so many technological improvements that need to be done (Sharma & Chen, 2009). The efficiency of SWH systems should be sufficiently high, and system cost must be low. Research is underway to improve the thermal efficiency of SWH. It has been shown through several studies that the thermal efficiency of SWH systems can be improved by the application of energy storage mediums like phase change materials (PCMs). Canbazog˘lu et al. (2005) have identified sodium thiosulfate pentahydrate as an effective PCM for improving the thermal efficiency of the storage tank of SWH systems. The efficiency increment will help in delivering the maximum amount of solar energy as heat. The selection of the optimum tilt angle of the collector around a year also plays an important role in the generation of heat.

Another big obstacle to the implementation of SWH systems in the SACs in the financial support given by these countries. The SACs are developing countries, and the majority of their funds are allocated to social security schemes. Therefore, the maximum RE projects in these countries suffer from a shortage of funds.

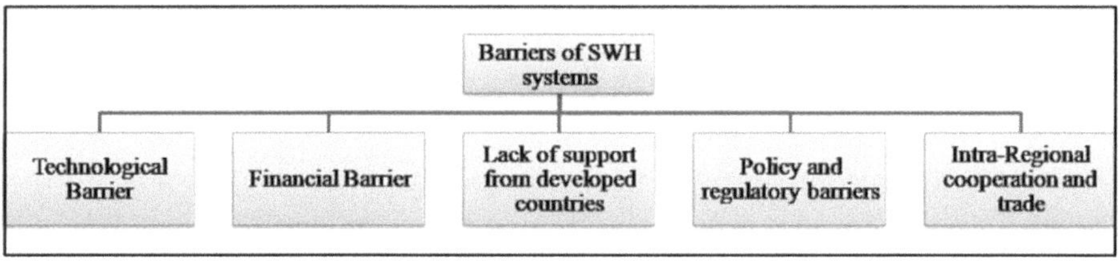

FIGURE 9.10 Barriers to the SWH system.

The developed countries should come forward to support SWH projects by considering them under the clean development mechanism (CDM). It is also advisable that SACs should help each other by making separate financial organizations to foster SWH systems in their countries. The forecast growth in energy demand is very high at SACs, so it is highly desirable to implicate the energy-efficient projects of RE like SWH systems for long-term investments in these countries. In this regard, the infrastructure also plays an important role in holding SWH systems on the rooftop so that the planning of implication of sustainable infrastructure for holding up SWH systems may also be equally enviable (Liu, Sheng, & Azhgaliyeva, 2019). The lack of support from developed countries in the form of finance and new technology is also a big impediment to nurturing SWH systems in SACs. This obstacle can be removed effectively by offering a favorable mechanism to support the SWH system by developed countries in the form of green incentives for SACs. The SACs are in the primary stage of making RE-based policies and regulations. This is not a good sign for these RE projects. These countries should increase the speed of making policies and regulations so that the framework will be on the stage of the application. These countries should consult the developed countries through the combined effort. The intraregional cooperation and trade among the SACs in the field of RE are negligible, which is also a barrier to increasing the number of installations at these locations.

9.8 THE POLICIES AND FINANCIAL SCHEMES

It is revealed by the studies that SACs are lagging in making effective policies for installing SWH systems. They are also not financing energy projects effectively. The awareness about the benefit of installing SWH systems in the public domain is very limited in these countries. This is not a good sign for the success of these projects. The survey revealed that only a few organizations are available for trading SWH systems, and most local manufactures are not promoted by governments because of price competition from the international market. In the international market, China is the world leader in making cheaper SWH systems, and it is very difficult for a local manufacturer to compete with Chinese products. The governments of these countries should make policies in favor of local manufacturers to promote the manufacturing hub in their own country. Such type of policy will save electrical energy, reduce pollution, and generate local employment. To boost the installation of SWH projects, there is a need for the advertisement of national policies along with an awareness campaign. The government should also offer the maximum amount of subsidy for such kinds of projects. The impact of installation will be high if the joint effort is taken by all SACs in this regard.

9.9 THE CLEAN DEVELOPMENT MECHANISM (CDM)

The CDM is an arrangement under the Kyoto Protocol to allow developed countries to purchase carbon credits from the RE projects of developing countries. It facilitates the developed countries to invest in new technology in developing countries. It is cheaper to invest in these RE projects in comparison to investing in their own countries for the mitigation of CO_2 emissions. The SWH systems have a great potential for mitigation of GHGs as they utilize the majority of solar energy for heating water apart from some percentage of fuel from electricity, generators, and combustible fuels, etc., as a surplus. The SWH systems are an eligible candidate for CDM projects because they have the capability of displacing GHG emissions directly. The installation of these systems will create employment due to the use of locally manufactured SWHs and promote additional savings. Moreover, the introduction of these systems will replace coal-generated electricity that will reduce emissions of suspended particulate matter, i.e., SO_x and NO_x (Purohit & Michaelowa, 2008). The application of the SWH system could be in three different sectors residential, commercial, and industrial, but one thing that is notable here is that the individual projects are very difficult to cover by CDM projects because of the low value of GHG emission reduction potential throughout the life span of the energy project. It is a major hurdle for CDM projects in SACs and even around the globe. The CDM Executive Board recognized this problem and adopted simplified CDM modalities and procedures for qualifying small-scale projects defined as (i) RE project activities with a maximum output capacity equivalent of up to 15 MW; (ii) energy efficiency improvement project activities that reduce energy consumption by an amount equivalent to 60 GWh per year; and (iii) other project activities whose emission reductions are less than 60 kt CO_2 per year. It is clear that in SACs, there is a large potential for the SWH system, but at present, these countries have a fewer number of CDM projects covered by SWH systems. In these situations, there is a requirement for a combined effort to increase the number of projects of the SWH system covered by the CDM projects. The registered CDM projects in the SACs are listed in Table 9.5.

TABLE 9.5 Registered CDM Projects in South Asia

Countries	Number	Percentage	Register	CER (Percentage)*
India	2,065	29	1,510	16
Pakistan	42	0.6	33	0.4

TABLE 9.5 *(Continued)*

Countries	Number	Percentage	Register	CER (Percentage)*
Sri Lanka	28	0.4	19	0.1
Nepal	8	0.1	6	0.05
Bangladesh	8	0.1	4	0.1
Bhutan	6	0.1	3	0

*CER (certified emission reductions) as a percentage of total CERs (1,786,252) issued to South Asian projects until 2012.

Source: UNEP DTU CDM/JI Pipeline Analysis and Database, cdmpipeline.org.

9.10 CONCLUSION

It is evident that SACs are in an energy crisis, and at the same time, these countries also have great potential for RES. But it is at present underutilized considering the vast potential. The SWH systems are not so much popular in these countries as they should be. The climate resource assessment of these countries shows that all countries have daily solar radiation greater than 3.5 kWh/m^2/d, which is extremely favorable for SWH systems. The cumulative cash flow graphs of each SAC with the help of RETScreen Clean Energy Management Software are also studied here, and it is found that the range of SPP is from 6.3 years to 10.7 years, which shows the economic feasibility of SWH systems in these countries. Despite having all favorable situations in these countries, the number of installations is very little and neglected by the respective governments of countries as well as by the international community. There is also a need to make a separate international committee for SAC to facilitate RE projects so that these projects can get certified easily and implemented effectively in these regions. Through this certification, the SWH projects could be of great interest and facilitate the purchase of carbon credits by developed countries, and it is expected that more projects will be covered by CDM.

ACKNOWLEDGMENT

The author (Abhishek Anand) is highly obliged to the University Grants Commission (UGC) and Ministry of Education, Government of India, New Delhi, for providing the Senior Research Fellowship (SRF).

KEYWORDS

- **certified emission reductions**
- **clean development mechanism**
- **energy-saving potential**
- **Kyoto protocol**
- **renewable energy**
- **solar energy**
- **solar water heating system**
- **South Asian countries**

REFERENCES

Ahmed S., Mahmood A., Hasan A., Sidhu G. A. S., & Butt M. F. U., (2016). A comparative review of China, India and Pakistan renewable energy sectors and sharing opportunities. *Renew. Sust. Energ. Rev.* , *57,* 216–225. https://doi.org/10.1016/j.rser.2015.12.191.

Bhuiyan, A. N., (n.d.). *Environmental Protection and Sustainable Development within South Asian Economies,* 1–13.

Canbazoğlu, S., Şahinaslan, A., Ekmekyapar, A., Aksoy, Ý. G., & Akarsu, F., (2005). Enhancement of solar thermal energy storage performance using sodium thiosulfate pentahydrate of a conventional solar water-heating system. *Energy Build, 37*, 235–242. doi: 10.1016/j.enbuild.2004.06.016.

Iftikhar, M. N., Najeeb, F., Mahazzam, S., & Khan, S. A., (2015). Sustainable energy for all in South Asia: Potential, challenges, and solutions. *Sustainable Development Policy Institute,* 1–33.

International Institute for Energy Conservation, UNEP Report. (2011). *Solar Water Heating Applications: Assessment of Country Successes – Draft.*

Kalogirou, S., (2009). Thermal performance, economic and environmental life cycle analysis of thermosiphon solar water heaters. *Sol. Energy, 83*, 39–48. doi: 10.1016/j.solener.2008.06.005.

Liu, Y., Sheng, Z., & Azhgaliyeva, D., (2019). *ASEAN: Impacts of Regional Asian Development Bank Institute.* ADBI Working Paper 1041, https://www.adb.org/publications/toward-energy-security-asean.

Natural Resources Canada, (2005). *Clean Energy Project Analysis.*

Panwar, N. L., Kaushik, S. C., & Kothari, S., (2011). Role of renewable energy sources in environmental protection: A review. *Renew. Sustain. Energy Rev., 15*, 1513–1524. doi: 10.1016/j.rser.2010.11.037.

Parks, N., Sea, R., Arabia, S., International T, Agency E, Lia, R. A., et al., (n.d.). *Australian Energy Resource Assessment – Chapter 10 – Solar Energy.*

Pranti, A. S., Iqubal, M. S., Saifullah, A. Z. A., & Ahmmed, K., (2013). Current energy situation and comparative solar power possibility analysis for obtaining sustainable energy security in South Asia. *Int. J. Sci. Technol. Res., 2*, 1–10.

Purohit, P., & Michaelowa, A., (2008). CDM potential of solar water heating systems in India. *Sol. Energy, 82*, 799–811. doi: 10.1016/j.solener.2008.02.016.

Sharma, A., & Chen, C. R., (2009). *Solar Water Heating System with Phase Change Materials, 1*, 297–307.

Shukla, A. K., Sudhakar, K., & Baredar, P., (2017). Renewable energy resources in South Asian countries: Challenges, policy, and recommendations. *Resour. Technol.,* 1–4. doi: 10.1016/j.reffit.2016.12.003.

Singh, S., Anand, A., Shukla, A., & Sharma, A. (2020). Technical, financial, and environmental feasibility of solar water heater for residential, commercial, and industrial application: A theoretical approach. *Mater. Sci. Energy Technol., 3*, 648–671. doi: 10.1016/j.mset.2020.07.001.

Sudhir, S., Ram, K., & Alison, D. (2014). *Low-Carbon Development in South Asia*, 60.

CHAPTER 10

Energy Sustainability Through Solar Dryer in South Asian Countries

ANKIT SRIVASTAVA,[1] ABHISHEK ANAND,[1] AMRITANSHU SHUKLA,[2] and ATUL SHARMA[1]

[1]*Non-Conventional Energy Laboratory, Rajiv Gandhi Institute of Petroleum Technology, Jais, Amethi, Uttar Pradesh, India*

[2]*Department of Physics, University of Lucknow, Lucknow, U.P., India*

ABSTRACT

In South-Asian countries (SAC), agricultural products are dried under the open sun. However, owing to interference from external contaminants and uneven drying rates, this method of drying affects the quality of the dried goods. Solar dryers of many sorts have been created and developed in different regions of the world. Numerous types of solar dryers have been designed and developed in various parts of the world. The solar drying system is utilized to remove moisture from food and other crops, which require quality maintenance and preservation over a long period. This sustainable way of drying saves the consumption of valuable conventional fuels in the SAC region, where energy is costly and highly imported. In this chapter, the authors provide information about the various solar drying techniques with due focus on the status of solar drying in the SAC region.

10.1 INTRODUCTION

The SAC region includes developing countries like India, Sri Lanka, Pakistan, Bhutan, Nepal, Afghanistan, and the Maldives. The speedy increase

The South Asian Energy Corridor: Toward Sustainability and Security. Amritanshu Shukla, PhD, Atul Sharma, PhD & Saurabh Mishra, PhD (Eds.)

in social and economic development has augmented the demand for energy in this region. In these countries, there is a huge mismatch between the demand and supply, and therefore, it is very difficult to attain the new social and economic development target for 2020 set by the respective governments (Shukla, Sudhakar, & Baredar, 2017). The use of large-scale fossil-based energy resources has caused some serious threats like greenhouse gas emissions, ozone layer depletion, and climate change. To decrease overdependence on fossil fuels and to manage the growing energy demand, the SAC countries have initiated several efforts to switch to renewable-based energy resources. Because of large-scale solar energy availability, renewable energy (RE) is a conspicuous choice to manage their electricity supply and to minimize greenhouse gas emission. In solar PV/T technology, the part of radiation falling on the collector can be used to generate both heat and electricity. The heat energy generated can be applied for space heating and other industrial processes. The electricity generated can be either utilized on hand, stored, or exported to the grid (Palit, 2013). Wind energy can also be used to generate electricity using kinetic energy. It has two major disadvantages, i.e., high initial costs and technology's immaturity. Besides that, the construction of the turbines is extremely expensive. Other disadvantages include esthetic impact, the remoteness of the location, noise, etc. Due to these drawbacks, solar energy is convenient in comparison to wind energy. There are two ways to use solar energy, i.e., solar thermal and photovoltaic (PV) (Mustayen, Mekhilef, & Saidur, 2014).

Solar energy can also be used for the drying of agriculture products. Drying is done to extract moisture content from the agricultural product, which helps in preserving the agricultural products for a longer time. Post-harvest losses in perishables are projected to reach 30–50% in SACs where cold storage facilities are insufficient. Drying is a traditional way of preserving food. It makes the transportation process easier as the dried product becomes lighter due to the loss of moisture. Natural sun-drying is quick and economical but suffers from some serious difficulties, such as uncontrolled drying rate, non-uniform drying, food degradation due to fungi and bacteria, and harm due to birds, insects, rodents, etc. The solar drying technique eliminates the drawbacks of natural sun drying, making the process of preservation much simpler, increasing the availability of certain commodities, increasing storage space, and ensuring cost-effective transport.

This chapter aims to provide information on solar drying in SACs. As a result, solar energy may be used to dry various agricultural products while preserving the drying product quality, which can also reduce the use of other drying energy sources in the South Asian region.

10.2 ENERGY SCENARIO IN THE SOUTH ASIAN REGION

Energy generation and energy security are the wheels of a country's economic development. Table 10.1 shows the statistics of electricity use per capita of SACs.

TABLE 10.1 Comparison of Energy Use per Capita in SAC (Palit, 2013)

Country	Per Capita Electricity Consumption in kWh
India	597
Sri Lanka	416
Pakistan	451
Afghanistan	35
Bangladesh	228
Nepal	91
Maldives	–
Bhutan	–

Table 10.2 shows the statistics of various energy sources utilized by SACs for the fulfillment of their energy needs.

The requirement of energy is going to increase in SACs in the near future due to increasing population and economic development. While meeting future energy demands, energy security and sustainability should also be considered. Solar energy, as we all know, may be employed in a number of ways to reduce the usage of nonrenewable resources. Agriculture product drying is done for food preservation and sometimes for consumption. As a result of using a solar dryer system to dry agricultural goods, the usage of non-renewable resources is reduced, and the quality of dried products is improved.

10.3 DRYING TECHNOLOGY

The majority of the population around the globe is undernourished. The widespread reduction in soil fertility caused by improper agriculture and excessive fertilizer usage reduces production. The massive rate of deforestation, lack of sophisticated equipment, and shortage of high-grade seeds further add to the misery. The lack of distribution outlets, inappropriate way of transportation, high post-harvest losses due to pests and spoilage, and the

TABLE 10.2 Energy Generation from Various Sources of Energy in major South-Asian countries (Salam et al. 2020)

Country	Oil (Unit in TWh)	Natural Gas (Unit in TWh)	Coal (Unit in TWh)	Nuclear (Unit in TWh)	Hydro (Unit in TWh)	RERs (Unit in TWh)
India	339.5	407	82.5	20.9	81.4	9.3
Pakistan	2583	539.6	4931	98.8	357	253.5
Bangladesh	87.2	266.3	26.7	Nil	2.35	1.16

lack of adequate drying and Storage facilities aggravate the existing situation even further. As a result, the focus should be on creating efficient drying processes that would allow farmers to produce high-quality commercial goods. A dryer's main function is not just to dry faster but to enhance the quality of the product. The type of dryer used and the way it is handled have a major impact on product quality. Solar drying appears to be a potential alternative for addressing present post-harvest issues in poor nations such as the SACs region (Visavale, 2012).

10.3.1 NATURAL SUN-DRYING

Outdoors and uncontrolled open sun drying (OSD) is still the most widely used technique to save and process agricultural items. The uncontrolled drying experiences several issues like breeze conceived dust, invasion by creepy-crawly, and the items sometimes get degraded to such a level that occasionally become less market esteem. Over-drying, lack of drying, and pollution from remote materials are also important drawbacks of normal solar drying. As a result, the nutritious value of the agricultural product is reduced. The decline in quality produces a loss of value in international markets, resulting in financial losses.

10.3.2 SOLAR DRYING

Solar drying is a more efficient method of preserving farm products. In this case, the utilization of solar energy for controlled drying resulted in an overall improvement in product quality. The device used to dry agricultural products is known as the solar dryer.

10.3.3 CLASSIFICATION OF SOLAR DRYERS

Different types of drying equipment are available. The two most significant categories are based on (1) the heat transmission mechanism of wet solids and (2) the handling qualities and physical features of the wet substance. The first technique shows variations in dryer design and function, but the second way is more effective in identifying a set of dryers for preliminary evaluation in a specific drying situation. The drying equipment heat transfer categorization diagram is given in Figure 10.1.

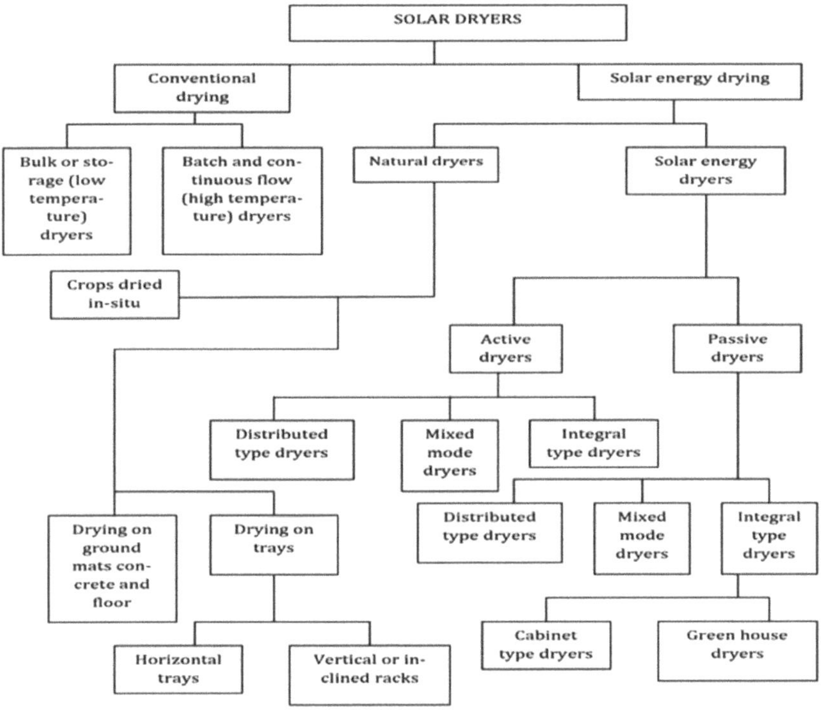

FIGURE 10.1 Classifications of solar dryers and drying modes (Sontakke & Salve, 2015).

There are three distinct sub-classes of active or passive solar drying systems:

- Direct-type solar dryers;
- Indirect-type solar dryers; and
- Hybrid solar dryers.

Figure 10.2 shows the main characteristics and the standard designs of the various solar-energy dryers.

10.3.3.1 OPEN SUN DRYING (OSD)

The operating concept of solar-powered open-air drying is depicted in Figure 10.3. The crops are often spread on the ground, sheet, or cement floor, where they get short solar wavelength radiation as well as natural air circulation for most of the day. A portion of the energy is reflected back, while the remainder is absorbed by the surface, depending on the color of the crops. The absorbed radiation is transformed into thermal energy, raising the temperature of the substance. As a result of the moist air blowing over the crop field, there are certain losses, such as the loss of long wavelength radiation from the crop surface to ambient air as well as convective heat loss. The OSD process does not usually meet the requisite quality standards, and it is often difficult to market the goods internationally. A more sophisticated method of utilizing solar energy in crop drying has come out called solar drying due to insufficiencies involved in OSD (Ekechukwu & Norton, 1999).

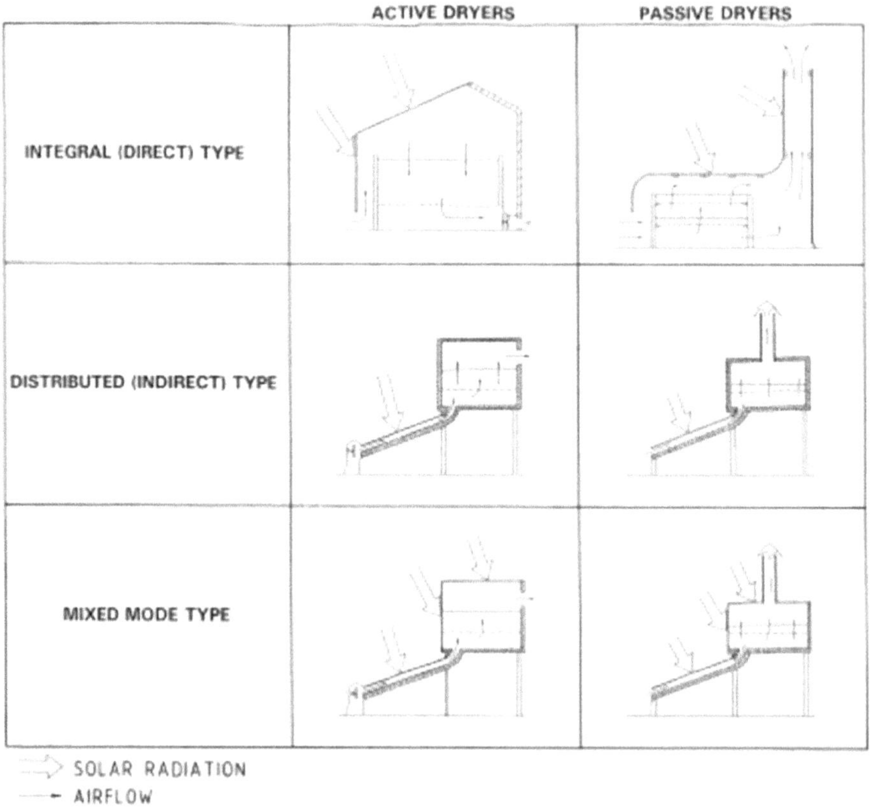

FIGURE 10.2 Typical configuration of solar energy dryers (Prakash & Kumar, 2013).

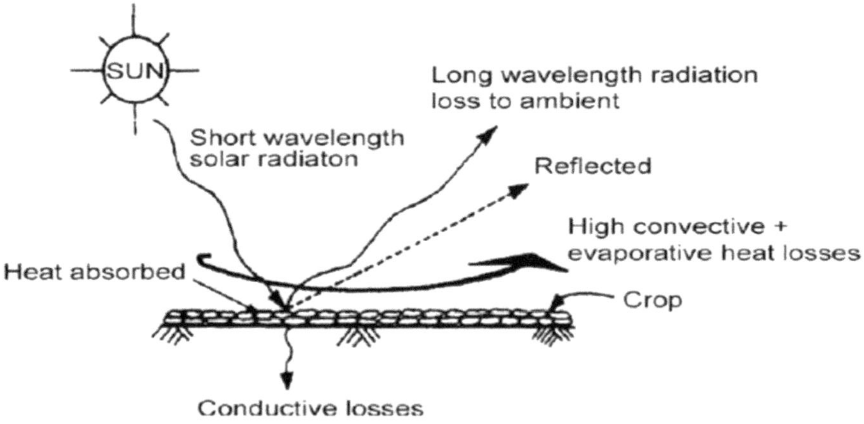

FIGURE 10.3 Open sun drying working principle (Ekechukwu & Norton, 1999).

10.3.3.2 *DIRECT SOLAR DRYING (DSD)*

The working principle of direct drying is shown in Figure 10.4. As indicated in the diagram, air from below enters the cabinet and exits through the top exit. It is separated from the moisture here. There is some amount of radiation that is reflected in the atmosphere, and the remainder is transferred through a cabinet dryer. The transmitted radiation is reflected from the cultivated surface, and the rest is absorbed by the plant's surface, which raises its temperature while releasing long wavelengths of radiation. Due to its glass cover, this radiation cannot escape into the atmosphere. The temperature within the cabinet is often greater than the temperature outside. The glass cover in the dryer cabinet therefore limits direct convective losses to the outside, which plays a significant role in boosting crop and cabinet temperatures.

Solar drying has the following benefits over open-air drying:

- Construction is simple and less costly than the indirect type;
- Provide rain, dew, debris protection, etc.
- The following limitation applies to a cabinet dryer:
- Responsible for overheating and crop damage;
- Poor steam removal rates that lead to relatively slow drying rates;
- Low power limits and thus applications of small size;
- Crop discoloration by the direct sun radiation exposure;

- Condensation of humidity within the covering of glass decreases transmissibility;
- The inadequate temperature rise influences the removal of moisture.

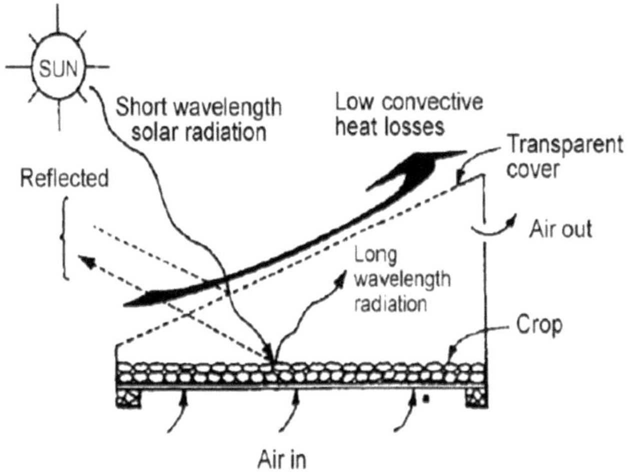

FIGURE 10.4 Direct solar drying working principle (Ekechukwu & Norton, 1999).

10.3.3.3 INDIRECT SOLAR DRYING (ISD)

They differ in terms of heat transfer and vapor removal from direct dryers. Figure 10.5 describes indirect solar drying's (ISD) working theory. The plants in these indirect solar panels are placed inside an opaque drying case on a rack or shelves, and a single device is used to warm the air in the cabinet known as the solar panel. The heated air will flow through and through the wet crop providing heat for humidity evaporation by convective thermal transmission between the hot air and the wet crop. The difference in humidity between drying air and the air present near the crop surface causes the drying of the crop.

ISD has the following advantages:

- Provides better drying control and a higher quality of the product obtained than drying with the sun.
- No localized heat damage and caramelization occur, as the crops are covered and open to direct radiation.
- Can be worked at higher temperatures for deep drying layers.

- Recommended for the crops that are photosensitive.
- High power than direct drying. However, these structures are comparatively complex, requiring additional capital investment and a higher maintenance price than direct drying units (Sharma, Chen, & Vu Lan, 2009).

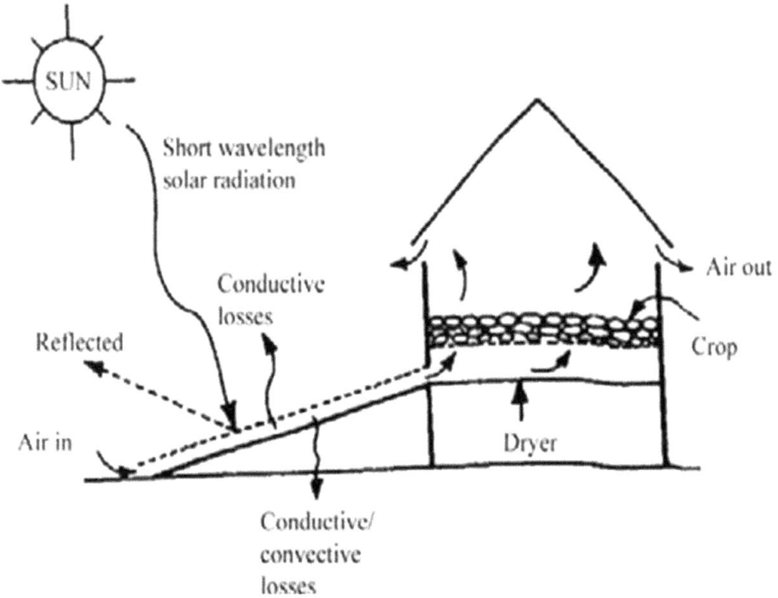

FIGURE 10.5 Indirect solar drying system working principle (Sharma, Chen, & Vu Lan, 2009).

10.3.3.4 HYBRID SOLAR DRYING (HBD)

Direct and indirect forms of solar panels are combined in hybrid solar dryers, resulting in a unique product. Thermal energy for drying is generated by combining direct solar radiation to dry the product with preheated air in a collector heater.

10.3.4 TYPES OF SOLAR DRYER (METHOD)

Direct or indirect solar dryers are categorized into two types based on drying methods: (i) passive dryers, which dry directly from the sun's rays with or without natural air circulation; and (ii) active solar dryers, which circulate hot air using a fan (forced convection).

10.3.4.1 PASSIVE SOLAR DRYING SYSTEMS

Using buoyancy, wind pressure, or a mix of both, passive solar dryers heat and circulate air naturally. Passive mode is used by basic dryers and greenhouse dryers alike. This method is still used in many Mediterranean, tropical, and subtropical regions of the world, including Africa and Asia, as well as tiny farming communities in the United States of America. These materials are basic, inexpensive, easy to install, and may be used in remote regions without access to electricity. The materials are easily accessible in the surrounding region. When it comes to drying little amounts of fruit and vegetables such as bananas, mangoes, and potatoes using a passive dryer, you can't go wrong.

10.3.4.2 ACTIVE SOLAR DRYING SYSTEMS

External methods such as fans or pumps are used in active solar drying systems to transfer warm air from the collector to the drying beds. Because they employ forced convection, all active solar dryers use forced convection. A traditional active solar dryer only relies on solar energy for the heat source, while motorized fans or ventilators are used for air circulation. These dryers are widely used in big commercial drying operations, where they are used in conjunction with standard fossil-fuel dryers to improve drying control by reducing the influence of variations. Active solar dryers are said to be good for drying foods with greater moisture content, such as kiwi fruit, papaya, brinjal, cauliflower slices, and cabbage. Primary, indirect, and hybrid solar-energy dryers are the three types of active solar-energy dryers (Umesh & Karale, 2013).

10.4 ROLE OF SOLAR DRYER IN SAC

Open-air and uncontrolled sun-drying is still the most frequent way of preserving and processing agricultural goods in SACs. Uncontrolled drying, however, suffers from a serious problems such as windborne dust insect infestation, a product can be severely degraded to such an extent that it sometimes becomes less market worth and, resulting in loss of food quality that can have adverse economic effects on the domestic and international market. Thus, the use of solar dryers is beneficial for maintaining the quality of agricultural products as well as requires less maintenance. The source of energy is sunlight, which is free, and so increases the wealth of farmers in SACs.

The main advantages of the solar dryer are listed below:

- Improved quality of products is achieved;
- It reduces losses and lowers the market price of the products;
- Products will be protected against flies, rain, and dust;
- Prevents fuel dependence and reduces environmental impact;
- It is more efficient and cheaper.

A wide range of agricultural goods can be dried in SACs. India is a large producer of dry agricultural goods such as preserved onions, cucumbers, and gherkins. Provisionally preserved items include mushrooms and truffles, dried truffles, dried mangoes, dried ginger, asparagus dried, dehydrated garlic powder, dehydrated garlic flakes, dried garlic, dried potatoes, and dried spices, etc. (Sharma, Chen, & Vu Lan, 2009).

In Pakistan, mainly dry fruits (especially Oranges and Mangoes), vegetables, and various spices are to be dried. Bangladesh crop that needs drying are grapes, mangoes, chilly, ginger, garlic, and other spices. In Nepal, Bhutan, Sri Lanka, and Maldives, numerous crops need to be dried for preservation as well as for consumption. So, the uses of the solar dryer in SACs enhance the quality of agriculture products after drying and also the wealth of the farmers.

Many other uses of solar dryers include:

- Fruits (apples, apricots, grapes, pineapples, bananas, melons, plums, etc.), beets, mangoes, dates, and figs.
- Vegetables (cabbage, broccoli, peppers, herbs, onions, squash, tomatoes, asparagus, celery, etc., potatoes, peas, carrots, pepper, cassava, yams, red cedar, mahogany.
- Corn, maize, rice, cassava, and cocoa.
- Fish, meat, and mushrooms.
- Spices (dry chili peppers, garlic).
- Tea, coffee, cocoa, tobacco, cashew, and macadamia tree, milk, hay, coconut.

- Clothing, wool, lighting, and co-treatment of wood (Abu-Hamdeh & Alnefaie, 2019).

Meat drying necessitates extra caution due to its high protein content, which promotes microbial development, e.g., drying fish involves careful purification of each unit of the dryer. Solar dryer enables rapid drying and prevents the loss of nutrients. No overlap of two layers of food dried in a solar dryer can occur (Abu-Hamdeh & Alnefaie, 2019). Dryness reduces in rainy weather. The dryer is quickly cooled by rainy water, which causes the condensation to produce moisture on the cover film. It'll take some time to work properly again after a break.

- Typical drying times range from 1 to 3 days, for big slices may be more.
- Depending on water content, drying in smaller batches may be required.
- A rainy day can also be handled by using proper Ventilator systems.
- Proper drying and storage in the proper air will retain freshness for six to 12 months.

The number of agricultural products can be stored for six to 12 months while preserving the quality of food with the application of sun dryers in SACs. There are different designs available for the solar dryer that can be used depending on the product to be dried (Bal, Satya, & Naik, 2010). Different products have different drying times. The solar dyer should be accordingly chosen so that maximum efficiency can be achieved. Major agriculture products of SACs and their drying time are listed in Tables 10.3–10.7.

TABLE 10.3 Fruits Drying Time and Dry Status (Stefan, 2019)

Fruits	Dry-Time/h	Preparation	Degree of Dry-Status
Apples	8	About 10 mm thick disks	40%
Pears	10	Quartered	25%
Apricots	12	Halve	30%
Cherry	11	Completely	40%
Prunes	12	Halve	35%
Peaches	10	Thin slices	20%
Plantains	8	8 mm little slices	50%
Whortleberries	10	Completely	30%

TABLE 10.3 *(Continued)*

Fruits	Dry-Time/h	Preparation	Degree of Dry-Status
Brambles/ raspberries	12	Completely	25%
Strawberries	14	Halve	25%
Elder	12	Completely	30%

TABLE 10.4 Drying Time and Dry Status of Vegetables (Kant et al., 2016)

Vegetables	Dry-Time/h	Preparation	Degree of Dry-Status
Carrots	6	Small piece	25%
Pod pepper	6	10 mm strips	30%
Tomatoes	8	Little slices	25%
Cauliflower	6	Blanch in small pieces	50%
Beans	5	Small slices	60%
Onion-vegetables	5	Small pieces	50%

TABLE 10.5 Grains Drying Time and Dry Status (Pirasteh et al., 2014)

Grains	Dry-Time/h	Preparation	Degree of Dry-Status
Maize	6	Whole grains	80%
Grains	5	Without shell	90%
Seeds, kernels, nuts	4	Without shell	90%

TABLE 10.6 Herbs Drying Time and Dry Status (Rabha, 2017)

Herbs	Dry-Time/h
Solid, brittle herbal types	2–3 weeks in dry air
Smoothly leaves herbal types	In the dryer with moderate heat, 4–6 h

10.4.1 ROLE OF THE SOLAR DRYER IN INDIA

Solar drying is the amendment of the old sun-drying technique which is useful for all farm commodities. Solar dryers were originally used in the Republic of India to dry non-agricultural items before being expanded to agricultural products. Table 10.7 shows the chronology of the works in the solar drying system in India.

TABLE 10.7 Early Work Done in the Solar Drying System in India (Tiwari, 2016)

Year	Place	Product	Dryer Type
1954	National Physical Laboratory (NPL) New Delhi	Coal	Solar heated air
1955	National Physical Laboratory (NPL), New Delhi	Jaggery from sugar cane juice and palm juice	Plane glass mirror concentrators
1968	Khadi and Village Industries Commission (KVIC), Ahmedabad	Jaggery from palm Nira	Plane glass mirror concentrators
1972	Forest Research Institute (FRI), Dehradun	Timber	Solar Kiln
1972	Central Arid Zone Research Institute (CAZRI), Jodhpur	Fruits and vegetables	Solar cabinet dryer
1978	Annam alai University	Paddy	Solar paddy dryer
1980	National Industrial Development Corporation (NIDC)	Grain	Solar grain dryer

In India, around 80% of the farmers are poor and marginal producers. They suffer from unsteady market costs. Farmers are not in a position to preserve their produce even for a week owing to high electricity prices, poor infrastructure, and lack of funding for storage facilities. When market prices are poor, farmers normally have to be forced to throw away their produce, leading to substantial waste. Some farmers have even committed suicide as a result of the increasing debt (Masoom, Kosmopoulos, Bansal, & Kazadzis, 2020).

Farmers may boost their revenues by using low-cost technologies like solar dryers. Simultaneously, urges nations to decrease food losses and progress toward resource development goals, particularly national food loss and waste (Kumar, Singh, Prakash, & Ashutosh, 2014).

The first move was to establish a demand for dried goods. This secured return on investment made it possible for farmers to receive a loan from an organization, a self-help group, or a bank. Within 1 to 2 years, they have been able to buy the dryer. It is now possible for farmers to tailor their earnings to market prices. If the price levels on the market are poor, the goods will dry up and wait at least six months for a better price. Farmers can sell it immediately when the prices are high. There are no additional power costs due to using solar energy. Farmers can thus increase their income because the solar dryer uses available and free solar energy. Solar dryers of various sorts are now in use across the world to dry a variety of agricultural goods.

- **High Efficient Solar Dryer:** The solar drying device uses solar energy to heat the air and to dry any food material or an alternative source powered dryer, which is useful in minimizing the waste of agricultural products or other items and helps to conserve the product.

10.4.1.1 SOLAR CABINET DRYER

The Solar Cabinet Dryer appears to be a heated box that may be used to dry a broad variety of items. It comprises a parallelogram box insulated at the base and coated with a double-bedded clear roof, probably a strengthened glass. At the bottom of the box, ventilation holes are created to flow into the fresh air. The outlet is provided at the side and rear panel of the box. The box can be made of M.S. or wood, according to the requirements, and the inner rear panel can be coated with aluminum sheet, foil, or any other reflective material for increased efficiency. The hot air, step by step, removes the wet content of the product. This dryer has wide applications in food, herbs, and spices.

10.4.1.2 SOLAR TUNNEL DRYER

The Solar Tunnel Dryer is one of the solar dryer device models and is one of the best efficiencies because of direct conduction heating and solar radiation heating, and it can be used for limited drying capacities of approximately 10 kg per day to 500 kg per day. The dryer is suitable for drying garlic, ginger, cassava, jackfruit, carrot, etc.

10.4.1.3 SOLAR HERB DRYER

This dryer is made specifically for drying herbs. The size of a dryer might vary depending on the amount of drying capacity required. Temperature within the dryer is maintained as per the requirement for drying any particular herb.

10.4.1.4 INDUSTRIAL SOLAR DRYER

These solar dryers have a large capacity for drying various food products. In these solar dryers, the temperature is controlled according to the food to be dried. These sorts of dryers have a significant starting cost.

10.4.1.5 SOLAR TENT DRYER

This solar dryer is constructed with the use of a tent which makes its initial cost lower, but somehow, it takes larger time for drying than other efficient dryers.

10.4.1.6 FISH DRYING SOLAR DRYER

Traditionally, the drying of fish was achieved by drying it in the open sun. Depending on the type of fish, this typically needs around 5–10 days. In addition, it also takes a long time, induces pollution by dust and other contaminants, and is unhygienic, producing a lot of odors as well. Solar Dryer not only does it dry the fish in 6 to 10 hours, but it also keeps fish hygienic, eliminates the possibility of contamination with dust particles and harmful bacteria, and keeps it free of droppings of birds and rats, etc. As solar dried fish is not only very clean but can also be kept for a longer period and has a higher product value than usually dried fish, the use of solar dryers also gives fishermen good income (Lakhani & Patel, 2013).

10.4.2 ROLE OF SOLAR DRYER IN SRI LANKA

Sri Lanka is an island in the Indian Ocean with latitudes of 6° and 9°50’ and longitudes of 79°50’ and 82°50.’ The overall land area of the island is 6.5 million hectares, with an estimated 19 million people (Senadeera & Kalugalage, 2004). Sri Lanka is both a manufacturer of dry goods and a consumer of them. Today, agricultural crop production has been boosted by increased demand for dehydration and process raw materials. In many dynamic global markets for food, the Country would have a promising future if the right technology were adopted to boost food quality and diversity. Food dehydration locally acts as a preventive measure for food shortages as well as a means of generating high revenue through the development of value added. Cost and quality are key factors in the drying of food, as it is solely an energy method. On the contrary, in Sri Lanka’s rural areas, farmers are responsible for sun-drying perishables in the open air, which creates a number of issues for them. Open sun-drying can be time-consuming, with a risk of pollution, infestation, and attack by a microorganism. When compared to sun drying, solar dryers can have greater air temperatures and lower relative humidity, which can enhance drying speeds while also lowering the ultimate humidity content of the dried product. As a consequence, both throughout the drying process and during storage, the danger of spoiling is minimized. While the necessary quality levels can be achieved by the use of electrical dryers, they are not economically feasible due to the high cost and lack of availability of national grid connections in the rural sector (Weerasena & LA, 2020). Although the use of petroleum and electricity in rural agriculture-based businesses has risen in Sri Lanka, the utilization of these energy sources for dehydration at the community level is currently not practical. Due to climatic conditions and soil fertility, Sri Lanka developed the following crops and fruits. The major crop of Sri Lanka, which needs to be dry, is listed in Tables 10.8–10.10.

TABLE 10.8 Major Crops, Their Extents, Production, and Trade (Weerasena & LA, 2020)

Crops	Area 1,000 Ha	Production 1,000 Mt	Export 1,000 Mt	Imports 1,000 Mt
Tea	195	284	268	–
Rubber	159	97	43	–
Coconut	439	2,808 mil. nuts	916.48 mil. nuts	–
Rice	781	2,868	1.53	214

TABLE 10.9 Production and Area of Major Fruit Crops in Sri Lanka (Weerasena & LA, 2020)

Crops	Area (ha)	Production (1,000 Fruits)
Banana	48,075	3,310,600 bunches
Cashew	21,218	1,898,326
Lime	6,955	1,17,663
Mango	25,800	4,31,214
Orange	3,464	23,998
Papaya	3,476	26,874
Passion fruit	425	6,202
Pineapple	4,774	32,626

TABLE 10.10 Production of Export Crops (Weerasena & LA, 2020)

Crop	Production (Mt)
Coffee	2,422
Cocoa	1,538

TABLE 10.10 *(Continued)*

Crop	Production (Mt)
Cinnamon bark	11,503
Leaf oil	139
Pepper	4,557
Cloves	1,902
Cardamom	70
Citronella oil	108

Solar drying of food crops is a crucial field for Sri Lankan rural economic development. The economic benefit, however, is negligible as a result of the lack of product consistency and quantities with the drying methods currently in use. Present solar dryers need several days of drying so that the moisture content of the perishable products can be reduced to safe levels, which produce unfavorable products. The dryer was well embraced as an alternative to conventional drying of the sun in the rural field. The dehydrated goods produced were of acceptable quality in terms of flavor characteristics and customer acceptance. Due to the use of solar dryers, farmers of Sri Lanka can increase their income, and the quality of the dried products is also good, which increases the market acceptability of dried products.

10.4.3 ROLE OF THE SOLAR DRYER IN BHUTAN

Bhutan agriculture plays a leading role in the economy of Bhutan. For most of the population, agriculture tends to be the main means of livelihood. Agriculture employs around 80% of Bhutan's population. More than 95% of women in the nation work in agriculture. The majority of the refugees in this Himalayan country work in agriculture (Lhendup, 2005). Maize and rice are the two main crops grown in Bhutan. Rice accounts for 43% of total domestic cereal cultivation, whereas maize accounts for 49%. Rice is the most important crop in the country. Wheat and other minor cereal crops are grown in the country's agricultural sector. In areas with adequate irrigation, paddy is the main crop. In addition to paddy, these fields are planted with wheat, barley, oilseeds, potatoes, and vegetables. In dryland areas, maize is typically cultivated at a lower height. In order for fertility to be created, foresters in the nation serve as a source of livestock feed and organic materials. Organic farming in Bhutan is encouraged by the governing bodies.

Bhutan is now generating 100% organic farming in 2020 (Department of Agriculture Ministry of Agriculture and Forests, 2016).

Solar thermal systems in the agricultural sector have been demonstrated to be practical, cheap, and ecologically beneficial in Bhutan, where they are used to preserve vegetables, fruit, and other products. Solar heating systems for drying food and other crops can enhance product quality while decreasing waste and traditional fuel use, therefore improving quality of life. Solar drying is a cost-effective and efficient method of preserving food and agricultural commodities. Bhutan produces organic agriculture products, so the demand for high-quality dried agriculture products is higher than in other countries (Raj, 1973). Major crops, vegetables, and fruits of Bhutan are listed in Tables 10.11–10.13, which required drying for food preservation while maintaining the quality of dried agriculture products.

TABLE 10.11 Cereal, Oilseeds, Spices, Legumes, and Pulses, and Roots/Tubers

Crop Type	Crop Name	Harvested Area (Acres)	Production (MT)	(Kg/acre)
Cereal	Paddy	53,055	85,090	1,604
	Maize	56,609	82,035	1,449
	Wheat	3,717	2,521	678
	Barley	2,451	1,702	694
	Buckwheat	6,897	3,705	537
Oilseeds	Mustard	2,395	892	370
	Groundnut	208	149	718
	Soya bean	544	254	466
	Sunflower	15	7	474
	Perilla/Naam	41	12	282
Spices	Cardamom	11,086	2,736	247
	Ginger	4,773	10,871	2,278
Legumes and pulses	Rajma Bean	1,565	994	635
	Mung Bean	952	482	506
Roots and tubers Tapioca	Sweet potato	31	29	940
	Tapioca	278	415	1,490

Source: Agriculture in Bhutan (2020).

TABLE 10.12 Vegetable Production in Bhutan

Crop Name	Cultivated Area (Acres)	Quantity Produced (MT)	Yield (Kg/Acre)
Asparagus	417	239	574
Chili	5,538	9,907	1,789
Cabbage	2,738	6,685	2,442
Cauliflower	1,512	2,082	1,377
Carrot	607	1,276	2,103
Radish	2,871	6,490	2,261
Turnip	1,603	10,499	6,551
Beans	3,385	4,409	1,302
Peas	795	1,014	1,275
Tomato	347	455	1,310
Broccoli	725	1,004	1,385
Eggplant	408	585	1,433
Lady's finger	43	42	964
Green leaves	1,458	1,937	1,328
Onion bulb	442	414	935
Garlic	1,409	1,176	835

Source: Bhutan: Organic by Country Profile and Land Use (2020).

TABLE 10.13 Fruit Production in Bhutan

Commodities	Total Trees (No's)	Bearing Trees (No's)	Production (MT)	Yield (Kg/Bearing Tree)
Apple	2,42,903	1,96,708	6,587	33
Mandarin	1,665,797	8,82,807	42,003	48
Areca nut	1,423,208	7,26,075	9,467	13
Mango	82,153	23,494	644	27
Pear	39,575	16,726	963	58
Peach	27,087	18,131	972	54
Plum	15,849	8,920	376	42
Walnut	24,072	7,984	181	21
Jackfruit	11,113	5,406	775	143
Guava	36,405	26,360	665	25

Source: Agriculture in Bhutan (2020).

The tables give information on the numerous crops, vegetables, and fruits grown by Bhutanese farmers. Solar dryer helps farmers to increase their income since Bhutan produces 100% organic food, so the demand for dried agriculture products is higher than in other countries, which are also good for Bhutan farmers.

10.4.4 ROLE OF SOLAR DRYER IN NEPAL

Most of Nepal's economy is based on agriculture. Rural areas are home to the bulk of the population, which is heavily reliant on agriculture for a living. Agriculture accounts for 33.9% of the GDP. The agricultural system is based on subsistence farming and mixed farming, with smallholder farmers dominating the landscape. About 80% of the population is employed in agriculture (Raj, 1973). A considerable proportion of post-harvest losses in fruit and vegetables may be of great economic concern in Nepalese agriculture. As a result of post-harvest loss, the horticultural product loses their quality and quantity at any point along the post-harvest chain, which includes the change in edibility and wholesomeness or quality of the food, which ultimately prevents its use. Fruit and vegetable post-harvest losses in Nepal are estimated to be between 20% and 50%. Within the harvesting and retail markets, goods are subject to a range of processes under varying environmental factors, including transport and storage. Post-harvest losses during handling, transportation, storage, and delivery are major concerns for the marketing of consumable fruits and vegetables. This loss, which leads to low per individual supply, not only indicates the loss of a significant amount of capital but also increases transport and marketing costs (Joshi, Pradhan, & Pathak, 2000).

However, given the low income of the rural population, the relatively costly investment in traditional energy-based dryers remains a barrier to the widespread adoption of these conventional energy-based dryers. Solar dryers are a viable alternative to open sun dryers and traditional sun dryers. Solar power is inexpensive, readily available, and safe, making it one of the most promising alternative energy sources.

In the near future, large-scale deployments of solar energy systems that directly convert solar radiation into heat are expected. The correct application of sun drying technologies can replace moderate amounts of fuel wood or fossil fuels now utilized in industrialized nations for food and agriculture processing (Ghimire et al., 2018).

In Nepal, the following agriculture products are majorly used for drying. Due to OSD's low quality of the dried product produced, various solar dryers are used to dry agriculture products now. Table 10.14 shows the agricultural products of Nepal with their price.

TABLE 10.14 Agriculture Products with Their Price in Nepal (Report W. Bank, 2020)

SL. No.	Product	Price/Kg (NRs)
1.	Gundruk (Mustard)	675
2.	Gundruk (Rayo)	509
3.	Chilly	350
4.	Akabare	1,125
5.	Masyoura (maas, Karkalo)	600
6.	Masyoura (mung)	600
7.	Mushroom	1,540
8.	Fish	1,700
9.	Kiwi	1,250
10.	Peach	870
11.	Orange	880
12.	Coconut	1,190
13.	Pineapple	900
14.	Mango	880

Nepal's government offers subsidies for solar dryers to facilitate solar drying, but the operation is cumbersome and cannot be afforded by farmers. Blanket's solution to the subsidy amount (same amount for all), irrespective of residence and economic status, is not appropriate. Priority should be given to people from the remote region and to those belonging to a lower economic group (Ghimire et al., 2018).

10.4.5 ROLE OF SOLAR DRYERS IN AFGHANISTAN

Afghanistan has a distinct culture of drying agricultural products. In the majority of the areas, the climate is extremely dry, so ideal for drying. Nearly everything is dried in the open, i.e., fresh cheese, meat, all kinds of vegetables and fruits, and a lot of raisins, tomato paste, herbs, and spices (FAO, 2013). Even honeydew melons are often dried to a delicious sweet. Farmers are used to drying portions of their crops in order to sell them dried on the market.

A lot of drying is done for personal use and survival. People who live in cities and are unable to cultivate their own food have the habit of purchasing vegetables in bulk during peak season and drying them for use throughout the winter. Afghan traders from the past are still well-known in India for their dried fruits. In today's international marketplaces, quality standards are quite high. However, from a hygiene standpoint, it is also recommended not to dry food in the open at Afghanistan's local marketplaces. Due to the elevation, solar radiation is quite intense, and the number of sunny days per year is rather high. Dried fruits are Afghanistan's main export to several nations across the world. Dry fruits of Afghanistan, which are exports, are listed in Table 10.15 with their production quantity.

TABLE 10.15 Export of Afghanistan Dried Fruits in Quantity (Rasoly & Chandrashekar, 2018)

Name	Quantity in Tons (2016–2017)
Raisins	19,172
Apricots	2,624
Cherry	25
Dried fig	7,585
Almonds	9,036
Walnuts	3,024
Pistachio	1,069
Dried berry	862
Dried plum	504
Pinecone	91
Other fruits	4,126

The new solar dryer dries fruits and vegetables faster and cleaner than the conventional drying method and protects them against damage to the open air. Within the dryer, the product is put on five layers which have a heat control system that can regulate the degree of heat produced by sunlight. The dryer also has tools such as a measuring device, washing basins, and slicing knives, all designed to improve product efficiency. Solar dryer increases the quality and quantity of dry fruits exported to other countries and satisfies the quality expected at the international level (Karim, 2020).

10.4.6 ROLE OF SOLAR DRYER IN THE MALDIVES

Approximately 10% of the area is suitable for farming in the Maldives. More than a few household gardens have been able to successfully cultivate millet and other grains, as well as a variety of tropical vegetables and fruits. Coconut palms produce copra and coir, which are the second and third most significant exports after fish. Rice, a basic meal for the people, must be imported in large quantities. Breadfruit, mangoes, pumpkins, taro, bananas, papayas, melons, limes, and chili peppers are among the other significant crops in the region. There were small amounts of sorghum, millet, and maize in 1999. In 1999, coconuts and copra accounted for 12,000 tons of production (Shyamala, n.d.). Fruit and vegetable post-harvest losses are significant, ranging from 20% to 44%. Fruit and vegetable post-harvest losses range from 20% to 44%. Using a solar dryer reduces post-harvest losses and provides food for the Maldives' population (FAO, 2013).

10.4.7 ROLE OF THE SOLAR DRYER IN PAKISTAN

Pakistan's agriculture sector is vital to the country's economy, accounting for 18.9% of GDP and employing 42.3% of the population. It is also a significant source of foreign exchange profits, as well as a catalyst for expansion in other industries. To encourage growth in this sector, the government is concentrating on helping small and marginalized farmers and promoting small-scale innovative technology. One of the methods used to dry agricultural products is the solar dryer. Solar dryers preserve grain, fruits, and vegetables, decrease losses, dry faster and more evenly, and provide a higher-quality product than open-air techniques.

Under hygienic circumstances, solar-drying technology gives an option for processing vegetables and fruits in accordance with national and international standards while using no energy. It saves resources, time, and space, improves product quality, makes the process more cost-effective, and preserves the environment.

Solar dryers are currently being used to dry large amounts of fruit transport and ultimately sell them on the urban market, resulting in a positive impact on this area's economy. NGOs are actively working to make the use of such dryers more popular. Solar dryers might be utilized equally successfully in the provinces of Punjab and Sindh to dry agricultural goods for increased market value and to generate local jobs. The main crops, vegetables, and fruits dried using solar dryers and exported to other counties are listed in Table 10.16.

TABLE 10.16 Agricultural Commodities of Pakistan (Qadir, 2020)

SL. No.	Commodity
1.	Oilseeds
2.	Chilies and peppers, dry
3.	Cottonseed
4.	Mangoes, mangosteens, guavas
5.	Pulses
6.	Roots and tubers
7.	Spices
8.	Apricots
9.	Spinach
10.	Dates
11.	Tobacco, unmanufactured
12.	Onions, dry
13.	Rice, paddy
14.	Sunflower seed
15.	Indigenous cattle meat
16.	Almonds with shell

Solar drying has the potential to minimize post-harvest losses while also improving agricultural product quality. Five sun-drying systems have been erected in the Sindh province's Khairpur and Sukkur regions at Pano Aquil, Pryalo, Pir Jo Goth, Therhi, and Hayyat Solangi in date-producing zones. At these sites, the government also conducted training for local farmers to use and maintain these systems to benefit from this technology to increase their production as well as to produce quality dry products. The farmers' capacity-building initiative would assist in increasing farmers' competence and understanding of suitable drying procedures.

Farmers may increase their output and profitability by using this technology to produce high-quality dates and other fruits. This system would also be an addition to the system in order to achieve the country's national aim of food security.

10.5 CONCLUSION

The SACs are very populated, so the use of solar dryers for drying various agricultural products can fulfill the requirement of nutrition in the future. As the solar dryer uses solar energy for the drying process, it is possible to save the energy needed for drying that is used by traditional energy sources.

- Solar drying is a low-cost food drying technique that can be quickly applied to contribute positively to fruits and vegetables in the food processing industry.
- In view of the increased cost and lack of non-renewable energy (RE) sources, this can be used to encourage the use of RE sources as an income-generating alternative.
- Performance research is conducted, and solar drying is found to be stronger, quicker, and more hygienic.
- The purpose of using these effective drying technologies is to minimize spoilage, and increase the quality of the product and overall processing hygiene in order to greatly improve farmers' agricultural returns in appreciation of the hard work they have put into crop cultivation.
- The use of phase-change materials increases solar dryer performance.

KEYWORDS

- **drying technology**
- **energy scenario**
- **energy sustainability**
- **hybrid solar drying**
- **open sun drying**
- **passive solar drying systems**
- **solar dryer**
- **South-Asian countries (SAC)**

REFERENCES

Abu-Hamdeh, N. H., & Alnefaie, K. A., (2019). Assessment of thermal performance of PCM in latent heat storage system for different applications. *Sol. Energy, 177*, 317–323. https://doi.org/10.1016/j.solener.2018.11.035.

Bal, L. M., Satya, S., & Naik, S. N., (2010). Solar dryer with thermal energy storage systems for drying agricultural food products: A review. *Renew. Sustain. Energy Rev., 14*, 2298–314. https://doi.org/10.1016/j.rser.2010.04.014.

Department of Agriculture Ministry of Agriculture & Forests Royal Government of Bhutan Thimphu : Bhutan, Agric. Stat., Department of Agriculture, MoAF, Thimphu, pp. 2–81, 2016.

Dorji, T. (2015). "Bhutan : Organic By Country Profile & Land Use," Natl. Soil Serv. Cent., Bhutan: Organic By 2020, pp. 1–42.

Ekechukwu, O. V., & Norton, B., (1999). 99/02111 Review of solar-energy drying systems II: An overview of solar drying technology. *Fuel Energy Abstr., 40*, 216. https://doi.org/10.1016/s0140-6701(99)97881-5.

FAO, (2013). *Food and Agriculture Organization Maldives Country Programming Framework Office of the FAO Representative for Sri Lanka and Maldives COLOMBO, Sri Lanka*. Gov. Mlad.

Ghimire, D., Lamsal, G., Paudel, B., Khatri, S., & Bhusal, B., (2018). Analysis of trend in area, production, and yield of major vegetables of Nepal. *Trends Hortic.*, 1. https://doi.org/10.24294/th.v1i2.914.

Joshi, C. B., Pradhan, B. D., & Pathak, T. P., (2000). Application of solar drying systems in rural Nepal. *World Renew. Energy Congr. VI 2000*, 2237–2240. https://doi.org/10.1016/b978-008043865-8/50483-9.

Kant, K., Shukla, A., Sharma, A., Kumar, A., & Jain, A. (2016). Thermal energy storage based solar drying systems: A review. *Innov. Food Sci. Emerg. Technol., 34*, 86–99. https://doi.org/10.1016/j.ifset.2016.01.007.

Karim, S. A., (2020). *Solar Dryers Save Time, Increase Income for Afghan Farmers*, 2015–2016.

Kumar, A., Singh, R., Prakash, O., & Ashutosh (2014). Review on global solar drying status. *Agric. Eng. Int. CIGR J., 16*, 161–177.

Lakhani, J., & Patel, V., (2013). *Design and Development of Solar Dryer Submitted by 2013*, 1–46. https://doi.org/10.13140/RG.2.2.28719.94880.

Lhendup, T., (2005). Technical and financial feasibility of a solar dryer in Bhutan. *Energy Sustain. Dev., 9*, 17–24. https://doi.org/10.1016/S0973-0826(08)60496-2.

Masoom, A., Kosmopoulos, P., Bansal, A., & Kazadzis, S., (2020). Solar energy estimations in India using remote sensing technologies and validation with sun photometers in urban areas. *Remote Sens., 12*, 1–25. https://doi.org/10.3390/rs12020254.

Mustayen, A. G. M. B., Mekhilef, S., & Saidur, R., (2014). Performance study of different solar dryers: A review. *Renew. Sustain. Energy Rev., 34*, 463–470. https://doi.org/10.1016/j.rser.2014.03.020.

Palit, D., (2013). Solar energy programs for rural electrification: Experiences and lessons from South Asia. *Energy Sustain. Dev., 17*, 270–279. https://doi.org/10.1016/j.esd.2013.01.002.

Pirasteh, G., Saidur, R., Rahman, S. M. A., & Rahim, N. A., (2014). A review of the development of solar drying applications. *Renew. Sustain. Energy Rev., 31*, 133–148. https://doi.org/10.1016/j.rser.2013.11.052.

Prakash, O., & Kumar, A., (2013). Historical review and recent trends in solar drying systems. *Int. J. Green Energy, 10*, 690–738. https://doi.org/10.1080/15435075.2012.727113.

Qadir, S., (2020). Enhancing Pakistan's agricultural sector exports to India. *SSRN Electron J.* https://doi.org/10.2139/ssrn.3545767.

Rabha, D. K. (2017). *Development and Performance Investigation of a Solar Dryer Integrated with Latent Heat Storage*. Doctor of Philosophy by. Indian Inst Technol Guwahati Guwahati.

Raj, P. A., (1973). *Development in Nepal at a Glance*, 1–17.

Rasoly, M., & Chandrashekar, H. M., (2018). *Export Performance of Dried Fruits from Afghanistan: A Study in Afghanistan, 5*, 16–22.

Report W. Bank, (2020). *Agriculture in Nepal*, pp. 1989–1991.
Salam, Rashiqa Abdul, Khuram Pervez Amber, Naeem Iqbal Ratyal, Mehboob Alam, Naveed Akram, Carlos Quiterio Gómez Muñoz, and Fausto Pedro García Márquez. 2020. "An Overview on Energy and Development of Energy Integration in Major South Asian Countries: The Building Sector." *Energies* 13 (21): 5776. https://doi.org/10.3390/en13215776.
Senadeera, W., & Kalugalage, I. S., (2004). *Performance Evaluation of an Affordable Solar Dryer for Drying Crops*. Bienn. Conf. Soc. Eng. Agric., Dubbo.
Sharma, A., Chen, C. R., & Vu Lan, N., (2009). Solar-energy drying systems: A review. *Renew. Sustain. Energy Rev., 13*, 1185–210. https://doi.org/10.1016/j.rser.2008.08.015.
Shukla, A. K., Sudhakar, K., & Baredar, P., (2017). *Renewable Energy Resources in South Asian Countries: Challenges, Policy, and Recommendations*, *3*, 342–346. https://doi.org/10.1016/j.reffit.2016.12.003.
Shyamala, R., (2018). "Solar energy for food processing," *Soc. Energy, Environ. Dev. (SEED), 1*(1), 1–37.
Sontakke, M. S., & Salve, P. S. P., (2015). *Solar Drying Technologies: A Review, 4*, 29–35.
Stefan, S., (2019). Solar drying of food. *Altern. Technol., 11*, 10.
Suresh Chandra Babu, A. G. (2020). "Agriculture in Bhutan," *Econ. Reforms Food Secur.* Impact Trade Technol. South Asia, vol. 1, pp. 4–5.
Tiwari, A., (2016). A review on solar drying of agricultural produce. *J Food Process Technol., 7*. https://doi.org/10.4172/2157-7110.1000623.
Umesh, T., & Karale, S. R., (2013). A review paper on solar dryers. *Int. J. Eng. Res. Appl., 3*, 896–902.
Visavale, G. L., (2012). *Solar Drying: Fundamentals, Applications, and Innovations*. https://doi.org/ISBN: 978-981-07-3336-0.
Weerasena SSBDGJ and LA, (2020). *Crop Diversification in Sri Lanka* (pp. 1–14). Dir. Gen. Agric. Dep. Agric. Perad. Sri Lanka.

CHAPTER 11

Biofuels an Answer to Roaring Energy Demand: Technology to Implications

SHIVANI CHATURVEDI,[1] AJAY KUMAR CHATURVEDI,[2] JYOTI SINGH,[3] PRASANTA KUMAR ROUT,[4] LATA NAIN,[3] and SUNIL KUMAR KHARE[1]

[1]*Enzyme and Microbial Biochemistry Lab, Department of Chemistry, Indian Institute of Technology, Delhi, India*

[2]*AWHO Colony, Tyagi Vihar, Lucknow, Uttar Pradesh, India*

[3]*Division of Microbiology, ICAR–Indian Agricultural Research Institute, New Delhi, India*

[4]*Process Chemistry and Technology Department, Central Institute of Medicinal and Aromatic Plants, Lucknow, Uttar Pradesh, India*

ABSTRACT

With the increase in urbanization, industrialization, fast-track communication, and high demand refrigeration are prime requirements of smart cities. Rising demand in mechanization of agriculture and automation in the industries need energy. To meet the energy requirement, the Government of India, in its energy platter, has set a target for achieving 227 gigawatts (GW) of energy by the year 2022. Within this target, the RE share was 10 GW, which has already been achieved in 2018 due to systematic efforts in this sector. Sources of renewable clean energy are divided as; biomass (8.9%), thermal energy (4.2%) (geothermal and solar heat), hydroelectricity (3.9%), and the remaining are electricity from wind and other forms of biomass (2.2%). This data illustrated the potential of biomass to convert liquid transport fuel through suitable technologies. Transformation of biomass into energy is

The South Asian Energy Corridor: Toward Sustainability and Security. Amritanshu Shukla, PhD, Atul Sharma, PhD & Saurabh Mishra, PhD (Eds.)

carried out in three major technological interventions: mechanical methods, thermo-chemical conversions, and biotransformation.

On these lines, India has also committed 'Intended Nationally Determined Contribution' (INDC) towards the Paris Agreement of the 'United Nations Framework Convention on Climate Change' (UNFCCC). For achieving these targets, the effective use of agro-residues in biofuel plays a major role in the current trends to curtail the issue of food crop residue burning in Northern India. In line with the biofuel policy of India, the push for 2nd generation of biofuels is the right step to extract energy from biomass without carbon credit in the form of liquid biofuels such as bioethanol and biodiesel. In the futuristic sense, we need to come up with cutting-edge technologies to realize the idea of 4th generation carbon-neutral biofuel.

11.1 INTRODUCTION

Conventional energy is limited and is going to be depleted soon, so natural alternative sources that are sustainable and safer for the environment have to be explored (Adams et al., 2018). Global warming, a dearth of fossil fuel, and increasing environmental concern initiated the search for alternative RE resources. RE sources like wind, geothermal, hydroelectricity, solar, and biomass have been used according to the geographical and environmental conditions of an area. According to the 2015 records of global investment scenario, solar and wind energy are much higher than biomass and hydropower energy. It contributes to around 89.3% of total investment in the RE sector (REN21, 2016). According to a rigorous study on the use of RE resources by G-7 countries, it was found that Canada utilizes wind energy, Italy and France use solar energy, the United Kingdom depends on hydroelectricity as a RE source, and France utilizes biomass energy and reduces CO_2 emissions (REN21, 2016; Destek & Aslan, 2019). France, Japan, and Germany were working on renewable alternative sources like wind, solar, and biomass, but they found biomass energy to be the most efficient and eco-friendly source (Destek & Aslan, 2019; Mehmet et al., 2019; Tgarguifa et al., 2018). In India, there are varied geographical variations, so alternative energy sources like wind are appropriate for the eastern and western Ghats; and solar for Rajasthan. Gujrat and Maharashtra, whereas biomass-based biofuel for good agro-practice states like Karnataka, Andhra Pradesh, Tamil Nadu, Punjab, Haryana, and Uttar Pradesh. In this regard, a lot of work has been done so far. Still, there is a big gap between production capacity and consumption of energy sources which needs to be narrowed down (Dimzon et al., 2018).

In Southeast Asian countries, the focus on the development of RE technologies has increased (Hu et al., 2017). RE resources not only lower the carbon footprint but also reduce the emission of greenhouse gases (GHGs) (Sarkodie & Strezov, 2018; Ahmed et al., 2017). In India, a lot of policies and targets have been set to develop sustainable RE policies.

11.2 GOVERNMENT TARGETS AND PRESENT SCENARIO

As of 2013, the total installed capacity for electricity production in India is around 2,666 GW. Out of this total energy generation, 10.5% is contributed by RE, from which 12.83% is being produced through biomass. India has 500 million metric tons (MMT) of biomass availability per year and has 17,500 MW total biomass power generation capacity (Kumar et al., 2015). The Indian Government has set the target for achieving 227 gigawatts (GW) in the energy platter, including 10 GW from RE up to the year 2022. Renewable clean energy consumption is divided; into biomass (8.9%), thermal energy (4.2%) (geothermal and solar heat), hydroelectricity (3.9%), and the remaining is electricity from wind and other forms of biomass (2.2%). Due to better Government policy to boost the agricultural sector, India has already achieved this target of 10-GW energy in 2018. This shows the potential to realize greater energy conversion from biomass through technological interventions such as mechanical methods, thermo-chemical treatment, and biotransformation/microbial conversion. In thermal transformation, direct combustion and incineration can be done to convert biomass into energy, while a few other chemical processes like plasma treatment, pyrolysis, torrefaction, and gasification are used for transformation to energy. Anaerobic digestion (AD), composting, and ethanol fermentation were biochemical ways to deal with biomass and convert it into cleaner energy (Gumisiriza et al., 2017).

11.2.1 WAYS TO HANDLE THE CURRENT SITUATION

The basic reason behind the energy crises is the increasing population, which in turn, increases the consumption rate, creating a severe dearth of resources. Besides this, the use of outdated, poor infrastructure, along with the mass wastages of energy due to lack of insulation for leakages, and less alertness of individuals for not switching off lights, fans, and televisions, further create energy crises. To handle these situations, dependency on renewable options should be enhanced by the consumer. The distribution system should also be taken care of as far as major breakdown or tripping are concerned. Frequent strikes and damage to public properties like electric poles, wires, and transformers also contribute to the crises. These energy crises can be well tackled with the support of RE resources according to the availability of area-specific natural sources. Other than this, using highly energy-efficient products like CFLs and LEDs can lower the use of energy. Newer technologies like touch dimmers and preset or slide lighting systems have helped to curtail the electricity need. Alternatively, energy simulation software supports the reduction of running business energy costs and the creation of energy-efficient building infrastructure needs of the hour. Encouraging energy policies for reducing energy bills can also play an important tool not only for the common person but also for the corporate consumer to develop a general attitude towards a reduction in energy use. Developing countries like India can play a key role in lowering carbon emissions. Utilizing the biomass generated from various sources could be a reliable and sustainable way to go forward for India.

11.3 CONTRIBUTION OF BIOMASS IN ENERGY DEMAND

Biomass derived from agro-industries is a sustainable and safe green solution for the production of useful chemicals and fuels without carbon emission (Zhou et al., 2011).

This biomass can be procured from agricultural, forestry, and agro-based industries. However, biomass generated from lignocellulosic material has tremendous benefits over other sources due to their non-edibility and, therefore, no issue of food source supply security (Ye & Cheng, 2002). European Union (EU) produced approximately 67% of primary energy from biomass. This biomass calorific value is approximately 14–19 MJ/kg. In the year 2012, around 370 TWh of bio-energy electricity was obtained, which according to the International Energy Academy, corresponds to 1.5% of world electricity generation (Kucharska et al., 2018). The renewable nature of lignocellulosic material also draws the special interest of researchers for their utilization in energy production (Asgher et al., 2013; Ofori-Bateng & Lee, 2013).

Ligno cellulosic materials contribute approximately 48% of energy originating from biomass (Balat, 2011).

Ligno cellulosic material mainly consists of cellulose, hemicelluloses, and lignin, with a smaller contribution of phenolic compounds, some minerals, and acetyl groups. These wastes have a variable composition of proteins, minerals, and sugars. Mostly this lignocellulosic matter has 40–50% cellulose, 10–25% lignin, and 20–30% hemicelluloses (Iqbal et al., 2011).

The lignocellulosic biomass is a potential substrate for the production of bio-hydrogen (Wulf & Kaltschmitt, 2013), bioethanol (Saini et al., 2015), biogas (Sarvari et al., 2016), synthetic biofuels (Hannula, 2015) and methane (Liew et al., 2012). Hydrogen is supposed to be the basic substrate for many petrochemical processes (olefins hydrogenation, aniline from nitrobenzene, methanol, integrated circuits, optical fibers, synthetic gas, synthesis of ammonia, methane, and production of plastics) (Atif et al., 2005). Hydro conversions, hydrafination, and hydrocracking also use hydrogen (Kucharska et al., 2018). Rocket engines, grid storage and conversion, and automotive engines have a high application of compressed hydrogen (500–700 Ba) (Tani et al., 2017; Kucharska et al., 2018). Due to this nutritionally rich status, it can be used for several useful product formations (Sadh et al., 2018). This biomass makes high-value products like enzymes, fine chemicals, and substrates for microbial fermentation (Asgher et al., 2013), fillers for thermoplastic composite materials (Waligorska & Laniecki, 2009), bioethanol (Grabarczyk et al., 2009), beauty masks and many pharmaceutical products.

There are mainly two kinds of wastes generated from agricultural sources, one is purely direct agricultural field wastes, and another is industrial processing wastes. In agricultural waste, the direct field residue consists of stems, seeds, stalks, leaves, roots, and barks, while after processing, molasses, bagasse, husks, stubble, peels, shells, and pulps were produced (Figure 11.1). These residues are underutilized and need more extensive research on them (Sadh et al., 2018). Seed cakes of *Pongamia pinnata*, *Jatraopha*, and neem cake also contribute to renewable biomass and can be utilized in many ways (Chaturvedi et al., 2009a, b).

Wastes generated from processing industries are mainly juices, pulps, shells, husks, bran, dust, and straws (Figure 11.2). Oil industries like soybean, neem, Jatropha, sesame, mustard, canola, sunflower, coconut, peanuts, etc., also made huge piles of seed cakes in the surroundings (Figure 11.2). They have different types of compositions of cellulose, hemicelluloses, lignin, moisture, etc. Non-edible oil seed cake of Jatropha can be utilized after bioaugmentation as a feasible and sustainable compost in agricultural fields (Chaturvedi et al., 2012). In India, every year, approximately 20% of the production of vegetables and fruits is reported to be waste (Rudra et al.,

2015). India produces a huge amount of soybean, wheat, cotton, and apple, and its high production increased the waste generation. This accumulated, untreated, and underutilized waste may cause serious problems for the environment but also can affect negatively the health of both humans and animals (Sadh et al., 2018) (Figure 11.3).

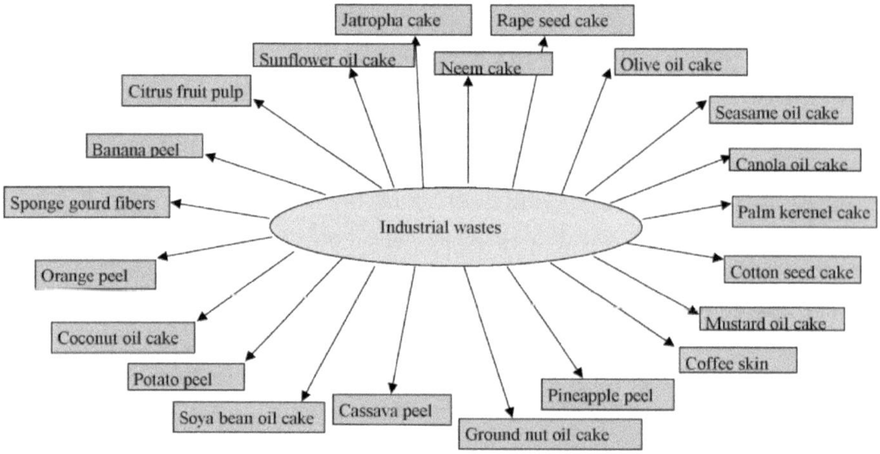

FIGURE 11.1 Various kinds of industrial processing waste were underutilized and need more extensive research.

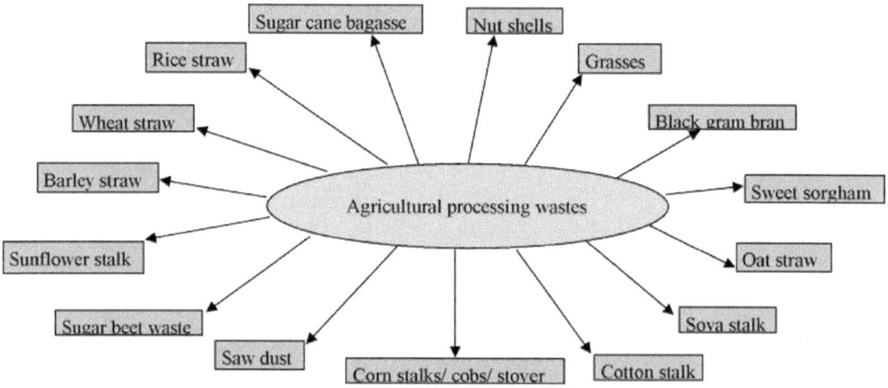

FIGURE 11.2 Wastes generated from processing agriculture are mainly juices, pulps, shells, husks, bran, dust, and straws. Oil industries like soybean, neem, jatropha, sesame, mustard, canola, sunflower, coconut, peanuts, etc., also made huge piles of seed cakes in the surroundings.

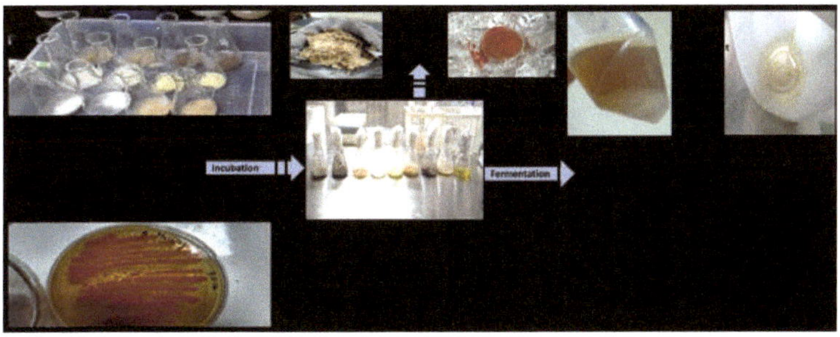

FIGURE 11.3 Pictorial representation of the whole processing mechanism for fatty acid production from starchy biomass (including incubation, fermentation, and pigment formation of production media).

Molecules of cellulose are held together by intermolecular hydrogen bonds in normal conditions, but they have a high tendency to build up intra-molecular as well as intermolecular hydrogen bonds, which in turn make them insoluble to organic solvents (Iqbal et al., 2011). Hemicelluloses are polymers of hexoses and pentoses like glucomannan, glucuronoxylan, and a small amount of other polysaccharides (Anwar et al., 2014). Lignin consists of mainly three types of aromatic alcohols, it acts as a binding matter for a gap between and around hemicelluloses and cellulose (Anwar et al., 2014). Altogether, lignocellulosic material composition mainly depends on the source from which it is derived from (Anwar et al., 2014).

Due to the complex structure of lignocellulosic materials, it is difficult to manage. Along with this, if its strong cell wall has been pretreated, some of the resultant by-products hinder the whole process of saccharification (Kumar & Sharma, 2017). It is also resistant to chemical and enzymatic degradation (Barakat et al., 2013). Despite the low cost and abundance of lignocellulosic material, its conversion to value-added fine chemicals with economical costs is a big challenge (Alanso et al., 2013).

There should be an appropriate selection of pretreatment techniques suitable for the specific biomass type. In this regard, the energy consumption should be minimal, the hemicelluloses' part should be preserved, the low-cost catalyst of pretreatment or recyclable pretreatment catalysis should be taken, biomass particle size should not be reduced after the process, and minimal production of degraded products (Wyman, 1999). To recover cellulose from lignocellulosic material, lignin barriers should be broken down, and this pretreatment is necessary (Anwar et al., 2014).

11.4 A MAJOR ROLE OF MICROBIAL FUEL GENERATED FROM BIOMASS

In order to explore the economic and RE sources (Kricka et al., 2015), the development of second-generation biofuel from using agro-industrial waste (Zhang et al., 2011) without disturbing food supplies (Jin et al., 2015) was established (plant, animal, and microbial fuel).

11.4.1 PRODUCTION OF 2ND GENERATION BIOFUEL

Earlier, for biofuel production, agricultural crops like sugar beets, canola, corn, potatoes, sugar cane, and soybean were used, and so termed as the first generation of biofuels. This category of biofuel actually creates food vs fuel competition. Due to this problem, second generation of biofuels gets more attention in which agricultural wastes, lignocellulosic biomass, grass, crop residues, and vegetable peel wastes were transformed into biofuels (Hironi et al., 2018). Basically, the second generation of biofuel utilizes the non-food waste part to generate biodiesel or bioethanol form of biofuels.

11.4.2 PRODUCTION OF BIOETHANOL AND BIODIESEL FROM LIGNOCELLULOSIC BIOMASS

Lignocellulosic biomasses need pretreatment before enzymatic saccharification to glucose. It was reported that lignin inhibited the saccharification, and pretreatment helped to remove lignin and also provided improved cellulosic surface for enzymatic hydrolyzes (Rout et al., 2016). Both by solid-state fermentation and submerged fermentation, available sugars get converted into useful final products. Conversion of biomass is mainly associated with five steps; first is biomass selection, second is pretreatment of the biomass, third is saccharification of carbohydrate polymer, fourth is enzymatic fermentation of pentoses and hexoses, and the fifth step is isolation and purification of the final desired product (Hironi et al., 2018). Practically, the economical production of bioethanol is dependent on two key steps the pretreatment process and the cost of the enzymes for saccharification. The key step for bio-ethanol production is effective pre-treatment process for partial separation of lignin. A number of pretreatment processes are listed in our earlier review article as follows physical pretreatment, chemical pretreatment, physic-chemical, sonochemical, biological, and pulse-electric-field

(Rout et al., 2016). Generally, a combination of pretreatment processes is economically effective for the separation of major amounts of lignin from the agricultural residues. Similarly, the high value commercial enzyme cost might be curtailed using laboratory-produced enzymes, in this aspect, our work on *Trichoderma* spp. is practical and effective.

The global vegetable oil market is predicted to grow at a rate of 3.25% during 2019–2024 (National Academy of Agricultural Sciences, 2017). The demand grew at a compound annual growth rate (CAGR) of 4% during 2011–2018, with utilization volume reaching 200 million tons in the year 2018. Due to the high demand for vegetable oils, the non-conventional source of lipids is an attractive alternative. Oleaginous algae, heterotrophic bacteria, fungi, and yeast produce single cell oil (SCO) (Rottig et al., 2016) out of them yeast generated microbial lipid was found to be the better source (Sitepu et al., 2014). They have the ability to utilize cheap substrates like agro-industrial biomass (Qin et al., 2017). These biomasses could be from the olive mill (Arous et al., 2017); banana peel (Chaturvedi et al., 2018a); sweet sorghum (Matsakas et al., 2014); starchy waste (Chaturvedi et al., 2018b, c).

Oleaginous yeasts are popular because of their growth adaptability in both aerobic and anaerobic conditions and their simple cultural environment requirement. But for triglyceride production, they need a higher C/N ratio of more than 30 (Możejko-Ciesielska & Kiewisz, 2016; Dasguptab et al., 2017). These yeasts have the ability to produce more than 20% triglycerides than biomass weight, and lipid production can be enhanced up to 70% of the biomass in the conditions of limited nitrogen availability (Papanikolaou & Aggelis, 2011a, b). A few strains of oleaginous yeast, like *Saccharomyces pastorinus*, *Rhodotorula glutinis, and Rhodotorula mucilaginosa*, were explored and found to produce not only intracellular lipids but also extracellular lipids. The extracellular lipid formation makes the procurement process cost-effective as compared to intracellular lipid. As the cell wall of yeast is very tough to break, it needs lots of energy. By getting the lipid extracellularly, the effort to break the cell wall can be avoided, and so the energy can be saved. These yeasts were able to produce saturated lipids like octadecanoic acid, hexadecenoic, and tetradecanoic acids (Chaturvedi et al., 2018b).

These oleaginous yeasts, along with some more strains (*Trichosporon cutaneum*, *Cryptococcus curvatus*, and *Lipomyces starkeyi*), were screened on various real-life substrate as banana peel, yam peel, cassava peel, potato peel, rice husk, barley husk, wheat bran and corn residue. The results were very encouraging, these strains were able to utilize starchy waste as a carbon source and were capable of producing a good amount of lipid. Out of them, *C. curvatus* produces 16 g of lipid by using wheat bran waste and 14.3 g of lipid by consuming banana peel starch. These results suggest not only an environmentally friendly alternative to household and agro-industrial processing waste but also give a pathway for SCO production (Chaturvedi et al., 2019a).

India is the leading producer of bananas, so the peel of bananas has also been explored and found to be an excellent substrate for lipid production due to the fermentation process (Chaturvedi et al., 2018a). In southern parts of India, cassava is also a popular crop used in many forms. Peel of cassava is also being utilized as a cheap source of carbon and has been found to be a very good source of carbon for SCO production (Chaturvedi et al., 2019b).

In a country like India, where there is so much diversity in agricultural crops, these robust efforts towards biomass utilization not only solve the issue of biofuel generation but also provide waste management alternatives for meaningful utilization.

11.5 CONCLUSIONS

The depletion of fossil fuel and the increasing population has led to the search for alternative energy sources. The lignocellulosic biomass is available in huge amounts, and its utilization in the second generation of biofuel helps researchers to explore the alternative RE source. Indian economy is based on agriculture, but the direct burning of crop residue creates havoc-like situations in the Northern part of the country. For the future, the conversion of food waste and agro-residues to liquid biofuel, such as bioethanol and biodiesel, are the most promising alternative options in the hands of scientists and technologists. The most difficult task is the conversion of lignocellulosic biomass into sugars by saccharification, which is associated with an expensive process of pretreatment as well as the cost of enzyme production. Though India has been tested the preliminary success of the production of bioethanol, still awaiting the mass scale and disintegration mole of production. The processes of pre-treatment, saccharification, and in-house enzyme production through an integrative biochemical approach may lead to technological intervention in the near future. Alternatively, in bio-hydrogen production, pre-treated lignocellulosic biomass might be opted for the production of hydrogen by using renewable solar energy.

Biomass as carbon source using oleaginous yeast will provide important fatty acid methyl ester (FAME) produced under submerged condition as intracellular lipids were methyl 13 methyltetradecanoate ($C_{16}H_{32}O_2$), methyl tetra decanoate ($C_{15}H_{30}O_2$), hexadecanoic acid ($C_{34}H_{68}O_2$), pentadecanoic acid ($C_{15}H_{30}O_2$), and dodecanoic acid ($C_{14}H_{28}O_2$). Biomass-based substrate submerged condition produced extracellular lipid in the form of FAME. The key FAME were methyl 18 fluro octadecanoate ($C_{19}H_{37}FO_2$), methyl 13 methyltetradecanoate ($C_{16}H_{32}O_2$), 9 hexadecanoic acid methyl ester ($C_{17}H_{32}O_2$), and cyclopentane undecanoic acid methyl ester ($C_{17}H_{32}O_2$), All these standard fatty acids may result in better specific gravity of oil similar to vegetable oil/biodiesel. This strategy not only provides environmentally friendly biofuel but also suggests waste management alternatives.

In the World scenario, vegetable oil is very limited for non-food applications. The alternative futuristic sources with commercial potential, like oleaginous algae, heterotrophic bacteria, fungi, and yeast, are promising. The perfect balance among energy requirements, food security, greenhouse gas reduction, and socio-economic development factors conceptualized the appropriate strong policy design for sustainable energy.

In line with the biofuel policy of India, the push for the 2nd generation of biofuels is the step in the right direction. In the futuristic sense, we need to come up with cutting-edge technologies to actualize the idea of 4th generation carbon-neutral biofuel.

FUNDING

The first author (Shivani Chaturvedi) is thankful for financial support from the university grant commission post-doctoral fellowship program, UGC award letter-number F.15-1/2017/PDFWM-2017-18-DEL-3915 (SA-II) from UGC, India.

AUTHORS' CONTRIBUTIONS

Shivani Chaturvedi and Ajay K. Chaturvedi, Jyoti Singh, Prasanta K. Rout, along with Sunil K. Khare and Lata Nain, wrote the manuscript. All the authors read and approved the final manuscript.

CONFLICTS OF INTERESTS

All authors declare that they have no conflicts of interest.

KEYWORDS

- **biodiesel**
- **bioethanol**
- **biofuels**
- **biomass**
- **clean energy**
- **climate change**
- **energy demand**
- **renewable energy**

REFERENCES

Adams, S., Klobodu, E. K. M., & Apio, A., (2018). Renewable and non-renewable energy, regime type, and economic growth. *Renew. Energy, 125*, 755–767. https://doi.org/10.1016/j.renene.2018.02.135.

Ahmed, T., Mekhilef, S., Shah, R., Mithulananthan, N., Seyedmahmoudian, M., & Horan, B., (2017). ASEAN power grid: A secure transmission infrastructure for clean and sustainable energy for South-East Asia. *Renew. Sust. Energ. Rev., 67*, 1420–1435. https://doi. org/10.1016/j.rser.2016.09.055.

Alonso, D. M., Wettstein, S. G., & Dumesic, J. A., (2013). Gamma-valerolactone is a sustainable platform molecule derived from lignocellulosic biomass. *Green Chem., 15*, 584.

Anwar, Z., Gulfraz, M., & Irshad, M. (2014). Agro-industrial lignocellulosic biomass a key to unlock the future bio-energy: A brief review. *J. Radiation Res. & Applied Sci., 7*, 163–173.

Arous, F., Atitallah, I. B., Nasri, M., & Mechichi, T. (2017). A sustainable use of low-cost raw substrates for biodiesel production by the oleaginous yeast *Wickerhamomyces anomalus. 3 Biotech, 7*, 268.

Asgher, M., Ahmad, Z., & Iqbal, H. M. N., (2013). Alkali and enzymatic delignification of sugarcane bagasse to expose cellulose polymers for saccharification and bio-ethanol production. *Industrial Crops and Products, 44*, 488–495.

Atif, A. A. Y., Fakhru'l-Razi, A., Ngan, M. A., Morimoto, M., Iyuke, S. E., & Veziroglu, N. T., (2005). Fed batch production of hydrogen from palm oil mill effluent using anaerobic microflora. *Int. J. Hydrogen Energ., 30,* 1393–1397.

Balat, M., (2011). Production of bioethanol from lignocellulosic materials via the biochemical pathway: A review. *Energy Convers Manag., 52*, 858–875.

Barakat, A., Vries, H., & Rouau, X., (2013). Dry fractionation process as an important step in current and future lignocellulose biorefineries: A review. *Bioresour. Technol., 134*, 362–373.

Chaturvedi, S., Bhatacharya, A., & Khare, S. K., (2018c). Trends in oil production from oleaginous yeast using biomass: Biotechnological potential and constraints. *Appl. Biochem. Microbiol., 54,* 361–369.

Chaturvedi, S., Bhattacharya, A., Nain, L., Prasanna, R., & Khare, S. K. (2019a). Valorization of agro-starchy wastes as substrates for oleaginous microbes. *Biomass Bioenerg., 127*, 105294.

Chaturvedi, S., Kumar, A., & Singh, B., (2009a). Utilizing composted jatropha & neem cake and tobacco waste to sustain garlic yields in Indo-Gangetic plains. *J. Phytology, 1*(6), 353–360.

Chaturvedi, S., Kumar, A., Singh, B., Nain, L., Joshi, M., & Satya, S., (2012). Bioaugmented composting of jatropha de-oiled cake and vegetable waste under aerobic and partial anaerobic conditions. *J. Basic Microbiol., 52*, 1–9.

Chaturvedi, S., Kumar, V., & Satya, S., (2009b). Composting effects of *Pongamia pinnata* on tomato fertilization. *Arch Agron. Soil Sci., 55*(5), 535–546.

Chaturvedi, S., Kumari, A., Bhatacharya, A., Sharma, A., Nain, L., & Khare, S. K. (2018a). Banana peel waste management for single cell oil production. *Energy Ecol. Env., 3*, 296–303.

Chaturvedi, S., Kumari, A., Nain, L., & Khare, S. K., (2018b). Bioprospecting microbes for Single cell oil production from starchy wastes. *Pre. Biochem. Biotechnol., 48*, 296–302.

Chaturvedi, S., Singh, B., Nain, L., Khare, S. K., Pandey, A. K., & Satya, S., (2010). Evaluation of hydrolytic enzymes in bioaugmented compost of *Jatropha* cake under aerobic and partial anaerobic conditions. *Annals Microbiol., 60*(4), 685–691.

Chaturvedi, S., Tiwari, R., Bhattacharya, A., Nain, L., & Khare, S. K. (2019b). Production of single-cell oil by using cassava peel substrate from oleaginous yeast *Rhodotorula glutinis*. Biocatal. *Agric. Biotechnol., 21*, 101308.

Dasgupta, D., Ghoshab, D., Bandhu, S., & Adhikaria, D. K. (2017). Lignocellulosic sugar management for xylitol and ethanol fermentation with multiple cell recycling by *Kluyveromyces marxianus* IIPE453. *Microbiol. Res., 200*, 64–72.

Destek, M. A., & Aslan, A., (2019). Disaggregated renewable energy consumption and environmental pollution nexus in G-7 countries. *Renewable Energ.* https://doi.org/10.1016/j.renene.2019.11.138.

Dimzon, I. K. D., Morata, A. S., Müller, J., Yanela, R. K., Lebertz, S., Weil, H., Perez, T. R., et al., (2018). Trace organic chemical pollutants from the lake waters of San Pablo City, Philippines, by targeted and non-targeted analysis. *Sci. Total Environ., 639*, 588–595. https://doi.org/10.1016/j.scitotenv.2018.05.217.

Goryunov, A. G., Goryunova, N. N., Ogunlana, A. O., & Manenti, F., (2016). Production of energy from biomass: Near or distant future prospects? *Chem. Eng. Trans., 52*, 1219–1224.

Grabarczyk, J., Paukszta, D., & Borysiak, S., (2009). Kompozyty polimerów termoplastycznych z materiałami lignocelulozowymi. *Tech Trans., 106*, 93–97.

Hannula, I., (2015). *Synthetic Fuels and Light Olefins from Biomass Residues, Carbon Dioxide and Electricity* (pp. 15). Performance and cost analysis; Teknologian tutkimuskeskus VTT Oy: Otaniemi, Finland.

Hu, L., Shi, X., Qiao, S., Lin, T., Li, Y., Bai, Y., Wu, B., Liu, S., Kornkanitnan, N., & Khokiattiwong, S., (2017). Sources and mass inventory of sedimentary polycyclic aromatic

hydrocarbons in the Gulf of Thailand: Implications for pathways and energy structure in SE. *Asia Sci. Total Environ., 575*, 982–995. https://doi.org/ 10.1016/j.scitotenv.2016.09.158.
Iqbal, H. M. N., Ahmed, I., Zia, M. A., & Irfan, M., (2011). Purification and characterization of the kinetic parameters of cellulase produced from wheat straw by *Trichoderma viride* under SSF and its detergent compatibility. *Advances in Bioscience and Biotechnology, 2*(3), 149–156.
Iqbal, H. M. N., Asgher, M., & Bhatti, H. N., (2011). Optimization of physical and nutritional factors for the synthesis of lignin degrading enzymes by a novel strain of *Trametes versicolor*. *BioResources, 6*, 1273–1278.
Jin, M., Slininger, P. J., Dien, B. S., Waghmode, S., Moser, B. R., Orjuela, A., Da Costa, S. L., & Balan, V., (2015). Microbial lipid-based lignocellulosic biorefinery: Feasibility and challenges. *Trends Biotechnol., 33*, 43–54.
Kricka, W., James, T. C., Fitzpatrick, J., & Bond, U., (2015). Engineering *Saccharomyces pastorianus* for the co-utilisation of xylose and cellulose from biomass. *Microb. Cell Fact., 14*, 61.
Kucharska, K., Rybarczyk, P., Hołowacz, I., Łukajtis, R., Glinka, M., & Kaminski, M., (2018). Pretreatment of lignocellulosic materials as substrates for fermentation processes. *Molecules 23*, 2937.
Kumar, A. K., & Sharma, S., (2017). Recent updates on different methods of pretreatment of lignocellulosic feedstocks: A review. *Bioresour. Bioprocess, 4*, 7.
Liew, L. N., Shi, J., & Li, Y., (2012). Methane production from solid-state anaerobic digestion of lignocellulosic biomass. *Biomass Bioenergy, 46*, 125–132.
Matsakas, L., Sterioti, A., Rova, U., & Christakopoulos, P., (2014). Use of dried sweet sorghum for the efficient production of lipids by the yeast *Lipomyces starkeyi* CBS 1807. *Ind. Crop Prod. 62*, 367–372.
Możejko-Ciesielska, J., & Kiewisz, R., (2016). Bacterial polyhydroxyalkanoates: Still fabulous? *Microbiol. Res., 192*, 271–282.
National Academy of Agricultural Sciences, (2017). Vegetable oil economy and production problems in India, *Strategy Paper*, 7.
Ofori-Boateng, C., & Lee, K. T., (2013). Sustainable utilization of oil palm wastes for bioactive phytochemicals for the benefit of the oil palm and nutraceutical industries. *Phytochem. Rev.*, 1–18.
Papanikolaou, S., & Aggelis, G., (2011a). Lipids of oleaginous yeasts. part I: Biochemistry of single-cell oil production. *Eur. J. Lipid Sci. Technol., 113*, 1031–1051.
Papanikolaou, S., & Aggelis, G., (2011b). Lipids of oleaginous yeasts. Part II: Technology and potential applications. *Eur. J. Lipid Sci. Technol., 113*, 1052–1073.
Qin, L., Liu, L., Zeng, A. P., & Wei, D., (2017). From low-cost substrates to single-cell oils synthesized by oleaginous yeasts. *Bioresour. Technol., 245*, 1507–1519.
REN21. Renewables Global Status Report, (2016). *REN21 Secretariate.*
Röttig, A., Hauschild, P., Madkour, M. H., Al-Ansari, A. M., Almakishah, N. H., & Steinbüchel, A., (2016). Analysis and optimization of triacylglycerol synthesis in novel oleaginous *Rhodococcus* and *Streptomyces* strains isolated from desert soil. *J. Biotechnol., 225*, 48–56.
Rout, P. K., Nannaware, A. D., Prakash, O., Kalra, A., & Rajasekharan, R., (2016). Synthesis of hydroxymethylfurfural from cellulose using green processes: A promising biochemical and biofuel feedstock. *Chem. Eng. Sci., 142*, 318–346.

Rudra, S. G., Nishad, J., Jakhar, N., & Kaur, C. (2015). Food industry waste: Mine of nutraceuticals *Intern. J. Sci. Environ. Technol., 4*(1), 205–229.

Sadh, P. K., Duhan, S., & Duhan, J. S., (2018). Agro-industrial wastes and their utilization using solid state fermentation: A review. *Bioresour. Bioprocess, 5*, 1.

Saini, J. K., Saini, R., & Tewari, L., (2015). Lignocellulosic agriculture wastes as biomass feedstocks for second-generation bioethanol production: Concepts and recent developments. *3 Biotech, 5*, 337–353.

Sarkodie, S. A., & Strezov, V., (2018). Assessment of the contribution of Australia's energy production to CO_2 emissions and environmental degradation using a statistical dynamic approach. *Sci. Total Environ., 639*, 888–899. https://doi.org/10.1016/j.scitotenv.2018.05.204.

Sárvári, H. I., Tabatabaei, M., Karimi, K., & Kumar, R., (2016). Recent updates on biogas production-A review. *Biofuel Res. J., 3*, 394–402.

Sitepu, I. R., Garay, L. A., Sestric, R., Levin, D., Block, D. E., German, J. B., & Boundy-Mills, K. L., (2014). Oleaginous yeasts for biodiesel: Current and future trends in biology and production. *Biotechnol. Adv., 144*, 360–369.

Tani, H., Umemura, Y., & Daimon, Y., (2017). *Interface Tracking Simulations of Liquid Oxygen/Gaseous Hydrogen Coaxial Combustions at Subcritical Pressures*. In 55th AIAA Aerospace Sciences Meeting; AIAA SciTech Forum: Grapevine, TX, USA, 0144.

Tgarguifa, A., Abderafi, S., & Bounahmidi, T., (2018). Energy efficiency improvement of a bioethanol distillery by replacing a rectifying column with a pervaporation unit. *Renewable Energ., 122*, 239–250.

Waligórska, M., & Łaniecki, M., (2009). Fotobiologiczna produkcja wodoru z mieszaniny alkoholi i lotnych kwasów tłuszczowych. *In'z Ap Chem., 48*, 121, 122.

Wulf, C., & Kaltschmitt, M., (2013). Life cycle assessment of biohydrogen production as a transportation fuel in Germany. *Bioresour. Technol., 150*, 466–475.

Wyman, C. E., (1999). Biomass ethanol: Technical progress, opportunities, and commercial challenges. *Ann. Rev. Energy Environ., 24*, 189–226.

Zhang, F., Rodriguez, S., & Keasling, J. D., (2011a). Metabolic engineering of microbial pathways for advanced biofuels production. *Curr. Opin. Biotechnol., 22*, 775–783.

Zhou, C. H., Xia, X., Lin, C. X., Tong, D. S., & Beltramini, J., (2011). Catalytic conversion of lignocellulosic biomass to fine chemicals and fuels. *Chem. Soc. Rev., 40*, 5588–5617.

CHAPTER 12

Energy Sustainability Issues in Southeast Asia: Population, Poverty, and Energy Access

SAURABH MISHRA

Associate Professor, Department of Science and Humanities, Rajiv Gandhi Institute of Petroleum Technology, Uttar Pradesh, India

ABSTRACT

In the current world Scenario, Southeast Asian countries are pacing up the podium to claim the developed nation status. The GDP of the emerging Asian region is growing almost at the rate of around 6%. This hefty economic growth projection lays forth the need for a mammoth rise in energy demand in the Southeast Asian region. Hence, energy sustainability and transition pose the primary challenges to be resolved upfront. These challenges have roots embedded in the finer issues such as – population, gender, poverty, and energy access.

The population growth rate of a nation is an important determinant in recognizing the increase in energy dependency and the need for sustainability. Southeast Asia is at loggerheads with the problem of rapid and ever-increasing cases of population. High GDP Rates in Southeast Asian countries are a clear indication of the growth rates and development happening and the scope for the same. This has also added to the increased demand for energy. Another viable factor is the pressure of electronic facilitation and transition, which extends an ever-increasing burden on the demand for more power due to the increase in consumption using multifarious gadgets by an individual.

The South Asian Energy Corridor: Toward Sustainability and Security. Amritanshu Shukla, PhD, Atul Sharma, PhD & Saurabh Mishra, PhD (Eds.)

Since the renewable energy (RE) sector is fast growing with a lot of untapped potential, companies from across the globe have shown their interest in investing in the sector. Special tax rebates and exemptions have further fueled the investors in the sector. Yet the challenge remains to obtain the right energy mix for varied regions the world over. Hydropower has been the major growing alternative to conventional sources of energy in the Southeast Asian region. Whereas wind, geothermal, bio, and solar have shown lesser contributions to supplement the energy mix. Thus, it is of prime importance that energy sustainability in Southeast Asia is addressed in the context of the twin-faced problem-1. Energy Poverty and related issues, and 2. Development of energy infrastructure.

12.1 INTRODUCTION

Southeast Asia, with its impressive diversity in religion, culture, and history, is home to almost 10% of the world's population. The quantum of population, when set against the fact that the Southeast Asian part of the world is also known to be a vibrant hub of the world's most dynamic economic activities, delivers conclusively that – the region would also be one of the fastest growing regions in terms of energy demand in the world. The increase in population and the motivated urbanization process look to harness more towards attaining better services and facilities in the region. The energy consumption in Southeast Asia has grown two folds in the last couple of decades, and the annual economic growth of the region is exceeding 4%; Adnan Z. Amin, Director General, IRENA has estimates for the region that – "energy consumption is expected to more than double by 2040" (IRENA, 2018). Also, since big share of the population of the Southeast Asian region is hosted in developing countries like Thailand and Indonesia, this all the more swells the fact that 'Urbanization and Population' are the major driving force behind the region's growing energy demand.

Since much of the energy access has been burdened on traditional sources like – coal and other fossil fuels, energy sustainability has been a core issue of concern. All countries bounded by the expanse of their economy and population seek a continuous, adequate, and reliable source that is economically viable and feasible. International Energy Agency (IEA) mentioned in its Southeast Asia Energy Outlook 2017 that demand for oil and coal is expected to swell in the coming years. Until the prime years of the 21st century, the trade of fossil fuels (oil, gas, and coal) in the Southeast Asian region was known to be in a good balance, but the high dependency on oil, gas, and coal reserves and increasing population pressure in the region (Figure 12.1) has exposed the fact that the regional import and export balance is at stake. It is also a major pointer that in the coming years, the region may shift towards being the net importer of traditional fuel resources (IEA, 2019) (Figures 12.1–12.2).

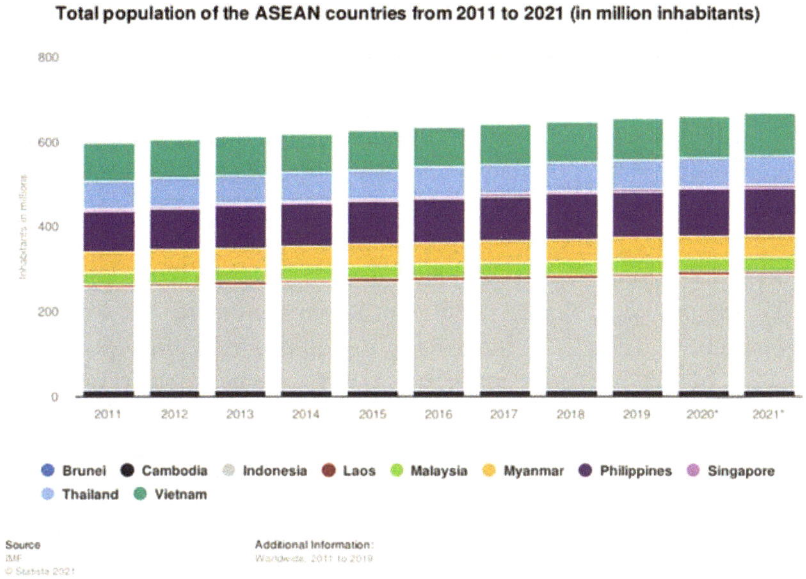

FIGURE 12.1 Total country-wise population distribution.
Source: Statista (2021).

12.2 POPULATION AND ENERGY SCENARIO

Growth and development have been the primary thrust area of Homo sapiens throughout the history of their existence on the face of this earth. Human anthropological records aver that the growth of human civilization and its need for energy have always been the two sides of the same coin. It is true that changes in energy consumption patterns are observed as an increase in population happens as it requires – more transport, per person, of resources, goods, and people; or if population-related growth in material consumption (Holdren, 1991). The threadbare fact is that people need energy; energy is the basis of entire development. However, it is important to know how the energy and population are affecting each other. This is a pertinent factor as its understanding of true nature could be a game changer in achieving energy sustainability.

Inquiries into the area have affected varied results. Counterintuitively, it has been observed that energy demand/consumption in densely populated urban areas in developed or developing societies tends to show economies of scale in various terms, thereby registering lower energy demand. Also, considering the case between the growing and populous countries like ones from the African continent and India and the United States of America have contrasting results in terms of population and energy consumption relationship. While the USA accounts for almost 20% of the world's energy, India, on the other hand, reaps only 05% of the world's energy. Dispersed settlements in thinly populated countries have shown to be more energy-expensive. Also, another important factor may be the share of the population in terms of – age, which may impact energy consumption. As per a study conducted in China, it was shown that – "population size, while it affects energy consumption, has no significant impact on energy production or real energy prices… for a 20% increase in population size, energy consumption increased by 43.6%; conversely, for a 20% decrease in population size, demand for energy decreased by 44.6%" (Huang, 2018). Still, at large, the case can be concluded in favor of population, urbanization, and growing needs which pull on a vicious nexus in terms of the everlasting need for energy consumption (Figure 12.2). Thus, population pressure and the energy needs are entwined in constructing the woes which add up to the struggle for achieving energy sustainability. Increased exposure and accessibility to the gadgets and life aiding solutions in day-to-day life have remarkably affected the way that people look at energy reliability. Walks of life are no more just dependent on human endeavors; Human endeavors are in turn capacitated by availability of various energy sources. High GDP rated are becoming more and more dependent on energy demand which is set up in growing nations as not satiable. Too much dependency on energy imports hangs the balance in negative returns.

12.3 ENERGY POVERTY AND SUSTAINABILITY

The absence of adequate modern sources of energy impedes the presence of decent living conditions. This very lack of modern sources of energy is Energy Poverty (Khanna et al., 2019). The concept of Energy Poverty was developed by Boardman (1991). He expressed – 'energy poverty' based on a unitary marker of the relationship between household income and energy consumption; the case being – if the energy consumption was more than 10% of the household income, the said household was considered to be energy poor. IEA has discussed energy poverty through two attributes: (i) lack of energy for clean cooking; and (ii) deficient access to electricity. It was elaborated in further context to the unavailability of modernity, reliability, and sustainability of energy resources (IEA, 2017). Energy Poverty solution is pertinent for ensuring the sustenance of modern-day lifestyle, employment creation, and overall, a comfortable livelihood (Pereira et al., 2018). Dependency on traditional energy resources for harnessing energy, such as wood, animal waste, and crop residues, is a burden on modern societies, switching to new renewable and non-renewable energy sources (RES) is the need of the hour (Goldemberg, 2000).

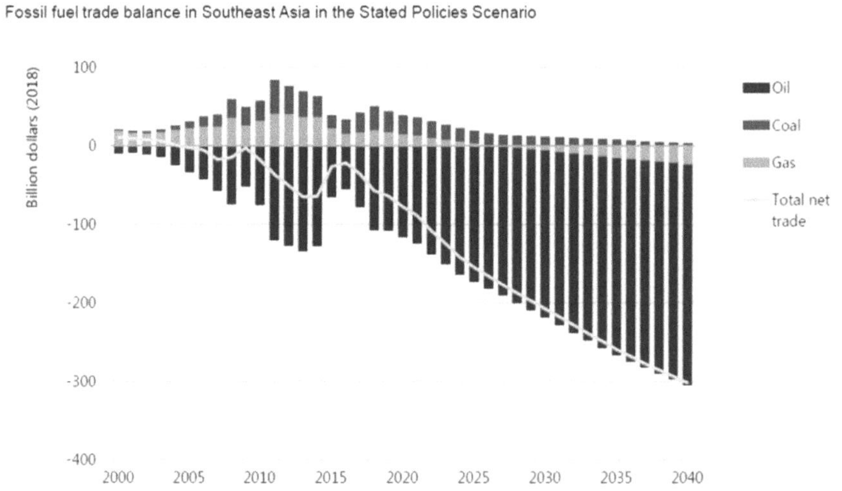

FIGURE 12.2 Southeast Asian fossil fuel trade.
Source: IEA (2019), Southeast Asian Energy Outlook.

Holdren & Smith (2000) argued that energy poverty is related to the energy ladder. The concept of five rung 'Energy Ladder' was given by Hosier & Dowd (1987) as: (i) gathered fuel wood; (ii) purchased fuelwood; (iii) transition fuels; (iv) kerosene; and (v) electricity. Initially conceived by Hosier and Dowd to mark the fuel phases, the energy ladder has come to mature as depicting a transition in energy resources caused by shift from traditional to modern fuel. The consumer will tend to choose more of better fuel as the pay capacity increases. Thus, as the income of people increases, it would primarily dictate the rise in the energy ladder. A fuel's rung is dictated primarily by its cost, a reflection of its cleanliness and efficiency (Goldemberg, 2000). As per the definition of poverty, 15% of 200 million people living in developed nations are energy-poor (IEA, 2017).

The situation of energy sufficiency throughout the world is in a grim state; as per IEA (2017), around 1.1 billion people in the world are not privileged with electricity (Higher rung of the ladder). The condition in Southeast Asian countries is no different 107 million people are electricity deprived. The bulk of the ASEAN population resides in Indonesia, the Philippines, and Myanmar (Figure 12.1), where the challenge is extreme. Indonesia contributes about 5.2% of total electricity poverty in developing Asia (IEA, 2017). Developed/developing countries like Singapore, Brunei, Vietnam, and Malaysia are nearing 100% electrification. The lower rung of the ladder, wherein Biomass is an important component, still shares a huge burden, as per World Bank's Global Tracking Framework 2017 (World Bank, 2017), with around 3.04 billion people being dependent on it. Around 29% of people out of this are living in ASEAN countries. In fact, lack of access to energy is only a part of the problem. Energy affordability is another important reason that forms prelude to the situation of energy poverty.

The estimate of primary energy demand and Fossil fuel import burden by the IEA in its 2019 (Figure 12.3) study confirms that as per the stated policy scenario (STEPS), dependency on fossil fuels is the largest, hovering around 900 million tons of oil equivalent, followed by the harness traditional biomass for energy and the other renewables. This has visibly shot up the import bills of 300 billion dollars. The main reasons behind the rise of demand in the fossil fuel sector are seen as the boom in the automobile sector and the rapid addition to the urban population. In the year 2018, the sale of new cars in – Singapore, Indonesia, Malaysia, Thailand, Vietnam, and the Philippines had risen by 5%, amounting to a sale of nearly 3.4 million units. It is also estimated that the same trend will be in vogue till 2040 (Figure 12.3).

Southeast Asia supports a breathtaking, diverse scenario in terms of per capita GDP, while countries like Singapore and Brunei record a per capita GDP higher than the world and OCED and record an equally high per capita energy demand (Figure 12.4). Whereas other sets of countries like – Indonesia, Myanmar, Thailand, Philippines, Cambodia, and Vietnam are lagging far behind their brethren. Fuel subsidies have a huge impact in the region, ASEAN has been known for its fuel reserves and production, and oil producing countries like Brunei, Indonesia, Malaysia, Myanmar, Thailand, and Vietnam. Due to the severe pressure, most of the countries are now forced to turn into importers rather than importers of the oil. It could be seen as an unfavorable contrast that according to the IEA (2107) reports, the demand for oil in the region is expected to rise from 4.7 million barrels per day to 6.6 million barrels per day, whilst the production of oil is going to take a dip.

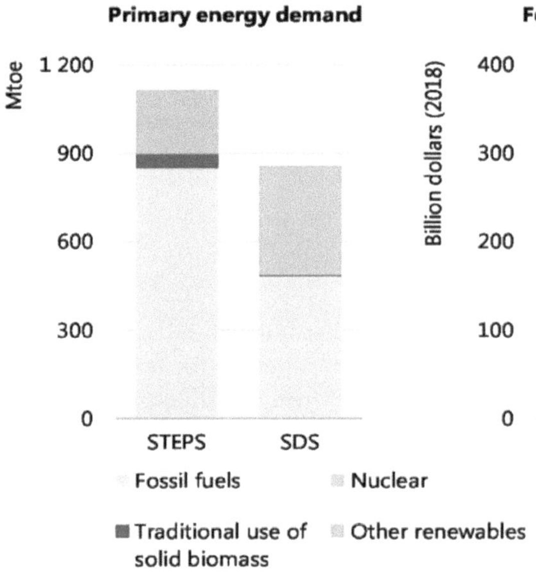

FIGURE 12.3 Comparison of selected energy indicators in the stated policies and sustainable development scenarios, 2040.
Source: Southeast Asia Energy Outlook (2019).

Much of the ASEAN is deficient with a profound public transport system, hence the individual reliance on private transport is high same is true for other industrial endeavors. Recurrence of the fact has forced the nations to help the citizens with fuel subsidies. Countries like Thailand and Vietnam had, of late, decided to use their funds to bear the subsidies. In Thailand, the State Oil Fund was put to purpose to help consumers cope with higher fuel prices to the tune of 50% of any increase in retail prices. While in Vietnam, the – Petroleum Stabilization Fund was used to reduce the price effect of the changes in international oil prices (Figure 12.4).

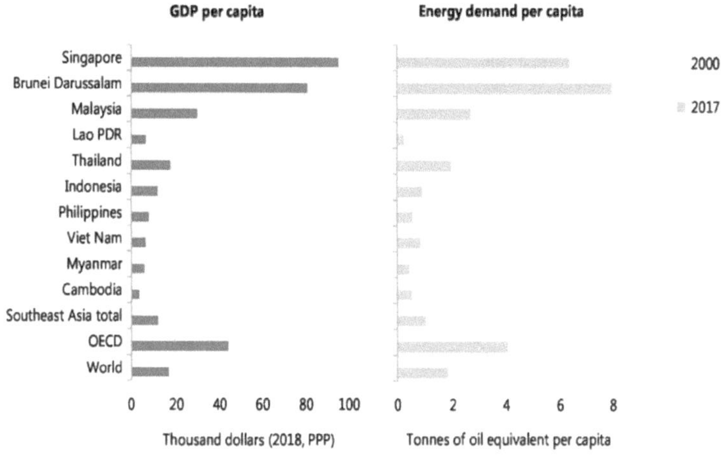

Notes: GDP = gross domestic product; ASEAN = Association of Southeast Asian Nations; OECD = Organisation for Economic Co-operation and Development. PPP = purchasing power parity. Lao PDR = Lao People's Democratic Republic. Energy demand per capita of Lao PDR is 2015 data.

FIGURE 12.4 GDP and energy demand per capita in Southeast Asia.
Source: Southeast Asia Energy Outlook (2019).

It is clear that the fuel subsidies have been a burden on the nation's reserves. Also, the subsidized fuel prices act as a motivation for the general mass to procure more vehicles which in turn would further add to the import expense and require more shelling from the national fund in the form of subsidies. As per the latest IEA (2019) data, the total subsidies in the Southeast Asian countries have been totaled to be around 35 million dollars (Figure 12.5). The bulk of the subsidy is spent on the oil, estimated to be approximately 28 million dollars. As much of the subsidies are provided for fossil fuels and energy generation by them, much of the subsidy is exhausted in the rich urban sectors and fails to reach the underdeveloped. Subsidies have been instrumental in absorbing the burden on citizens to some extent, but at the same time, impress a circumvention in path to achieving sustainability.

Another major player in the energy poverty chain is the issue of energy diversification. Thailand, among the ASEAN countries, stands as the 21st most populous nation in the world. Its landmass makes it the 50th largest nation in the world. Given the fact that – Thailand is 99% electrified (Provincial Electrical Authority of Thailand, 2018), even energy poverty is an issue of concern because – as per IEA (2016) most ,of the population is largely dependent on natural gas as the main means of energy facilitation. The case is that Thailand has its own production of natural gas, but the increased pressure on the resource has forced the country to start to import it. We are living in an age that demands circumstantially more energy; we tend to consume more and more of it and generally choose not to change or diversify the energy resource (Figure 12.5).

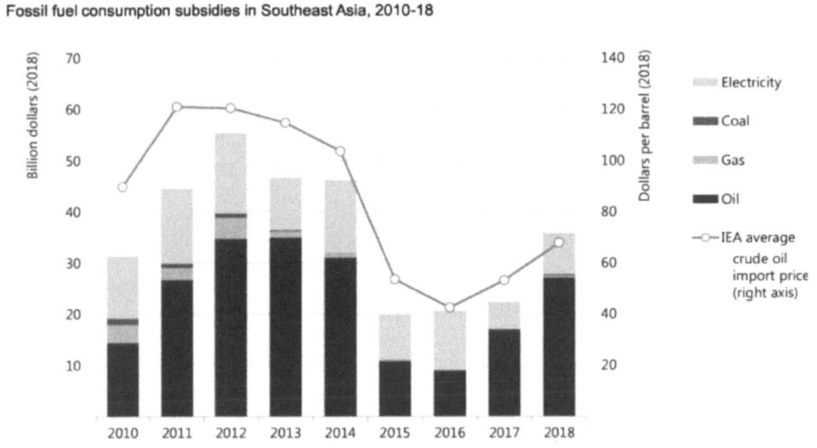

FIGURE 12.5 Subsidies on fossil fuel in Southeast Asia.
Source: Southeast Asia Outlook (2019).

12.4 CONCLUSION

In Southeast Asian countries, it is estimated that domestic coal reserves would continue to serve for another 50-odd years, whereas the natural gas reserves are estimated to be fast depleting and unproductive in the coming few decades, oil reserves, too, are sloth beneath the burden of excessive demand, may soon call it a day. This is bound to impact and slow down their progress and development. Aiming for a long-term sustainable and resilient development trajectory, ASEAN would have to look and transit towards a wider energy mix involving its rich RE possibilities like-Solar, wind, geothermal, and hydro. Investing national funds in energy infrastructure development is straight on the cards, this will help to alleviate the burden of energy poverty and also, in turn, burn out the subsidies. Also, subsidies are not evenly spread out and fail to branch to the areas where they are needed the most. Many ASEAN countries are facing challenges to electrify the rural and farfetched areas where terrain and geographical realities are hard to mitigate. Large portions of the people residing in Indonesia and the Philippines are hosted in island communities, which makes it prohibitively costly to connect to a national grid. Mini grids can cause the solution, but they face an upfront economic and technical challenge. The unavailability of infrastructure and investment, along with a lack to cater remotely located populations or low-density areas, act as impediments to accessibility and availability of energy to people (Khandker et al., 2012).

Action Plan of Action for Energy Cooperation (APAEC) which was ordained for 2016–2025, has a clear target of promoting clean energy and includes (Figure 12.6):

- Reducing energy intensity by 20% by 2020 and 30% by 2025 (based on 2005 level);
- Increasing RE to 23% of ASEAN's energy mix by 2025;
- Promoting clean coal technologies; and
- Encouraging civilian nuclear energy.

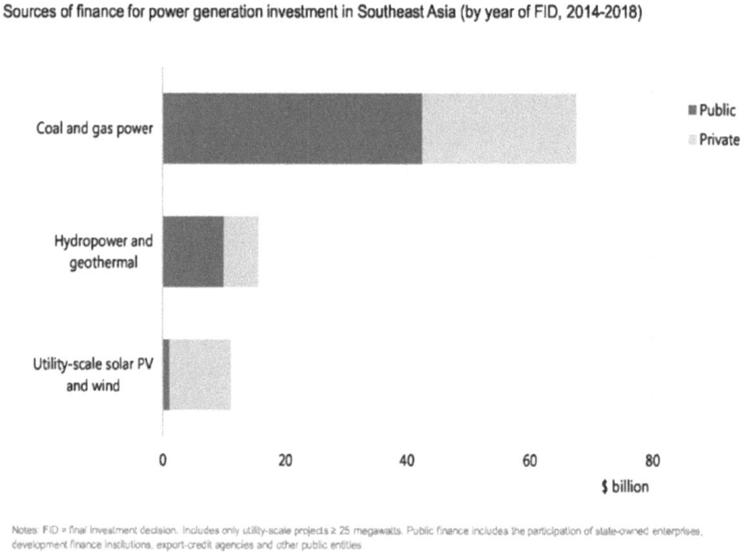

FIGURE 12.6 Finances for power generation in ASEAN.
Source: Southeast Asia Outlook (2019).

ASEAN has made a big move towards addressing the multifaceted problem of achieving sustainability, the seeds have been sown to reach the aspirational target of 23% renewables in the region's primary energy mix by 2025, Southeast Asian countries will have to substantially scale up their deployment of renewables in the power sector, as well as in heating, cooling, and transport (IEA, 2018). The exigent need to invest in the energy infrastructure requires the mobilization of investment and broad participation from the private sector, as well as the targeted use of public funds (IEA, 2019). The odd distribution of the investments in the energy infrastructure (Figure 12.6) poses a grim situation much of the public fund is dedicated to coal and gas power, with little scope left for others. Probably a large part of this investment comes in the form of imports and subsidies. Private players are seen to be encouraged to invest in alternate energy resources like – hydro, geothermal, solar, and wind. Collective effort and positive policies are the need of the hour to keep this momentum going, ASEAN, through its efforts in this direction, could well prove a role model and lead the way for other countries around the world to effect positive change.

KEYWORDS

- **ASEAN**
- **energy consumption**
- **energy poverty**
- **energy scenario**
- **fossil fuel**
- **population**
- **sustainability**
- **urbanization**

REFERENCES

Aaron O'Neill. (2021). *Total Population of the ASEAN Countries from 2011 to 2021*. Statista. https://www.statista.com/statistics/796222/total-population-of-the-asean-countries/ (accessed on 12 August 2023).

Barnes, D. F., Khandker, S. R., & Samad, H. A., (2011). Energy poverty in rural Bangladesh. *Energy Policy, 39*(2), 894–904.

Barnett, A., Bell, M., & Hoffman, K., (1982). *Rural Energy and the Third World*. Pergamon Press, Oxford.

Burke, P. J., (2013). The national-level energy ladder and its carbon implications. *Environment and Development Economics, 18*(4), 484–503.

Gaughan, A. E., Stevens, F. R., Linard, C., Jia, P., & Tatem, A. J., (2013). High resolution population distribution maps for Southeast Asia in 2010 and 2015. *PLoS One, 8*(2), e55882. https://doi.org/10.1371/journal.pone.0055882.

Goldemberg, J., (2000). Rural energy in developing countries. In: Goldemberg, J., Reddy, A. K. N., Smith, K. R., & Williams, R. H., (eds.), *World Energy Assessment: Energy and the Challenge of Sustainability*. United Nations Development Program, New York.

Goldemberg, J., Johansson, T. B., Reddy, A. K. N., & Williams, R. H., (1998). *Energy for a Sustainable World.* Wiley-Eastern Limited, New Delhi.

Grossman, G., & Krueger, A., (1995). Economic growth and the environment. *Quarterly Journal of Economics, 110*, 353–377.

Gwénaëlle, L., Ines, H., Nigel, B., & Sophie, B., (2009). *The Energy Access Situation in Developing Countries: A Review Focusing on the Least Developed Countries and Sub-Saharan Africa.* Published by UNDP and WHO, New York, USA.

Hausmann, R., Rodrik, D., & Velasco, A., (2005). *Growth Diagnostics.* John F. Kennedy School of Government, Harvard University, Cambridge, Massachusetts. http://ksghome.harvard.edu/-drodrik/barcelonafinalmarch2005.pdf (accessed on 12 August 2023).

Heindl, P., & Schussler, R., (2018). *Dynamic Properties of Energy Affordability Measures.* Discussion Paper No. 15-019. http://ftp.zew.de/pub/zew-docs/dp/dp15019.pdf (accessed on 12 August 2023).

Holdren, J. P., & Smith, K. R., (2000). Energy, the environment, and health. In: Kjellstrom, T., Streets, D., & Wang, X., (eds.), *World Energy Assessment: Energy and the Challenge of Sustainability* (pp. 61–110). United Nations Development Program, New York.

Holdren, J. P., (1991). Population and the energy problem. *Population Environment, 12*, 231–255. Springer.

Hosier, R. H., & Jeffrey, D., (1987). Household fuel choice in Zimbabwe: An empirical test of the energy ladder hypothesis. *Resources and Energy, 9*(4), 347–361.

Huang, H., Du, Z., & He, Y. (2018). The effect of population expansion on energy consumption in the canton of China: A simulation from computable general equilibrium approach. *International Journal of Engineering Sciences & Management Research, 5*(1), 19–26.

IEA, (2017). *South East Asia Outlook.*

IEA, (2017). *World Energy Outlook 2017.* OECD Publishing, Paris/International Energy Agency, Paris. https://doi.org/10.1787/weo-2017-en.

IEA, (2019). *South East Asian Energy Outlook.*

Indonesia Investments, (2017). *Electrification Ratio in Indonesia Rose to 92.8% in H1-2017.* August 2017. https://www.indonesia-investments.com/news/todays-headlines/electrification-ratio-indonesia-rose-to-92.8-in-h1-2017/item8067 (accessed on 12 August 2023).

International Energy Agency (IEA), (2017). *Energy Access Outlook 2017: From Poverty to Prosperity.* IEA, Paris.

IRENA, (2018). *Renewable Energy Market Analysis: Southeast Asia.* IRENA, Abu Dhabi.

Khanna, R. A., et al., (2019). Comprehensive energy poverty index: Measuring energy poverty and identifying micro-level solutions in South and Southeast Asia. *Energy Policy, 132*. Elsevier.

Pereira, M. G., Da Silva, N. F., & Freitas, M. A. V., (2018). Energy poverty and climate change: Elements to debate Chapter 17. In: *Oxford Handbook of Energy and Society* (pp. 317–339).

Tonn, B., & Eisenberg, J., (2007). The aging US population and residential energy demand. *Energy Policy, 35*, 734–745.

World Bank Report, (2017). *Global Tracking Framework: Progress Toward Sustainable Energy*. World Bank, Washington, DC.

CHAPTER 13

Role of Energy Education for Developing Secure and Sustainable Energy World

SAURABH MISHRA

Associate Professor, Department of Science and Humanities, Rajiv Gandhi Institute of Petroleum Technology, Uttar Pradesh, India

ABSTRACT

The increasing hunger for energy and relative industrialization and urbanization have been enacted without much thought into – the how and why of energy utilization. This has led to a silent development of threat to energy sustainability, as much of the world's energy has pooled in from traditional resources and thus been dependent on – coal, oil, and gas. At the same time, many solutions are at work to resolve ever-increasing vectors in the energy sustainability scenario. Educating about energy is a natural way to learn and appreciating energy related problems in a most general and effective manner. Technological development and deployment are the key areas that happen to be the cause bearers and are, in turn, dependent upon the energy professionals, well responded by energy-educated citizens. Southeast Asian countries have taken dedicated initiatives to start energy education programs at various levels of education, including higher education. Primary level education being the important foundation, has witnessed a change wherein the primary level wherein only elective courses or optional were offered to the students in the area of energy education. Education has a paramount role in moving towards a sustainable energy world. Energy education focuses on improving knowledge of – the importance and relevance of energy in our daily lives, which often goes unnoticed, it promotes awakening and consciousness towards the social, environmental, and development factors.

The South Asian Energy Corridor: Toward Sustainability and Security. Amritanshu Shukla, PhD, Atul Sharma, PhD & Saurabh Mishra, PhD (Eds.)

Structured efforts in the direction of energy sustainability will develop a trusted value chain which would help to better deal with problem factors being faced in the – logistics, operations, sales, and marketing areas in the energy sector. This would help and aid the stakeholders at all levels by making more comprehensive policies and cost-effective models, which in turn would benefit the consumers.

13.1 INTRODUCTION

Homo sapiens have been beholding the privilege of owning the Earth and its resources; as species that control the food chain. It has borne its consequences, eventually impacting the world in a threatening manner. After so many years of research and development (R&D), humanity still awaits a feasible solution to meet the energy demands of the clan. To most of us, energy has just happened – realizations of the creation of energy are still at naive stages for most of the world's inhabitants. This has resulted in using energy as a driver of daily needs and facilities without looking at its consequences. Since the shift from agrarian economic dependency to the dawn of the industrialization era, world has been looking at more and more energy resources to quench the thirst for energy as the cost for urbanizing and modernizing the world. The process of modernization and industrialization across the world has crossed the path with energy resilience in the modern world. The threat to energy sustainability with high dependency on conservative energy pool of- Coal and Oil created the need to think for energy alternatives and energy conservation. Energy sensitization and education have shown the way ahead to various stake holders across the world in working towards achieving energy security and sustainability.

Energy needs to be appreciated and understood by all in all its forms and galore. A formal introduction to this great vehicle of human development since ages has to be certainly more detailed and coaxial for – "People are familiar with but not educated on popular energy issues" (Mishra, 2016). Knowing energy and its drivers would help to create respect towards its usage and harness from various forms. It would also help to negotiate and mitigate challenges in challenges in energy transition and diversification. Energy transition and transformation has been an important aspect of harvesting change in the energy industry, with no room for any confusion or glitch. The entire activity is highly capital-intensive; a smooth transition is the only way out. Since transition is closely related to acceptability and awareness at the end user level, hence education with aim to sensitize the common masses towards energy sources and relative relevance is almost inescapable. UNESCO (1992) opined that – "Education is critical for promoting sustainable development and improving the capacity of the people to address environment and development issues. It is critical for achieving environmental and ethical awareness, values, attitudes, skills, and behavior consistent with sustainable development and for effective public participation in decision-making."

Most of the education done today happens at the primary levels, wherein middle level school going children are made to appreciate how energy affect human lives. Energy creation and forms are also seen to be discussed in brief in many curricula throughout the world. This thread of attaining knowledge is then left on its own to the students to explore further, which gets a backseat against the developed professional education curriculums and race to monetary survival in the future. There is a need for transformation in the existing education system in order to create the necessary skills for sustainable development (McGrath & Powell, 2016).

Education plays an important role in making citizens conscious about energy, the environment, and their relationship (Keser, 2003). Completely unaware and unabated with the new margins of constraints of energy which are drawn in by each batch of such social waves of new entrants in the world populace, without contributing and appreciating how gravely the energy world needs them to be a part of its rejig. Thus, faced with a low level of awareness, communicating the real essence and challenges of the energy industry is a pertinent aspect of dealing with energy problems (Philo, Happer, & Froggatt, 2012). The problem has been the core of many offshoots of energy dilemmas worldwide, and the situation is no different in the dynamic and vibrant economic and developing hub – the ASEAN.

Southeast Asian countries, in a press release (2012), had resolved to lay more stress on energy efficiency by including it as a part of study at the higher education level. Dr. P. Yoshida (Office of Policy and International Affairs at the U.S. Department of Energy) opined – "It is crucial to begin addressing tomorrow's regional energy issues today, and the way to do that is through education." While many solutions are at work to resolve ever-increasing vectors in the energy sustainability scenario. Energy education is certainly the best and the most cost-effective way to address the problem from the roots. Technological development and deployment are the key areas that can hold sustenance only when met out by successful deployment of energy professionals; well responded to by energy-educated citizens. The education sector is conceived as the poignant player which can host activities such as organizing training and offering new curricula at all levels (Win, 2014). UNESCO official development assistance has as an important part of the project – "Promotion of Energy Science Education for Sustainable Development in Asia."

In a study conducted in India, the Asian counterpart of ASEAN reg,arding the awareness and understanding of connotations of energy in people, it came out that 63% of the people understand and term Gasoline as a synonym to energy; only 5% of people recognized other forms of energy (Figure 13.1). It was also found that most of the respondents were aware of the energy-related terms but could not translate them into understanding further. Hence, there is a yawning gap that needs to be addressed in the area of energy education and awareness (Mishra, 2016). Southeast Asian countries have taken dedicated initiatives to starting energy education programs at various levels of education, including higher education. There has been a shift in the primary level wherein only elective courses or optional were offered to the students in the area of energy education. But much has undergone change, and a much more strengthened education framework has been put in place starting at school and undergraduate levels. Education thus would serve in its most primitive endeavor to enlighten the people as prime stakeholders in their role as consumers towards energy issues.

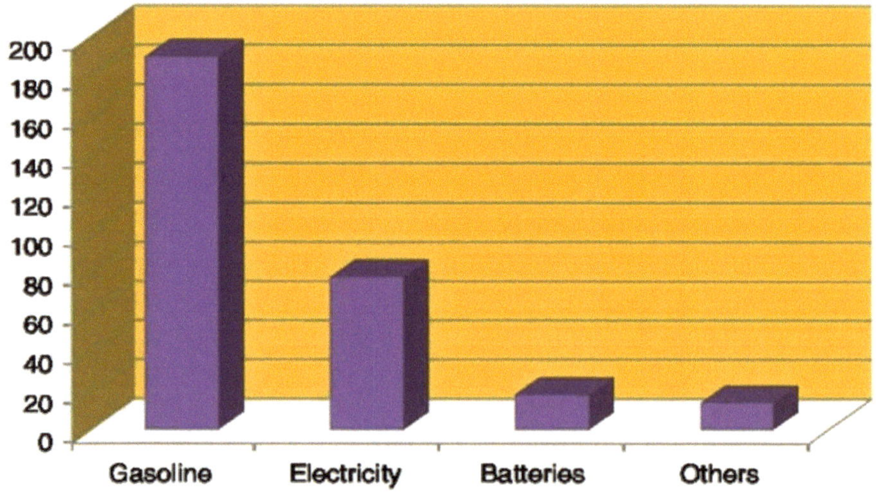

FIGURE 13.1 Connotations of energy (Mishra, 2016).

13.2 EDUCATION, EMPLOYMENT, AND GENDER

Education gives way to employability, and thus, the energy infrastructure and investments need to be overhauled and swelled to attract talented professionals as stakeholders of future gateways to energy excellence. In Southeast Asian countries, gender and employment are important factors for conceiving an energy-secure and sustained world. Energy consumerism holds no bar to the gender notion, but persistent inequalities in the wake of economic semblance and lack of autonomy have presented forth a shaded path. According to an estimate, in South Asia, women earn only 20%–40% in comparison to their male counterparts, even after the now improved education and employment venues. To ensure equitable access to energy, these differences must be factored in. World Bank's Gender at Work Report (2014) suggests that "on virtually every global measure, women are more economically excluded than men." Traditionally, women are tasked with domestic chores and minimal-level work; thus, women are subjected to energy poverty without much awakening and realization. Time spent by women is almost twice as much as men – on domestic work, and they work longer hours than men do when all work—paid or unpaid—is considered (UNDESA, 2010; Duflo, 2012). Releasing time for education and for economic, political, and social opportunities is a necessary first step (World Bank, 2012). Gender participation is an important participatory factor in achieving sustenance. It remains a fact that men and women have different energy consumption and access.

In Southeast Asian countries, the population is growing at a fast pace, and with a growing population, it is mandated that the generation of new resources and able professionals is required. The gender divide at any level is further going to be a barrier to – the expansion, generation, and deployment of new technologies. ASEAN being a hub of economic growth wo, would remain a pertinent question, how to manage a workforce that is literate and efficient in energy-related areas. Education stands as an important part in completing the cycle (Figure 13.2) of 4E's – education, employability, energy resources, and economic development, which holds the key to generic solutions to energy-related problems.

A ladder model based on the role of education in achieving sustainability (Figure 13.3) can be carved out by looking at various stakeholder and the operation of education on them and their contribution to riveting the energy sector of the Southeast Asian countries. The model aims to divide the stakeholders in the energy framework in terms of education and its impact and visualize how a strengthened and resilient energy system can be put in place for energy sustainability and development. It is proposed to visualize the model as a ladder. The first rung is – education for all (building awareness and knowledge propagation), the second rung is – education for consumers, the third rung is – education for policymakers, the fourth rung is education for investors, and the fifth rung is – education for professionals.

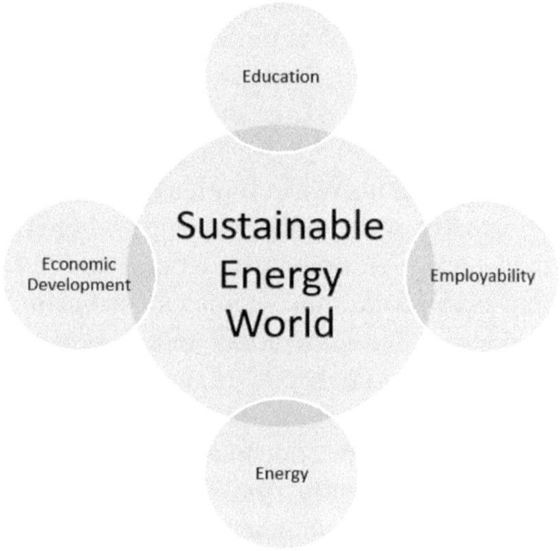

FIGURE 13.2 Four E's of the sustainable energy world.

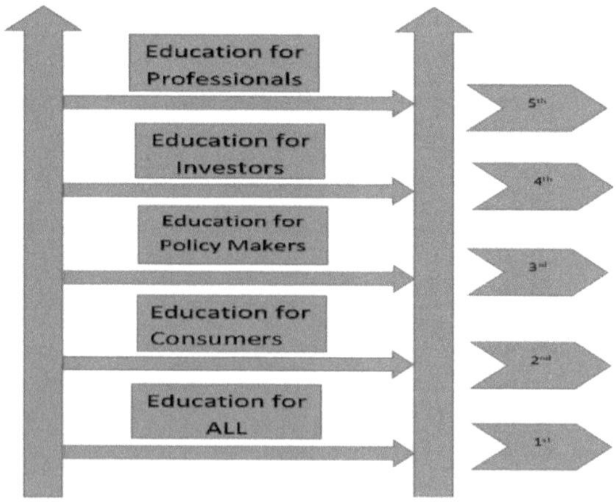

FIGURE 13.3 Ladder model for accessing the impact of education on various stakeholders.

Education stands as the foundation rung of the ladder, wherein the after factors of education are meant as the post rungs of the ladder. Education has a paramount role in moving towards a sustainable energy world. UNESCO (2012) has Education for Sustainable development (ESD) as one which aims to help people develop the attitudes, skills, and knowledge to make informed decisions for their and others benefit, now and in the future, and to act on these decisions. Thus, education with an aim towards sustainable development should have a continuity starting from the school level to the university level courses. It is also felt that learning through education is a process of – introspection, critical evaluation, and mutual dialog with one's own self to develop a consciousness (Van der Merwe & Albertyn, 2009). As per the second rung of the ladder, education helps to equip and empower the consumer towards ensuing consumer activities by honing the technology-related aspects and social behaviors of people (McLaughline, 2008). In a study, it has been observes that less educated people lack awareness and knowledge about the RE resources and options (Muresan et al., 2017), thus are potential obstacles in the path of sustainability achievement. It also noted that in terms of the third level of the ladder for policymakers, an ESD background in energy helps to assimilate the – awakening and consciousness towards the social, environmental, and development factors, and there is a background developed in terms of a connect between the higher-level education systems and the policy maker (McCowan, 2016). The generation of new knowledge and awareness with appropriate training in the areas can help to raise the leaders of tomorrow (Ciegis & Gineitienė, 2006; De Lange, 2013; Faham et al., 2017). Too much dependency on Fossil fuels in the Southeast Asian Region has led to money outflow, an evolving problem for most of the countries in the region (Figure 13.4). Decision-making and energy literacy are entwined endeavors toward creating a base of knowledgeable people who can support the implementation of smart and forward-looking policies (Yeh, Huang, & Yu, 2017).

IEA elaborated into context to unavailability of modernity, reliability, and sustainability of energy resources (IEA, 2017). The bulk of the ASEAN population resides in Indonesia, the Philippines, and Myanmar, where the challenge is extreme. Indonesia contributes about 5.2% of total electricity poverty in developing Asia (IEA, 2017). The fourth rung of the ladder corresponds to potential investors in the area of energy resource development. The road to achieving sustainability in the sector is untamable unless – a higher level of commitment investment in the energy sector in ASEAN has been dominated by the public sector, whereas it is high time (Figure 13.5) that the players other than the government start to take part in the infrastructural development of the energy sector. This cannot be realized unless educated investors are attracted with proper support from the equally responding policymakers. Finally, the fifth rung of the ladder prepares the employability options and opportunities that tend to attract the new talent pool, which promotes the strengthening, expansion, and reorienting of the energy base. Ploughing back benefits reaped from all the previous rungs of the ladder. In fact, in the proposed ladder model, each post rung of the ladder draws heavily upon the previous rung.

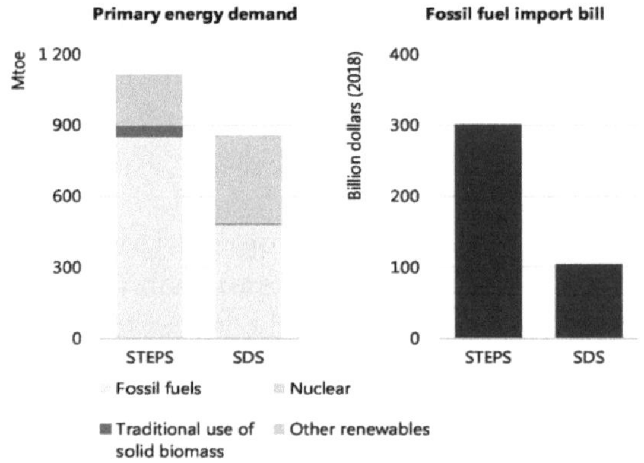

FIGURE 13.4 Fossil fuel imports burden due to dependency on traditional fuel.
Source: Southeast Asia Energy Outlook (2019).

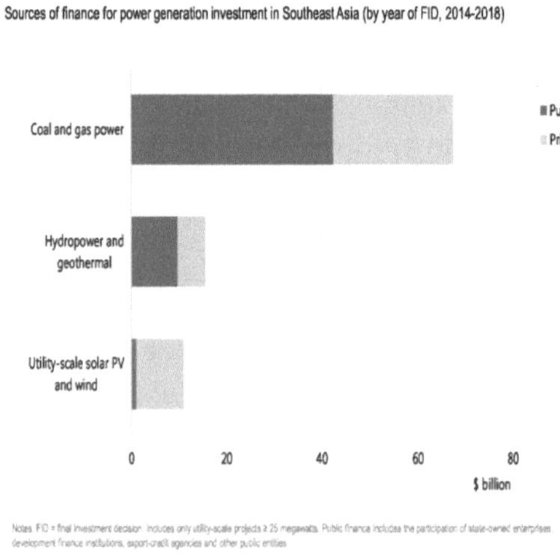

FIGURE 13.5 Finances for power generation in ASEAN.
Source: South–East Asia Outlook (2019).

13.3 CONCLUSION

Energy sustainability can be visualized in many ways and under plethora of conditions. Bonnett (2002) expressed that "sustainability" could be approached from the perspective of a frame of mind. Sustainability involves aiming to achieve a balance between generations in the distribution of energy, wealth, employability, and development (Kopnina, 2014). Energy education should focus on improving knowledge of – issues related to energy and allied factors such as – resource identification and creation, awareness, and transition, and developmental and environmental impacts. Structured efforts in the direction of energy sustainability will develop an ability to assimilate and resolve current problems relevant to energy, analyze and critically appreciate information relevant to energy issues (Lingga, Andri, & Swasti, 2019), and better able to create a business models and policies to the benefit of all stakeholders at various levels.

Hence education is essential to develop – policy frameworks that are key to determining a nation's ability to incentivize and deploy new technologies, attract private capital, modernize production, transmission, and distribution, and finally expand access to energy. Evolution of ESD in the energy sector as concept can be looked at by promoting literacy beyond general knowledge (Saadatian et al., 2009; Lozano et al., 2013). It helps to connect the multilevel and multidimensional stakeholders for better governance and growth (Kolleck et al., 2017). Thus, increasing public awareness through a better understanding of the concept of sustainable development (Lozano, 2006).

Energy-literate citizens who can play a role in the decision-making process and are committed to action help to realize a successful paradigm shift in terms of energy use (DeWaters & Powers, 2013). Energy literacy is not only about knowledge (cognitive understanding) but also considers aspects related to influence and behavior. There are three aspects of energy literacy, namely – cognitive, affective, and behavioral (DeWaters & Powers, 2013; Chen & Chen, 2015; Lee et al., 2015). A comprehensive approach to energy education is needed to develop a sustainable energy world. In the Southeast Asian context, it is conclusive that the region is burdened under the pressure of conventional fuel resources and hence seems derailed in the sense of energy sustainability. Issues ranging from – lack of – awareness, access, opportunity creation, funding, and favorable policies, along with disbursement of subsidies, have hit hard the sustainability program. The overall point in the case is that the proposed five-rung model can help to grove energy education and promote a sustainable and energy-proficient world along with the silver lining against the odds; ASEAN is a natural hub to be better able to harness the advantage courtesy of its geographical and situational location.

KEYWORDS

- **ASEAN**
- **decision-making**
- **employment**
- **energy education**
- **energy sources**
- **modernity**
- **sustainability**

REFERENCES

Acikgoz, C., (2011). Renewable energy education in Turkey. *Renew. Energy, 36*, 608–611.

Akhanda, M. A. R. (1998). Present status and future direction of renewable energy education in Bangladesh. Proc. Sixth International Symposium on Renewable Energy Education. 26–28 November 1998. Tata Energy Research Institute, New Delhi, 69–73.

Approach Paper for IRENA Capacity Building Strategy, (2012). International Renewable Energy Agency. www.irena.org (accessed on 12 August 2023).

ASEAN Promotes Energy Efficiency Through Education, (2012). ASEAN Secretariat News, Press Release. https://asean.org/asean-promotes-energy-efficiency-through-education/ (accessed on 12 August 2023).

Ball, A., & Grubnic, S., (2007). Sustainability accounting in the public sector. In: Unerman, J., Bebbington, J., & O'Dwyer, B., (eds.), *Sustainability Accounting and Accountability* (pp. 243–265). Routledge, Abingdon.

Berkovski, B., (1998). Global renewable energy education and training: World solar program 1996–2005. In: *Proceedings of the Sixth International Symposium on Renewable Energy Education* (pp. 11–23). Tata Energy Research Institute. New Delhi.

Berkovski, B., & Gottschalk, C. M., (1997). Strengthening human resource for new and renewable energy technologies of the 21st century: UNESCO engineering education and training program. *Renew. Energy, 10*, 441–450.

Bonnett, M., (2002). Education for sustainability as a frame of mind. *Environmental Education Research, 8*(1), 9–20. Available at: http://rrojasdatabank.info/sustbonnett.pdf (accessed on 12 August 2023).

Bourodinos, E. L. (1992). Renewable energy sources education and research as an education for survival. Progress in Solar Energy Education (Proceedings of The First International Symposium on Renewable Energy Education, 19–20 June 1991, Borlänge, Sweden). *1*, 12–16.

Broman, L., & Ott, A., (1988). Solar education: The way forward. *Sun. Work Eur., 6*, 245.
Broman, L., & Ott, A., (1992). On the need for solar energy education. *Prog. Sol. Energy Educ., 1,* 23–25.
Bryant, S., & Olson, J., (2009). Training carbon management engineers: Why new educational capacity is the biggest hurdle for geologic CO_2 storage. *Energy Procedia, 1*, 4741–4748.
Chaar, L. E., & Lamont, L. A., (2010). Nourishing green minds in the land of oil. *Renew. Energy, 35*, 570–575.
Charters, W. W. S., (1996). The solar challenge: Promoting effective education and training. *Prog. Sol. Energy Educ., 4*, 2–4.
Chen, K. L., Liu, S. Y., & Chen, P. H. (2015). Assessing multidimensional energy literacy of secondary students using contextualized assessment. *The International Journal of Environmental and Science Education, 10*(2), 201–218. doi:10.12973/ijese.2015.241a.
Chitauro, J. J. (1997). Strategies in the introduction of renewable energy studies in the education systems. *Progress in Solar Energy Education, 5,* 20–21.
Ciegis, R., & Gineitienė, D., (2006). The role of universities in promoting sustainability. *Inzinerine Ekonomika-Engineering Economics Journal, 48*(3), 56–62. Available at: http://inzeko.ktu.lt/index.php/EE/article/view/11373/6097 (accessed on 12 August 2023).
Cortese, A. D., (2003). The critical role of higher education in creating a sustainable future. *Plan High Educ.,* 15–22.
De Lange, D. E., (2013). How do universities make progress? Stakeholder-related mechanisms affecting adoption of sustainability in university curricula. *Journal of Business Ethics, 118*(1), 103–116. Available at: http://dx.doi.org/10.1007/s10551-012-1577-y.
DeWaters, J., & Powers, S., (2011). Energy literacy of secondary students in New York State (USA): A measure of knowledge, affect, and behavior. *Energy Policy, 39*, 1699–1710. doi: 10.1016/j.enpol.2010.12.049.
DeWaters, J., & Powers, S., (2013). Establishing measurement criteria for an energy literacy questionnaire. *J. Environ. Educ., 44*, 38–55. doi: 10.1080/00958964.2012.711378.
Dias, R. A., Mattos, C. R., & Balestievi, J. A. P., (2004). Energy education: Breaking up the rational energy use barriers. *Energy Policy, 32*, 1339–1347.
Douglas, B., (2001). RE education and training. *Refocus, 2*, 39–41.
El Fadel, M., Rachid, G., El Samra, R., Bou, B. G., & Hashisho, J., (2013). Knowledge management mapping and gap analysis in renewable energy: Towards a sustainable framework in developing countries. *Renew. Sustain. Energy Rev., 20*, 576–584.
Energy Training and Capacity Building: Programme 2011, International Energy Agency, 2010.
Faham, E., Rezvanfar, A., Mohammadi, S. H. M., & Nohooji, M. R., (2017). Using system dynamics to develop education for sustainable development in higher education with an emphasis on the sustainability competencies of students. *Technological Forecasting and Social Change, 123*(3), 307–326. Available at: https://doi.org/10.1016/j.techfore.2016.03.023.
Fara, V. L., Popa, M., & Creanga, I., (1993). On the creation of the solar energy section of the engineering physics department at the Polytechnic Institute of Bucharest. *Prog. Sol. Energy Educ., 2*, 13, 14.
Fridleifsson, I. B., (2010). *Capacity Building in Renewable Energy Technologies in Developing Countries*. World Energy Congress. Montreal, Canada.
Garg, H. P., & Kandpal, T. C., (1991). *Renewable Energy Education*. Paper presented at the Indian National Solar Energy Convention, Pune.

Garg, H. P., & Kandpal, T. C., (1993). *Global Trends in Renewable Energy Education.* Paper presented at the workshop on material science and physics of non-conventional energy sources, ICTP, Trieste, Italy.
Garg, H. P., & Kandpal, T. C., (1996). *Renewable Energy Engineering Education.* New Delhi: Omega Scientific Publishers.
Gender at Work: A Companion to the World Development Report on Jobs, (2014). Washington, DC.
Guiterrez-Martin, F., & Huttenhain, S. H., (2003). Environmental education: New paradigms and engineering syllabus. *J. Clean Prod., 11*, 247–251.
Gupta, C. L., (2003). Role of renewable energy technologies in generating sustainable livelihoods. *Renew. Sustain. Energy Rev., 7*, 155–174.
Hashim, H., & Ho Wai, S., (2011). Renewable energy policies and initiatives for a sustainable energy future in Malaysia. *Renew. Sustain. Energy Rev., 15*, 4780–4787.
Hasnain, S. M., Elami, U. A., Al- Awaji, S. H., Aba- Oud, H. A., & Smiai, M. S., (1995). Prospects and proposal for solar energy education programs. *Appl. Energy, 52*, 307–314.
Hower, J. C., (2002). Editorial: Coal education. *Int. J. Coal. Geol., 52*, 1, 2.
IAEA Capacity Building for Sustainable Energy Development. International Atomic Energy Agency (IAEA). Information Series 02-01566/FS Series 2/01/E.
IEA (2017). Southeast Asia Energy Outlook 2017, IEA, Paris, https://doi.org/10.1787/9789264285576-en.
IEA (2019). Southeast Asia Energy Outlook 2019, IEA, Paris. https://www.iea.org/reports/southeast-asia-energy-outlook-2019.
Jain, P. K., Lungu, E. M., & Buti, M., (2002). Renewable energy education in Botswana; needs, status, and proposed training programs. *Renew. Energy, 25*, 115–129.
Javed, F., (1996). Renewable energy education for vocational schools. *Progress in Solar Energy Education, 48–52.*
Jennings, P., (2009). New directions in renewable energy education. *Renew. Energy, 34*, 435–439.
Jennings, P., & Lund, C., (2001). Renewable energy education for sustainable development. *Renew. Energy, 22*, 113–118.
Kandpal, T. C., & Garg, H. P., (1999). Energy education. *Appl. Energy, 64*, 71–78.
Karabulut, A., Gedik, E., Kecebas, A., & Alkam, M. A., (2011). An investigation on renewable energy education at the university level in Turkey. *Renew. Energy, 36*, 1293–1297.
Karekezi, S., Turyareeba, P., & Ewagata, E., (1997). Renewable energy training in Kenya –the case of solar photovoltaics. *Prog. Sol. Energy Educ., 5*, 81–86.
Keser, O., (2003). Energy, environment, and education relationship, in developing countries' policies: A case study for Turkey. *Energy Sources, 25*(2)23–133.
Kiatsirioat, T., & Hinmlabh, J., (1996). Renewable energy education in Thai universities. *Progress in Solar Energy Education w4*, 56, 57.
Kiatsirioat, T., & Manurung, R., (1996). Renewable energy education in Thai universities. *Prog. Sol. Energy Educ., 4*, 56–57.
Kojima, N., (1997). Renewable energy in the educational arena in Japan. *Prog. Sol. Energy Educ., 5,* 71–73.
Kolleck, N., Jorgens, H., & Well, M., (2017). Levels of governance in policy innovation cycles in community education: The cases of education for sustainable development and climate change education. *Sustainability, 9*(12), 1966-1982. Available at: www.mdpi.com/2071-1050/9/11/1966 (accessed on 12 August 2023).

Koltun, M. M., (1992). Solar energy education in the USSR. *Prog. Sol. Energy Educ., 1*, 29–32.
Kopnina, E., (2014). Contesting 'environment' through the lens of sustainability: Examining implications for environmental education (EE) and education for sustainable development (ESD). *Culture Unbound: Journal of Current Cultural Research, 6*(5), 931–947.
Lalic, D., Popovski, K., Gecevska, V., Vasilevska, S. P., & Tesic, Z., (2011). Analysis of the opportunities and challenges for the renewable energy market in the Western Balkan countries. *Renew. Sustain. Energy Rev., 15*, 3187–3195.
Lawand, T. A., & Ayoub, J., (1996). Renewable energy activities in rural Argentina –educational aspects. In: *Proceedings of the World Renewable Energy Congress* (pp. 1194–1198).
Lee, L. S., Lee, Y. F., Altschuld, J. W., & Pan, Y. J., (2015). Energy literacy: Evaluating knowledge, affect, and behavior of students in Taiwan. *Energy Policy, 76*, 98–106. doi: 10.1016/j.enpol.2014.11.012.
Lingga, N. P., Andri, P. K., & Swasti, M., (2019). Energy literacy-based learning activities on female students. *International Journal of Scientific & Technology Research, 8*(8), 1711–1715.
Lozano, R., (2006). Incorporation and institutionalization of SD into universities: Breaking through barriers to change. *Journal of Cleaner Production, 14*(9–11), 787–796. Available at: https://doi.org/10.1016/j.jclepro.2005.12.010.
Lwin, K., & Torok, S. J., (1997). Hope for longer terms sustainable development and utilization of NRSE in Asia and the Pacific via training, capacity buildings, and educational efforts. *Program Sol. Energy Education, 5*, 27–30.
Mahmoudi, H., Abdellah, O., & Ghaffour, N., (2009). Capacity building strategies and policy for desalination using renewable energies in India. *Renew. Sustain. Energy Rev., 13*, 921–926.
Martinot, E., (1998). Energy efficiency and renewable energy in Russia: Transaction barriers, market intermediation and capacity building. *Energy Policy*, 905–915.
McCowan, T., (2016). Universities and the post-2015 development agenda: An analytical framework. *Higher Education Journal, 72*(4), 505–523. Available at: https://doi.org/10.1007/s10734-016-0035-7.
McGrath, S., & Powell, L., (2016). Skills for sustainable development: Transforming vocational education and training beyond 2015. *International Journal of Educational Development, 50*(10), 12–19. Available at: https://doi.org/10.1016/j.ijedudev.2016.05.006.
McVeigh, J. C., (1982). *Training in Solar Energy: Curriculum Development*. Report submitted to UNESCO. Paris.
Mishra, S., (2016). Renewable energy awareness and education in India. *Current and Sustainable/ Renewable Energy Reports, 3*(3, 4). Springer.
Mulugetta, Y., (2008). Human capacity and institutional development towards a sustainable energy future in Ethiopia. *Renew. Sustain. Energy Rev., 12*, 1435–1450.
Nahar, N. M., (1998). Role of renewable energy education in the development of arid regions. In: *Proceedings of the Sixth International Symposium on Renewable Energy Education* (pp. 221–226). New Delhi: Tata Energy Research Institute.
Naidu, B. S. K., (1998). India's leadership in renewable energy development for a sustainable future. In: *Proceedings of the Sixth International Symposium on Renewable Energy Education* (pp. 283–298). New Delhi: Tata Energy Research Institute.
Negro, S. O., Floortje, A., & Hekkert, M. P., (2012). Why does renewable energy diffuse so slowly? A review of innovation system problem. *Renew. Sustain. Energy Rev., 16*, 3826–3246.

Nicklas, M., (1993). Energy politics, can we achieve a sustainable energy path? *Solar Energy, 287–297.*
Olayiwola, F. R., (2001). National renewable energy policy objectives and programs in Botswana. *Renew Energy, 24*, 419–437.
Othman, M. Y., & Kamaruzzaman, S., (1999). Renewable energy education for ASEAN. *Renew Energy, 16*, 1225–1230.
Paltsev, S., Mehling, M., Winchester, N., Morris, J., & Ledvina, K., (2018). *Pathways to Paris: Association of Southeast Asian Nations (ASEAN) Technology and Policy Options to Reduce GHG Emissions*. MIT Joint Program Special Report. Pathways to Paris: ASEAN | MIT Global Change.
Pervez, I., & Hasan, M., (1996). Importance of renewable energy education in Malaysia. *Prog. Sol. Energy Educ., 4*, 31–33.
Philo, G., Happer, C., & Froggatt, A., (2012). Climate Change and Energy Security: Assessing the Impact of Information and Its Delivery on Attitudes and Behavior. In UK Energy Research Center.
Pinter, J., (1991). Solar energy in Hungary. In: *Proceedings of ISES Solar World Congress* (pp. 3842–3845). Denver, Colorado: Pergamon Press.
Scientific and Cultural Organization (UNESCO), Paris (France). *UNESCO Workshop on Training in Solar Energy*. Cadarache, France.
Shah, B., (1990). Energy *Education*. New Delhi (India): Northern Book Center.
Singh, Y. P., (1990). In: Shah, B., (ed.), *Energy Engineering and Education in 'Energy Education.'* New Delhi: Northern Book Center.
Singh, Y. P., & Ram, V. G., (1996). Renewable energy education: An imperative for our survival. In: Garg, H. P., & Kandpal, T. C., (eds.), *Renewable Energy Engineering Education* (pp. 109–120). New Delhi: Omega Scientific Publishers.
Survey on Education and Training in the Field of New and Renewable Sources of Energy. UNESCO report no. TER/ESR/395, United Nations Educational.
UNESCO, (2012). *2012 Report on the UN Decade of Education for Sustainable Development, Shaping the Education of Tomorrow*, Abridged, UNESCO, Paris.
United Nations Department of Economic and Social Affairs (UNDESA), (2010). *The World's Women 2010: Trends and Statistics.* New York.
Van Der, M. M., & Albertyn, R. M., (2009). Transformation through training: Application of emancipatory methods in a housing education program for rural women in South Africa. *Community Development Journal, 45*(2), 149–168. Oxford University Press. Available at: https://doi.org/10.1093/cdj/bsp001.
Wade, H. A., (1998). Human resource development for solar energy. In: *Proceedings of the Sixth International Symposium on Renewable Energy Education* (pp. 206–220). New Delhi: Tata Energy Research Institute.
World Development Indicators, https://datacatalog.worldbank.org/search?q=world%20development%20indicator.
World Development Report 2012: Gender and Development, (2012). Washington, DC.
Yeh, S. C., Huang, J. Y., & Yu, H. C., (2017). Analysis of energy literacy and misconceptions of junior high students in Taiwan. *Sustain., 9*. doi: 10.3390/su9030423.
Zografakis, N., Menegaki, A. N., & Tsagarakis, K. P., (2008). Effective education for energy efficiency. *Energy Policy, 36*, 3226–3232.

CHAPTER 14

Role of Energy in the Development of Agriculture Sector in South Asia

MUKUL SHARMA,[1] RAVEENA KARGWAL,[2] DEEPALI ATHEAYA,[1] and ANIL KUMAR[3,4]

[1]*Department of Mechanical Engineering, Bennett University, Greater Noida, Uttar Pradesh, India*

[2]*Department of Processing and Food Engineering, College of Agricultural Engineering and Technology, C.C.S. Haryana Agricultural University, Hisar, Haryana, India*

[3]*Department of Mechanical Engineering, Delhi Technological University, Delhi, India*

[4]*Center for Energy and Environment, Delhi Technological University, Delhi, India*

Abstract

Agriculture is the lifeline for South Asian countries, which provides more than 35% of employment in South Asian countries. This sector also actively participates in the GDP of nations. After industrialization, the machinery and implements which run on energy were deployed in the agriculture sector. This decreased the manual labor and time consumed during agricultural activities and resulted in a rise in production and profitability. At present, agriculture is energy-intensive, and in South Asian countries such as India and Pakistan, energy consumption in the agriculture sector is more than 12%, excluding the energy consumed during fertilizer and pesticide production. In this chapter, various agricultural activities involved during pre-harvesting and post-harvesting are discussed. The present role of conventional energies such as

The South Asian Energy Corridor: Toward Sustainability and Security. Amritanshu Shukla, PhD, Atul Sharma, PhD & Saurabh Mishra, PhD (Eds.)

electricity and diesel and non-conventional energies, viz. solar energy, wind energy, biomass energy, and biofuel in agriculture activities, are discussed. The current need of agriculture is to deploy sustainable agricultural practices with non-conventional energy sources to make this sector more productive and profitable for farmers.

14.1 Introduction

Agriculture is the primary employment of the citizens of South Asian countries. A high percentage of the total population is working in the agriculture sector across various South Asian countries (Nawaz & Farooq, 2021). As agriculture is the leading economic sector for people living in various South Asian countries, its share in GDP is high. Table 14.1 consists of the percentage of the labor force (Nawaz & Farooq, 2021) and the share of agriculture in GDP for various South Asian countries.

TABLE 14.1 Labor Force in Agriculture and Share of Agriculture in GDP of Various South Asian Countries

SL. No.	Name of Country	Percentage of Labor Force	Percentage Share of Agriculture in GDP*
1.	Afghanistan	59.0	25.8
2.	Bangladesh	44.0	12.7
3.	Bhutan	92.8	15.8
4.	India	54.0	16
5.	Nepal	92.9	24.3
6.	Pakistan	38.6	22
7.	Sri Lanka	43.0	7.4

*Data source World Bank (https://data.worldbank.org/indicator/NV.AGR.TOTL.ZS).

South Asian region consists of 11.71% of the land of Asia. It has a high variation in soil types, weather conditions, and agriculture systems (Srinivasarao et al., 2019; Nawaz & Farooq, 2021). An increase in population has disturbed farming practices and sustainability in farming across the region. After industrialization, agriculture practices developed, and the South Asian region's farmers adopted new modern agriculture practices. Before industrialization, the farmers depended on rainfall for agriculture. Land preparation, cultivation, and other pre- and post-harvesting processes depend on draft livestock and conventional implements. Draught animals such as bullocks draw these conventional implements. Although this practice was economical, it was time-consuming and resulted in low production due to a lack of irrigation technology and energy. The whole process needed a large extent of the workforce for achieving appropriate farm produce.

After industrialization, the agriculture sector developed to a high extent. The use of tractors instead of bullocks reduced the time needed for land preparation and transportation and decreased the workforce needed in agriculture. Electricity-based pumping system reduces the dependency of the farmer on rainfall. It turns into increasing agriculture production as well as farmer income. The use of chemical fertilizers and pesticides also helps in increasing crop production. Post-harvesting processes such as food processing and storage also help in increasing the farmer's income. Modern farming techniques helped increase production and profitability to some extent. Still, nowadays, the increased utilization of chemical fertilizers and pesticides results in low farmer income, soil infertility, and high GHG emissions (Nawaz & Farooq, 2021). Also, climate change, uneven rainfall trends, heatwaves, disease outbursts, insect attacks, etc., had increased, making agriculture more challenging and less profitable.

During agricultural activities, the farmer in the South Asian region depends upon various energies for crop production. These are electrical energy generated using various conventional and non-conventional energy sources and generated using crude petroleum oil. Electrical energy is generated using various sources such as coal, oil, natural gas, hydropower, etc. (Schnepf, 2004). Nevertheless, these sources are limited and cannot be renewed. Therefore, RE sources, viz. solar energy, wind energy, biomass, and biofuel energy, are encouraged for this purpose (Kumar et al., 2021; Sharma & Kumar, 2018). Several new solar energy technologies have been developed for the efficient generation of electricity. These technologies are solar PV (photovoltaics), hybrid PV/T, solar thermal, etc. (Saurabh, Atheaya, & Kumar, 2020). Apart from this, solar energy is also utilized for drying food crops and heating water.

This chapter focuses on various processes followed in the modern agriculture technique across South Asian countries, the use of various implements in those techniques, the role of conventional and non-conventional energies in the development of various agriculture activities, and sustainable agriculture techniques which are economical and can be followed to grow farmer profitability to a higher extent and make the agriculture sector more profitable.

14.2 VARIOUS PROCESSES INVOLVED IN AGRICULTURE

To get the product through modern agriculture technology, a farmer follows the following processes:

1. **Pre-Harvesting Process:** This involved the various techniques initialized from land preparation until the farmer harvested the farm produce. These techniques are as follows (Kargwal et al., 2019):
   i. **Tillage:** It is the most prevalent agricultural activity through which soil movement occurs. This process is broadly used throughout for producing the crop (Practice et al., 2017). Various tools, fertilizers, and equipment are used to physically manipulate soil that forms ideal conditions for seed germination, establishment, and plant growth (Chilur & Yadachi, 2017). The required power for this activity is generally provided by draft animals, human, and conventional energy-based machinery. The principal implements needed for this activity are plows, rotary hoes, basin lister, cultivators, harrows, planks, and rollers. Tillage process collectively has three different phases, which are followed simultaneously. First phase is known as primary tillage, which is performed for soil opening and loosening. Ploughs are the main implements for primary tillage. In ancient agriculture practices, the wooden plows were drawn with draft animals to perform the primary tillage process. In modern agriculture practices, iron soil-turning plows and rotary hoes were drawn by tractors for primary tillage. Next phase after primary tillage is secondary tillage which is performed using cultivators, harrows, planks, and rollers. In modern agriculture practices, these implements are drawn with a tractor. Final phase of the process is the layout of the seedbed. This process is generally followed after the sowing of the crop seeds. The implements used in this phase are country plow, ridge plow, and bund former (Dhiman & Dhiman, 2015).
   ii. **Sowing:** Agriculture activity in which the seeds are placed in the soil results in good germination in the field known as sowing or seeding. Perfect sowing results from an adequate seed per unit area, correct spacing between row to row and plant to plant, and placing seed to the correct depth. The implements needed for sowing are planter and seed drills. In the widespread agriculture practice of sowing, the seed is dropped by hand randomly on the surface of the prepared seedbed. This process depends on the farmer's skills. Seedbed is formed with the country plow. In another widespread technique, sowing is performed by seed drills (Dhiman & Dhiman, 2015). Both plows were made from

wood in ancient agriculture practice, and seed drills were made from wood and bamboo. These implements were drawn with animals. In modern agriculture practices, the implements are made from iron and drawn with tractors. The main advantages of this sowing process are the uniform amount of seed is fed in the seedbed, the uniform distance between crop plants is maintained, and proper depth of seeding is achieved.

iii. **Transplanting:** It is an agricultural activity that consists of two phases. First phase is seedlings preparation at the nursery, and the second phase is a plantation of plants in the prepared field. This process is highly time-consuming and generally performed for paddy crops, vegetables, and flowers. In modern agriculture practices, the implement used for this activity is known as a transplanter.

iv. **Interculture:** It is an agricultural activity of weed control and removal. Weeds grow together with the crop, absorb the soil nutrients, and compete with crop plants for light, moisture, and nutrients. Implements used for intercultural are dryland weeder, Cono weeder, engine operated weeder, sweep, and junior hoe.

v. **Irrigation:** It is a significant agricultural activity in which a controlled amount of water is provided to crop their growth and increase productivity. Irrigation is performed to provide water to crop plants artificially instead of counting on rainfall alone. Various ways of irrigating the crops are fetching water using canals, wells, gravitational flow, pipes, sprinklers, and other artificial means. In ancient agriculture practices, canals performed irrigation and drew water from wells through bullocks and windmills. In modern agriculture practices, crop irrigation is achieved by underground water fetched through pipes and sprinklers using pumps driven by diesel engines and electric submersible pumps.

vi. **Fertilizer and Pesticide Application:** Fertilizers help maintain the soil's fertility and allow the farmer to grow healthy and nutritious crops. The fertilizers are applied to the soil by farmer because these comprise major plant nutrients such as potassium, nitrogen, and phosphorus. Pesticides are needed for plant protection which helps in increasing the agriculture yield by protecting the crops from different pest attacks. Chemical pesticides are popular and widely used for pest control. In ancient agriculture techniques, farmers used organic fertilizers such as cattle dung, but since the process was slow, highly laborious and unavailability of large herds of cattle, this method became obsolete. In modern agriculture practices, chemical fertilizers and pesticides are popularly used and different equipment viz., hand, foot, and power-operated sprayers, compressors, and rockers are required to apply these fertilizers and pesticides.

vii. **Harvest and Threshing:** Harvesting and threshing is the final and significant activity of the pre-harvesting agriculture process. This can be accomplished either manually or by machinery operated by electric or diesel power. According to ancient agriculture practices, it is a highly labor-intensive activity in the agriculture system. Manual harvesting and threshing are time-consuming and demand a large workforce and require a hand chopper to harvest the crop. In modern agriculture practices, the harvest and threshing can be performed simultaneously within one or two days with a relatively more minor workforce. Devices like reapers, combine harvesters, etc., are used for harvesting and threshing purposes.

2. **Post-Harvesting Process:** This starts from the instant the crop is harvested from the field. The post-harvest technology maintains and extends the shelf life of crops and reduces food loss (Pedreschi, 2017). The post-harvest process comprises various unit stages, viz. sorting, drying, storage, milling, and packaging, to transport the harvested crop to the nearest market. The various stages of the post-harvest process are (Chilur & Yadachi, 2017; Kargwal et al., 2019; Kumar et al., 2021):

i. **Cleaning and Grading:** This activity is the most significant post-harvesting operation performed to remove the undesirable constituents from the threshed agriculture produce and separate food grains/products into various segments. The cleaning of agricultural products referred to the elimination of foreign materials from the desired products. During cleaning, grain washing, screening, and handpicking techniques are followed. Grading means classifying the cleaned agricultural produce into various batches depending on their quality, usage, and commercial values. Various energy-based machines viz. screen cleaner cum grader, air-screen cleaner cum grader, cellular surface separator, specific gravity separator, Pneumatic separator, magnetic separator, and color separator in modern agriculture practices cleaning and grading purposes.

ii. **Drying:** Crop drying is an essential activity that can increase the shelf life of a crop. This activity is prevalent among people living in South Asian countries since the ancient era. For crop storage for a more extended period, the drying activity is performed. In ancient agriculture practices, the food crop was dried using the sun's radiation in an open environment. This method is known as the open sun drying (OSD) method (Prakash, Kumar, & Sharaf-Eldeen, 2016). This process has several disadvantages like deterioration of food due to climate, dust, animals, birds, and microorganisms. Also, this method is very time-consuming. Various devices are developed for food drying in modern agricultural practices, which runs over conventional and non-conventional energy, preferably solar energy. These devices are known as dryers. Producing dried using these dryers has similar properties: color, texture, and taste as the original undried food. Several foods such as grains, fruits, leafy vegetables, fish, and various seafood are dried and stored for a more extended period using a drying process (Sharma, Atheaya, & Kumar, 2021; Udomkun et al., 2020).

iii. **Storage:** Drying food takes away all the moisture and juiciness of food, increasing its shelf life for a more extended period. However, if the farmer has to store the food for a shorter duration, cold storage is the better option. The food is kept at a low temperature which results in slowing the deterioration rate. It is kept stored inside an insulated room at optimum temperature and relative humidity (R_h), decreasing the production of the natural ripening agent, i.e., ethylene (Kumar & Mahajan, 2019). This method kept the food in its original form till it reached the market. Some food crops, viz. vegetables and fruits, are popularly stored in cold storage (Surendar, Babu, & Rao, 2020).

iv. **Milling:** A size reduction technology in which the raw material is ground, such as grinding food crop grains into flour, is popularly known as Milling (Kargwal et al., 2019). It is a vital food processing technique that influences the fundamental nutritional properties of food. The milling process is utilized for various purposes, including flour production, cocoa mass production, homogenization, ultra-grinding, and dehusking

(Kargwal et al., 2019). Fundamentally, four techniques are utilized for size reduction machines viz. compression, impact, shearing, and cutting. Grinder and crusher are commonly used for size reduction of farm produce. Various machines are used to perform the milling process, such as ball mills, attrition mills, hammer mills, dispersion, and colloid mills, disintegrators, oil expellers, etc.

v. **Material Handling and Transportation:** Material handling implements are needed to move the material from one place to another. In ancient practices, these operations were performed by manual labor. In modern technology, mechanical implements such as conveyors and elevators perform the material handling process. Furthermore, well-organized and efficient transport is a key to adequate farm management. It is an essential activity for the post-harvest agricultural process. Every farmer needs specific resources, i.e., seeds, manure, pesticides, packing materials, etc., due to which transportation plays an essential role in providing all the resources to a farmer (Kargwal et al., 2019). Along with this, transportation of harvested crop is needed either to warehouse where it is packed and stored or directly to the market. In ancient times, the harvested food crop was supplied to nearby markets only by bullock carts or hand carts. Therefore, farmers receive meager compensation for their produce, but transportation is advanced after developing IC engines in modern practices. The use of fuel-powered automobiles, trains, ships, and cargo airplanes increases a local farmer's range. The product from the field can be transported anywhere in the world in few days and helps develop farmers.

14.3 ROLE OF VARIOUS ENERGIES IN THE DEVELOPMENT OF AGRICULTURE

Production in an ample quantity of crop required a high amount of energy. For agriculture, energy can be categorized into two types viz. source-wise and operation-wise energy. Operation-wise energy is used during several operations performed in pre-harvest and post-harvest processes. Operation-wise energy used in pre-harvest during tillage, sowing, fertilizer application, interculture, harvesting, and threshing while in post-harvesting during cleaning and grading, drying, storage, packaging, and transportation.

From the preparation stage to the last stage of transportation, a hefty amount of energy is being used directly or indirectly. The direct form of energy is utilized during pre-harvest processes in machinery, irrigation, cultivation, and harvesting. During post-harvesting, the direct form of energy is stored, processed, and transported to the markets. Moreover, the indirect energy input is utilized in agriculture in chemical pesticides, mineral fertilizers, insecticides, and herbicides.

The equations to compute the direct and indirect energy inputs during pre-harvest and post-harvest are as follows:

1. **Direct Energy Inputs:** The direct energy inputs per hectare of agricultural produce consist of human effort, animal energy and fuel (mainly diesel). The energy can be computed using the following equations (Chilur & Yadachi, 2017; Chaudhary, Gangwar, & Pandey, 2006):

   i. **Human Energy:**

$$\text{Human energy}\left(\text{MJ / ha}\right) = \frac{\text{N}_{H} \times \text{T}_{\text{o}}}{\text{A}_{\text{o}}} \times EF_{H} \tag{1}$$

where; N_H is the number of humans; T_o is the operation time (in hours); A_o is the operating area (in hectare); and EF_H is the human energy equivalent factor (MJ/h).

   ii. **Animal Energy:**

$$\text{Animal energy}\left(\text{MJ / ha}\right) = \frac{N_{A} \times T_{o}}{A_{o}} \times \text{EF}_{A} \tag{2}$$

where; N_A is the number of animals; T_o is the operation time (in hours); A_o is the operating area (in hectare); EF_A is the animal energy equivalent factor (MJ/h).

   iii. **Fuel Energy:**

$$\text{Fuel Energy}\left(\text{MJ / ha}\right) = \frac{\text{F}_{\text{co}} \times \text{EF}_{\text{F}}}{\text{A}_{\text{o}}} \tag{3}$$

where; F_{CO} is the average fuel consumption (l/h); EF_F is the fuel energy equivalent factor (MJ/h); A_o is the operating area (in hectare).

2. **Indirect Energy Inputs:** This per hectare of agricultural produce comprise machines, seeds, fertilizers, and pesticides. These can be computed using the following equations (Chilur & Yadachi, 2017; Chaudhary, Gangwar, & Pandey, 2006):

   i. **Electrical Energy Consumed During Irrigation:**

$$\text{Electrical energy}\left(\text{MJ / ha}\right) = \frac{\text{kWh} \times \text{EF}_{\text{EE}}}{\text{A}_{\text{I}}} \tag{4}$$

where; kWh is the energy consumption reading from energy meter; A_I is the irrigated area (in hectare); and EF_{EE} is the electrical energy equivalent factor (MJ/kWh).

   ii. **Fertilizer Energy:**

$$\text{Energy of N (MJ/ha)} = \frac{\text{Share of N} \times \text{EF}_{\text{N}}}{A_o} \tag{5}$$

$$\text{Energy of P}_2\text{O}_5 \text{ (MJ/ha)} = \frac{\text{Share of P} \times \text{EF}_{\text{P}}}{\text{A}_o} \tag{6}$$

$$\text{Energy of K}_2\text{O (MJ/ha)} = \frac{\text{Share of K} \times \text{EF}_{\text{K}}}{A_o} \tag{7}$$

$$\text{Total Energy Input of Fertilizer} = \text{N} + \text{P}_2\text{O}_5 + \text{K}_2\text{O} \tag{8}$$

where; N is the nitrogen; EF_P is the nitrogen energy equivalent factor, MJ/ha; P is the phosphorus; EF_N is the phosphorus energy equivalent factor, MJ/ha; K is the potassium; EF_K is the potassium energy equivalent factor, MJ/ha; and A_o is the area.

iii. **Seed Energy:**

$$\text{Seed energy (MJ/ha)} = \frac{S \times EF_S}{A_o} \tag{9}$$

where; S is the seed, kg; A_o is the operating area, ha; EF_S is the seed energy equivalent factor (MJ/ha).

iv. **Pesticide Energy:**

$$\text{Chemical energy (MJ/ha)} = \frac{Q \times \text{EF}_{\text{P}}}{A_o} \tag{10}$$

where; Q is the quantity of pesticide, kg; A_0 is the operating area, ha; EF_P is the pesticide energy equivalent factor, MJ/ha.

v. **Machine Energy:**

$$\text{Machine energy}\left(\text{MJ / ha}\right) = \frac{W \times EF_M}{U_L} \times T \times R \tag{11}$$

where; W is the weight of machine, kg; U_L is the useful life, h; H is the time, h; R is the number of passes; and EF_M is the machine energy equivalent factor, MJ/ha.

3. **Energy Inputs During Post-Harvest:** During post-harvesting, various energy sources are utilized. These energies can be calculated using the following relations (Kargwal et al., 2019):
   i. **Manual Energy:** It was calculated by the multiplication of number of person involved in processing operation with time (Kargwal et al., 2019).

$$E_m = 0.075 \times N_H \times T \tag{12}$$

where; E_m is the manual energy, kW; N_H is the number of person involved in operation; T is the time taken to complete process, h; and 0.75: the average power of human labor in kW.

   ii. **Electrical Energy:** This used by the implement can be obtained via calculating the rated power of each motor and the number of hours of operation. A motor efficiency of 80% is anticipated to calculate the electric input (Kargwal et al., 2019):

$$E_p = \eta \times P_R \times T_o \tag{13}$$

where; E_p is the electrical power consumed, kWh; P_R is the rated power of motor, kW; T_o is the time of operation, h; η is the power factor (0.8).

4. **Energy Analysis:** After gathering data concerning different unit operations of agriculture crop production and processing, calculations must be completed regarding energy use efficiency, energy productivity, energy ratio and net energy gain (Kargwal et al., 2019).
   i. **Energy Use Efficiency:** Or energy ratio is the ratio of the input energy (direct and indirect energy) and the output (yield) from the agriculture crop production system. Output is determined once the yield is converted into energy by multiplying its energy equivalent into the energy units (Kargwal et al., 2019).

$$\text{Energy use efficiency} = \frac{\text{Energy output } (\text{MJ / ha})}{\text{Energy input } (\text{MJ / ha})} \tag{14}$$

ii. Energy Productivity: It is the total amount of harvested product divided by the total energy consumed by harvested product (Kargwal et al., 2019).

$$\text{Energy productivity}\left(\frac{\text{kg}}{\text{MJ}}\right) = \frac{\text{Grain yield}\left(\frac{\text{kg}}{\text{ha}}\right)}{\text{Total energy input}\left(\frac{\text{MJ}}{\text{ha}}\right)} \tag{15}$$

iii. Net Energy Gain: It is defined as the difference of output and input energy (Kargwal et al., 2019).

$$\text{Net Energy Gain } (\text{MJ / ha}) = \text{ Energy output } - \text{ Energy input} \tag{16}$$

iv. Specific Energy: It is the amount of energy spent to produce a unit of marketable product (Kargwal et al., 2019).

$$\text{Specific Energy } (\text{MJ / kg}) = \frac{\text{Energy Input } (\text{MJ / ha})}{\text{Agriculture crop Yield } (\text{kg / ha})} \tag{17}$$

In South Asian countries, agriculture and fish farming are prominent employment sources (Liu, Zhang, & Bae, 2017). The total energy consumption in agriculture for various South Asian countries is given in Table 14.2.

TABLE 14.2 Total Energy Consumption in Agriculture and Employment in Agriculture of Various South Asian Countries

SL. No.	Country Name	Employment in Agriculture (Percentage of Total Employment)	Electricity Consumption in Agriculture (%)	References
1.	Bangladesh	38.58	3.09	Das et al. (2020)
2.	Bhutan	55.31	2	Sharpe (2015)
3.	India	42.38	17.7	Energy Statistics India (2021)
4.	Nepal	65	1.3	ADB (2017)
5.	Pakistan	36.66	11.9509	Lin & Raza (2021)
6.	Thailand	31.16	3	Ko et al. (2017)

5. **Conventional Energy Involved During Pre- and Post-Harvesting Processes:** Pre-harvesting processes consist of highly energy-intensive activities, which are essential for getting quality farm produce. These processes involve the consumption of energy viz. electricity, petroleum products, and human efforts. A large extent of these energies is produced from conventional sources of energy. In South Asia, electricity is majorly produced using conventional coal, petroleum, nuclear, hydropower, etc. Electricity generation in various South Asian countries is given in Table 14.3. Apart from this, tractors and other agricultural implements also utilize conventional energy to a great extent. The tractors draw several agricultural implements, i.e., cultivators, harrows, threshers, etc., during various agricultural activities performed during the pre-harvest process. For irrigation purposes, electrical energy and diesel are used to run the pumps for lifting the underground water. Chilur & Yadachi (2017) had audited the maize production system in villages of Karnataka state in India. It was reported that 4631.26 MJ/ha conventional energy is consumed during pre-harvest processes, excluding the energy spent during chemical fertilizer production. Kargwal et al. (2019) surveyed the pearl millet production system in Haryana, India. It was concluded in the study that the total energy consumed during pre-harvesting processes of pearl millet production was 4785.52 MJ/ha.

TABLE 14.3 Electricity Generation Using Conventional and Non-Conventional Energy Sources in Various South Asian Countries

SL. No.	Name of Country	Electricity Generation Using Conventional Sources (%)	Electricity Generation Using Non-Conventional Sources (%)	References
1.	Afghanistan	53.2	46.8	World Bank Report (n.d.)
2.	Bangladesh	99.9	0.1	Bangladesh Power Review (n.d.)
3.	Bhutan	99	1	IRENA Report (n.d.)
4.	India	75.5	24.5	Energy Statistics India (2021)
5.	Nepal	100	0.0	ADB (2017)
6.	Pakistan	91	9	Unwin (n.d.)

TABLE 14.3 *(Continued)*

SL. No.	Name of Country	Electricity Generation Using Conventional Sources (%)	Electricity Generation Using Non-Conventional Sources (%)	References
7.	Sri Lanka	99.8	0.2	Sri Lanka Energy (n.d.)
8.	Thailand	79	21	Energy Statistics (n.d.)

Post-harvesting processes include the activities carried out after harvesting the crop. These activities consume energy in high amounts. Cleaning and grading process is machine-based which needs electrical power for its operation. Drying and cold storages are also energy-intensive processes (Jain et al., 2019; Surendar, Babu, & Rao, 2020). Conventional food dryers utilize electrical energy, and these are highly energy-intensive types of equipment. The heat energy for drying farm produce is generated by utilizing electrical power, primarily generated using non-conventional sources. In cold storage, a specific low temperature is needed to store the food products to slow down the ripening process of farm produce. Cold storage also draws electrical energy for the operation. Packing and transportation activity uses both electrical energy and diesel for their operation. Packed crop is transported by trucks, tractors, small loading vehicles to the market for selling. Kargwal et al. (Kargwal et al., 2019) audited energy consumption in the post-harvesting activity of pearl millet. During the processing of pearl millet in biscuit and cake making, it was observed that consuming 6.048 and 6.011 kWh/kg electrical energy, respectively. Chilur & Yadachi (2017) reported that 274.54 MJ/ha energy is required during transportation, which is contributed by diesel power. Surendar et al. (2020) had reported energy viability in cold storage. It was reported that 72% of total expenditure is made for the electricity consumed during cold storage. Hnin et al. (2019) studied various technologies of conventional food dryers and reported air-drying method involves specific energy consumption of 177.12 MJ/kg. Kumar et al. (2021) had reported a feasibility assessment for tea plantation and industry and observed that electrical energy used in the various plantation crops viz. rubber, tea, and coffee are 0.596, 0.679, and 0.264 kWh/kg, respectively, and the heat energy needed for them are 2.019, 28.39, and 27.78 MJ/kg, respectively.

6. **Non-Conventional Energy Involved During Pre- and Post-Harvesting Processes:** Conventional power is majorly available and utilized during various agricultural activities, but it has certain disadvantages. The major disadvantage of using conventional energy sources is that they produce high GHG (greenhouse gases) emissions (Sharma & Kumar, 2018). Apart from that, it is costly, and the conventional energy sources are depleting day by day. Therefore, non-conventional energy sources are now being used to generate electricity in South Asian countries as it is sustainable, renewable, economical, and environment friendly. Various RE sources utilized in South Asian countries are discussed as:

   i. **Solar Energy:** It is majorly used among South Asian countries for electricity generation, irrigation, and food drying. Electricity is being developed using solar photovoltaic power plants established across several South Asian countries. These countries are increasing power plant generation day by day to reducc dependencies on conventional energy sources. Solar irrigation pumps are also being arranged at subsidized prices for farmers in developing South Asian countries such as India to reduce agricultural costs and provide a cheap and sustainable solution (MNRE Annual Reports, 2021). The various solar dryers were developed for drying the farm produce sustainably and economically. The disadvantage of solar energy is that it cannot be available during night hours by embedding PCM (phase change materials) with the solar dryers (Lingayat et al., 2020). Sharma et al. (2021) studied various PCM-based solar dryers used in crop drying and are very economical. Singh et al. (2018) had studied various greenhouse solar dryers for drying food crops. Chauhan & Kumar (2018) developed a north wall insulated greenhouse dryer for gooseberry drying and concluded that the developed dryer reduced 41% more moisture content than OSD. Jain et al. (2019) simulated and studied energy analysis of a multi-shelf solar dryer for fenugreek drying. Kumar et al. (2021) reported solar PV plants and solar thermal collectors in tea plantation activities across South Asian countries such as Sri Lanka, Indonesia, and India. Solar PV is usually used to generate electricity and cover the consumption throughout the tea farms for tea plantation and processing activities, whereas solar thermal collectors fulfill the tea industry's heat demand.

ii. **Biomass and Biofuel Energy:** Biomass energy is energy generated from biomass viz. dry leaves, cow dung, garden waste, kitchen waste, etc. Biomass is used to generate heat and electricity, used during various pre- and post-harvest activities. Heat can be used during the drying of food. Kumar et al. (2021) discussed biomass for drying tea leaves and generating electricity using cogenerators in South Asian countries such as Sri Lanka and Northeast states of India.

Biofuel such as ethanol and biodiesel are blended with petrol and diesel, respectively. Biodiesel is extracted from jatropha, Karanja, neem, etc. (Raheman & Phadatare, 2004). This blend will reduce petroleum usage and are renewable sources that can be income sources for the farmers. These plants need less water for cultivation. The 20% biodiesel is blended with the diesel in India, which is used for agricultural activities (Joshi et al., 2017; Sharma & Kumar, 2018). Also, this blended diesel can be used to run diesel pumps for irrigation purposes. In Sri Lanka, biodiesel is produced from rubber and neem oil using transesterification. This biodiesel is blended with diesel, and in the future, 100% use of biodiesel is proposed (Murshed et al., 2020). In Pakistan, biodiesel is produced using mahua, neem, jatropha, sunflower seeds (Rauf, Wang, Yuan, & Tan, 2015). Shrivastava et al. (2020) analyzed the biodiesel development using palm oil by pyrolysis method for Thailand.

iii. **Wind Energy:** It is utilized for generating electricity using wind turbines which can be used in various agricultural activities. The wind farms generate electricity for the tea industry in coastal states such as Kerala and Tamil Nadu in India (Kumar et al., 2021; Sharma et al., 2012). Jahangiri et al. (2019) assessed various solar wind power plants installed across Afghanistan for electricity generation. Safiullah et al. (2016) stated that 1.9 MW of wind energy power plants were installed across Bangladesh for electricity production that will help in generating sustainable power, which is also used in the agriculture sector.

14.4 SUSTAINABLE AGRICULTURE PRACTICES

South Asian countries cover 11.71% of the land with high soil, crop, and climate diversity. Sustainable agriculture practices comprise direct seeding, zero tillage, bed sowing, highly efficient irrigation system, mixed farming with legumes, and organic pesticides and fertilizers (AFA, 2019). The use of organic pesticides and fertilizers is eco-friendly and economical. Cow dung-based fertilizer can be developed using earthworms known as vermi-compost, which is highly rich in nutrients and discouraged using chemical fertilizers (Joshi, Singh, & Vig, 2015). These fertilizers are free of cost as these are based on biowastes such as plant leaves, kitchen waste, etc., and dung of livestock available locally (Rakshit et al., 2017). The subsidized solar PV-based irrigation pumps can be used for pumping underground water for irrigation. Zero tillage can be used for crops such as wheat. Zero tillage technique also helps the soil sustain nitrogen, nutrients, and bio-organisms, increasing soil fertility (Feng & Balkcom, 2017). The mixed farming with legumes helps to sustain nitrogen in the soil, which can be used as organic nitrogen content for other non-leguminous crops, reducing urea usage in farming, which is also economical for farmers. These sustainable techniques have several advantages, such as they are free of cost, eco-friendly, helps to reduce the usage of water, preserve soil health, reduce GHG emission, increase crop productivity, and, hence, the profitability of farming (Aryal et al., 2021; Diakosavvas, 2017; Nawaz & Farooq, 2021).

14.5 SUMMARY

Agriculture is the leading employment sector for the citizen of South Asian countries. These countries have diversity in soil conditions, weather conditions, and crops. Before industrialization, the farming techniques were conventional, and farming depended on rainfall and draft animals. The pre-harvesting process in that era was laborious and time-consuming. Also, the agricultural production was more petite. Due to all these factors, the agriculture sector was non-profitable. After industrialization, modern machinery and implements were deployed in the agriculture sector, reducing the manual labor and time consumed during agricultural activities and increasing farming productivity. However, these machines are energy-intensive. They run over the electrical and diesel power, which is conventional energy. Presently, there is a shortage of conventional energy sources and energy generation from these sources, resulting in high GHG emissions. Moreover, in the future, a shift over non-conventional energy sources and sustainable agricultural practices is needed to overcome the disadvantages of modern agricultural practices and increase farmers' profits.

ACKNOWLEDGMENT

There is no direct funding for this work. The authors are highly thankful to Bennett University, Greater Noida, India, and Center for Energy and Environment, Delhi Technological University, for providing basic infrastructure for compiling this work.

KEYWORDS

- **agriculture**
- **energy**
- **post-harvesting**
- **pre-harvesting**
- **South Asia**
- **sustainable agriculture practices**

REFERENCES

ADB, (2017). *Nepal Energy Sector Assessment, Strategy, and Roadmap*. https://www.adb.org/sites/default/files/publication/356466/nepal-energy-assessment-road-map.pdf (accessed on 12 August 2023).

AFA, (2019). *Sustainable Family Farming Agriculture in South Asia Through Partnerships, 9*, 1–12.

Aryal, J. P., Rahut, D. B., Thapa, G., & Simtowe, F., (2021). Mechanization of small-scale farms in South Asia: Empirical evidence derived from farm households survey. *Technol. Soc. 65*, 101591. doi: 10.1016/j.techsoc.2021.101591.

Bangladesh Power Review. IEEFA Reports. https://ieefa.org/wp-content/uploads/2020/05/Bangladesh-Power-Review_May-2020.pdf (accessed on 12 August 2023).

Chaudhary, V. P., Gangwar, B., & Pandey, D. K., (2006). Auditing of energy use and output of different cropping systems in India. *Agricultural Engineering International: The CIGR e-Journal*, ManuscriptbEE05 001,8.

Chauhan, P. S., Kumar, A., & Nuntadusit, C., (2018). Thermo-environomical and drying kinetics of bitter gourd flakes drying under north wall insulated greenhouse dryer. *Sol. Energy 162*, 205–216. doi: 10.1016/j.solener.2018.01.023.

Chilur, R., & Yadachi, S., (2017). Energy audit of maize production system of selected villages of North Karnataka, India. *Int J Curr Microbiol Appl Sci., 6*, 3564–3571. doi: 10.20546/ijcmas.2017.608.427.

Das, N. K., Chakrabartty, J., Dey, M., Gupta, A. K. S., & Matin, M. A., (2020). Present energy scenario and future energy mix of Bangladesh. *Energy Strateg. Rev., 32*, 100576. doi: 10.1016/j.esr.2020.100576.

Dhiman, M., & Dhiman, J., (2015). Infusion of farm mechanization technologies in Indian agriculture: Progress and impact. *Indian J. Econ. Dev., 11*, 125. doi: 10.5958/2322-0430.2015.00014.1.

Diakosavvas, D., (2017). Improving energy efficiency in the agro-food chain. *Jt. Work Party Agric. Environ.*, 102.

Energy Statistics – Energy Policy and Planning Office – Thailand. https://www.eppo.go.th/epposite/images/Energy-Statistics/energyinformation/Energy_Statistics/Electricity/T05_02_03-2.xls (accessed on 12 August 2023).

Energy Statistics India, (2021). https://mospi.gov.in/publication/energy-statistics-india-2021 (accessed on 12 August 2023).

Feng, Y., & Balkcom, K. S. (2017). Chapter 11: Nutrient Cycling and Soil Biology in Row Crop Systems under Intensive Tillage. In: Al-Kaisi, M. M., Lowery, B. (eds.). *Soil Health and Intensification of Agroecosytems.* Academic Press, pp. 231–255

IRENA Report. *Renewable Readiness Assessment: Kingdom of Bhutan.* https://www.irena.org/-/media/Files/IRENA/Agency/Publication/2019/Dec/IRENA_RRA_Bhutan_2019.pdf (accessed on 12 August 2023).

Jahangiri, M., Haghani, A., Mostafaeipour, A., Khosravi, A., & Raeisi, H. A., (2019). Assessment of solar-wind power plants in Afghanistan: A review. *Renew. Sustain. Energy Rev., 99*, 169–190. doi: https://doi.org/10.1016/j.rser.2018.10.003.

Jain, A., Sharma, M., Kumar, A., Sharma, A., & Palamanit, A., (2019). Computational fluid dynamics simulation and energy analysis of domestic direct-type multi-shelf solar dryer. *J. Therm. Anal. Calorim.* doi: 10.1007/s10973-018-7973-5.

Joshi, G., Pandey, J. K., Rana, S., & Rawat, D. S., (2017). Challenges and opportunities for the application of biofuel. *Renew. Sustain. Energy Rev., 79*, 850–866. doi: 10.1016/j.rser.2017.05.185.

Joshi, R., Singh, J., & Vig, A. P., (2015). Vermicompost as an effective organic fertilizer and biocontrol agent: Effect on growth, yield, and quality of plants. *Rev. Environ. Sci. Bio/Technology, 14*, 137–159. doi: 10.1007/s11157-014-9347-1.

Kargwal, R., Garg, M. K., Singh, V. K., & Yadav, Y. K., (2019). Energy auditing of pearl millet production system in dryland region of Haryana Agricultural University in Hisar, India. *IJAE, 12*, 191–195. doi: 10.15740/HAS/IJAE/12.2/191-195.

Kargwal, R., Yadvika, Singh, V. K., Garg, M. K., Vinod, Kumar, M., & Mathur, M. (2019). Energy consumption pattern of value-added products of pearl millet. *Curr. J. Appl. Sci. Technol., 35*, 1–5. doi: 10.9734/cjast/2019/v35i130160.

Khaing, H. K., Zhang, M., Mujumdar, A. S., & Zhu, Y., (2019). Emerging food drying technologies with energy-saving characteristics: A review. *Dry Technol., 37*, 1465–1480. doi: 10.1080/07373937.2018.1510417.

Ko, C. H., Chaiprapat, S., Kim, L. H., Hadi, P., Hsu, S. C., & Leu, S. Y., (2017). Carbon sequestration potential via energy harvesting from agricultural biomass residues in Mekong River basin, Southeast Asia. *Renew. Sustain. Energy Rev., 68*, 1051–1062. doi: 10.1016/j.rser.2016.03.040.

Kumar, K. R., Dashora, K., Krishnan, N., Sanyal, S., Chandra, H., Dharmaraja, S., & Kumari, V., (2021). Feasibility assessment of renewable energy resources for tea plantation and industry in India: A review. *Renew. Sustain. Energy Rev., 145*, 111083. doi: 10.1016/j.rser.2021.111083.

Kumar, M., & Mahajan, B., (2019). *Design of Cold Storage for Fruits & Vegetables*. doi: 10.13140/RG.2.2.14335.82082.

Lin, B., & Raza, M. Y., (2021). Analysis of electricity consumption in Pakistan using index decomposition and decoupling approach. *Energy, 214*, 118888. doi: 10.1016/j.energy.2020.118888.

Lingayat, A. B., Chandramohan, V. P., Raju, V. R. K., & Meda, V., (2020). A review on indirect type solar dryers for agricultural crops – dryer setup, its performance, energy storage and important highlights. *Appl. Energy, 258*, 114005. doi: 10.1016/j.apenergy.2019.114005.

Liu, X., Zhang, S., & Bae, J., (2017). The impact of renewable energy and agriculture on carbon dioxide emissions: Investigating the environmental Kuznets curve in four selected ASEAN countries. *J. Clean Prod., 164*, 1239–1247. doi: 10.1016/j.jclepro.2017.07.086.

MNRE Annual Reports, (2021). https://mnre.gov.in/knowledge-center/publication (accessed on 12 August 2023).

Murshed, M., Mahmood, H., Alkhateeb, T. T. Y., & Bassim, M., (2020). The impacts of energy consumption, energy prices and energy import-dependency on gross and sectoral value-added in Sri Lanka. *Energies, 13*. doi: 10.3390/en13246565.

Nawaz, A., & Farooq, M., (2021). *Agricultural Practices and Sustainable Management in South Asia,* 36–48. doi: 10.1007/978-3-319-95981-8_112.

Pedreschi, R. (2017). Chapter 1: Postharvest Proteomics of Perishables. In: Colgrave, M. L. (ed.). Proteomics in Food Science. Academic Press, pp. 3–16. doi: https://doi.org/10.1016/B978-0-12-804007-2.00001-1.

Prakash, O., Kumar, A., & Sharaf-Eldeen, Y. I., (2016). Review on Indian solar drying status. *Curr. Sustain. Energy Reports, 3*, 113–120. doi: 10.1007/s40518-016-0058-9.

Raheman, H., & Phadatare, A. G., (2004). Diesel engine emissions and performance from blends of Karanja methyl ester and diesel. *Biomass and Bioenergy, 27*, 393–397. doi: 10.1016/j.biombioe.2004.03.002.

Rakshit, S., Chikkappa, K. G., Jat, S. L., Dhillon, B. S., & Singh, N. N. (2017). Scaling-up of proven technology for maize improvement through participatory approach in India. In: Pandey, P. R., Koirala, K. B. (eds.). *Best Practices of Maize Production Technologies in South Asia.* Published by the SAARC Agriculture Centre (SAC), 36–60.

Rauf, O., Wang, S., Yuan, P., & Tan, J., (2015). An overview of energy status and development in Pakistan. *Renew. Sustain. Energy Rev., 48*, 892–931. doi: 10.1016/j.rser.2015.04.012.

Saifullah, A. Z. A., Karim, A., & Karim, R., (2016). Wind energy potential in Bangladesh. *Am. J. Eng. Res., 5*, 85–94.

Saurabh, A., Atheaya, D., & Kumar, A., (2020). Study of hybrid photovoltaic thermal systems. In: *IOP Conference Series: Materials Science and Engineering*. Institute of Physics Publishing.

Schnepf, R., (2004). *CRS Report for Congress Energy Use in Agriculture*. https://nationalaglawcenter.org/wp-content/uploads/assets/crs/RL32677.pdf (accessed on 12 August 2023).

Sharma, A., Srivastava, J., Kar, S. K., & Kumar, A., (2012). Wind energy status in India: A short review. *Renew. Sustain. Energy Rev., 16*, 1157–1164. doi: 10.1016/j.rser.2011.11.018.

Sharma, M., & Kumar, A., (2018). Promising biomass materials for biofuels in India's context. *Mater Lett*. doi: 10.1016/j.matlet.2018.03.034.

Sharma, M., Atheaya, D., & Kumar, A., (2021). Recent advancements of PCM based indirect type solar drying systems: A state of art. *Mater Today Proc.* doi: 10.1016/j.matpr.2021.04.280.

Sharpe, T., (2015). *WEENTECH Proceedings in Energy RESD–2015*. Royal University of Bhutan (RUB) Bhutan and Sustainable Development.

Shrivastava, P., Khongphakdi, P., Palamanit, A., Kumar, A., & Tekasakul, P., (2020). Investigation of physicochemical properties of oil palm biomass for evaluating potential of biofuels production via pyrolysis processes. *Biomass Convers Biorefinery*. doi: 10.1007/s13399-019-00596-x. https://www.researchgate.net/publication/333433430_Soil_Health_Issues_for_Sustainability_of_South_Asian_Agriculture (accessed on 12 August 2023).

Singh, P., Shrivastava, V., & Kumar, A., (2018). Recent developments in greenhouse solar drying: A review. *Renew. Sustain. Energy Rev., 82*, 3250–3262. doi: https://doi.org/10.1016/j.rser.2017.10.020.

Sri Lanka Energy. https://www.export.gov/apex/article2?id=Sri-Lanka-Energy (accessed on 12 August 2023).

Srinivasarao, Ch., Sumanta Kundu, C., Subha Lakshmi, Y., Sudha Rani, K. C., Nataraj, B., Gangaiah, M., Jaya Laxmi, M., Vijay Sankar Babu, & Usha Rani, S. N., (2019) Soil Health Issues for Sustainability of South Asian Agriculture. *EC Agric 5,* 310–326.

Surendar, G., Babu, R., & Rao, N., (2020). Effect of electricity consumption on the financial viability of cold storage units: Recommendations for reducing energy cost. *Int. J. Adv. Sci. Technol., 29,* 6106–6115.

Udomkun, P., Romuli, S., Schock, S., Mahayothee, B., Sartas, M., Wossen, T., Njukwe, E., et al., (2020). Review of solar dryers for agricultural products in Asia and Africa: An innovation landscape approach. *J. Environ. Manage, 268*, 110730. doi: 10.1016/j.jenvman.2020.110730.

Unwin, J. What does Pakistan's energy mix look like and what is its future? https://www.power-technology.com/features/pakistan-energy-mix/ (accessed on 12 August 2023).

World Bank Report. *Afghanistan-Renewable-Energy-Development-Issues-and-Options*. https://documents1.worldbank.org/curated/en/352991530527393098/pdf/Afghanistan-Renewable-Energy-Development-Issues-and-Options.pdf (accessed on 12 August 2023).

Index

D

F

H

T

Z